WITHDRAWN

PARTIAL DIFFERENTIAL EQUATIONS
WITH MATHEMATICA®

PHYSICS SERIES

Consulting Editor **Paul Davies**
University of Adelaide

Introductory Statistical Mechanics   David S. Betts and Roy E. Turner

PARTIAL DIFFERENTIAL EQUATIONS
WITH MATHEMATICA®

Dimitri Vvedensky
Professor of theoretical solid state physics, Imperial College, London

Addison-Wesley Publishing Company
Wokingham, England · Reading, Massachusetts · Menlo Park, California · New York
Don Mills, Ontario · Amsterdam · Bonn · Sydney · Singapore
Tokyo · Madrid · San Juan · Milan · Paris · Mexico City · Seoul · Taipei

© 1993 Addison-Wesley Publishers Ltd.
© 1993 Addison-Wesley Publishing Company Inc.

All rights reserved. No part of this publication may be reproduced, stored in a retrieval system, or transmitted in any form or by any means, electronic, mechanical, photocopying, recording or otherwise, without prior written permission of the publisher.

The programs in this book have been included for their instructional value. They have been tested with care but are not guaranteed for any particular purpose. The publisher does not offer any warranties or representations nor does it accept any liabilities with respect to the programs.

Many of the designations used by manufacturers and sellers to distinguish their products are claimed as trademarks. Addison-Wesley has made every attempt to supply trademark information about manufacturers and their products mentioned in this book. A list of the trademark designations and their owners appears below.

Cover designed by Designers & Partners, Oxford
and printed by Erhedo Press, High Wycombe, Bucks.
Camera-ready copy prepared by the author.
Printed and bound in Great Britain by The University Press, Cambridge.

First printed 1992. Reprinted 1993 and 1994.

British Library Cataloguing in Publication Data
A catalogue record for this book is available from the British Library

Library of Congress Cataloging in Publication Data
Vvedensky, Dimitri D. (Dimitri Dimitrievich)
    Partial differential equations with Mathematica/Dimitri D. Vvedensky
    p. cm.
    Includes bibliographical references and index
    ISBN 0-201-54409-1
    1. Differential equations, Partial—Data processing.
    2. Mathematica (Computer file) I. Title.
    QA377.V84 1992
    515'.353—dc20   92-36876
                                                                  CIP

Trademark Notice
Mathematica is a trademark of Wolfram Research, Inc.
PostScript is a trademark of Adobe Systems, Inc.

Preface

This book provides an introduction to linear and nonlinear partial differential equations. The material is taken from two courses I taught at Imperial College to final-year physics undergraduates. The first course, given during 1986–1990, included many of the standard methods used for solving linear partial differential equations. This material forms the contents of Chapters 2–8. The second course, given in 1991–1992, provided an introduction to nonlinear partial differential equations. The main topics of this course are included in Chapters 9 and 10.

Although much of the material contained in this book may also be found in many standard textbooks, particularly that part dealing with linear partial differential equations, the treatment here offers a different twist in that extensive use is made of *Mathematica* (a product of Wolfram Research Inc.) to illustrate techniques and solutions and to provide examples that in many cases would not otherwise be practical. The goal of this approach, where a traditional presentation is interwoven with one based on mathematical computation, is to provide a more in-depth and detailed introduction to linear and nonlinear partial differential equations than that permitted by a conventional textbook. Specifically, by alleviating the necessity of carrying out complicated but mechanical and repetitive operations, the fine points of the methodologies used for solving differential equations can be demonstrated explicitly, the relationships among different ways of representing solutions to differential equations can be examined to whatever detail desired, and the reader can reconstruct, modify, and explore further

v

any of the examples or discussions in the book. This last point has the particularly attractive feature of making the reader a more active *participant* in the discussions, rather then simply a *spectator*.

Nevertheless, it must be stressed that neither knowledge of nor access to *Mathematica* is required to use this book. The parts of the text where results from *Mathematica* sessions are used and the method of presentation have been chosen to avoid alienating readers without access to *Mathematica*. For these readers, *Mathematica* can be regarded simply as a generator of different types of data. However, readers with access to *Mathematica* can benefit from the approach taken here by using the Notebooks that contain the sessions used in this book. These Notebooks can be obtained either by FTP from Imperial College (`ftp@cooker.sst.ph.ic.ac.uk`) or from Wolfram Research, Inc. through *MathSource* (FTP: `mathsource.wri.com`; email: `mathsource@wri.com`). The Imperial College system will also contain files with answers to the problems whose solutions do not appear in the book. Instructors who adopt this book for courses can contact their local Addison-Wesley representative for the detailed solutions to the problems. Most of the calculations in this book were performed with *Mathematica* version 1.2 on a Macintosh IIci with 8 megabytes of RAM, but for some of the larger calculations a kernel running remotely on an IBM R6000 workstation was used. The Notebooks have been prepared with *Mathematica* version 2.1.

The intended audience for this book falls into one of several categories. First are the students who have had a first course in ordinary differential equations and who may also have had an initial exposure to partial differential equations. This could include both advanced undergraduate and beginning graduate students. The second category encompasses those readers wishing to acquire a familiarity with nonlinear partial differential equations, particularly the modern approaches involving Bäcklund transformations and inverse scattering. Chapters 9 and 10 provide a self-contained introduction to these subjects, although several results from the earlier chapters are required at various points in the discussion. The material in Chapters 9 and 10 can be used as background before reading one of the more comprehensive treatments of the modern theory of nonlinear partial differential equations listed in the references at the end of those chapters. Finally, there are those who have already mastered the subject, but wish to become acquainted with the capabilities of *Mathematica* in a familiar context.

The problems at the end of each chapter are meant to be regarded as an integral part of the text. These problems are used variously to fill in the details of an argument presented in the main text, to apply and extend techniques that have been derived in the main text or the problems, and to provide practice on the pertinent background material. With few exceptions, the problems that require *Mathematica* are separated from those that do not. However, as discussed above, *Mathematica* opens up new realms for assigning problems both in terms of scope and the extent to which the detailed aspects of techniques can be examined without placing an undue burden on the reader. Thus, as in the main text, the problems that use *Mathematica* are often formulated so that the result can be appreciated by the general reader.

There are several people to whom I owe a debt of gratitude for their role in the preparation of this book. The editorial and production staff at Addison-Wesley have been helpful and encouraging (and patient!) from the start of this project. Particular thanks go to Michael Parkinson, Sarah Mallen, Susan Keany and Stephen Bishop. Paul Abbott of Wolfram Research provided many examples and suggestions of the uses of *Mathematica* in the early stages of writing and made available packages included in version 2.0 prior to release. An Educational Research Grant from Wolfram Research and a grant from Imperial College enabled me to use *Mathematica* in the second of my two courses. Thanks go to Jim Fallon of Wolfram Research and Professors Bryan Coles and John Archer of Imperial College for their generosity. Most of the figures in this book were generated with *Mathematica*. Most of those that were not were produced by David Marks using a pre-release version of his PostScript drawing package *SuperScript*. I extend my gratitude to him for this and for the many times I called upon his 'TeXpertise.' Finally, I would like to express my appreciation to the students who took my courses and who made many constructive comments and suggestions on the course notes that served as the basis of this book.

Dimitri D. Vvedensky
London, October 1992

Contents

Preface		v
1.	**Introduction**	**1**
	1.1 Introduction to Partial Differential Equations	2
	1.2 Outline of Book	6
	1.3 Introduction to *Mathematica*	8
	1.4 Mathematical Preliminaries	10
	Further Reading	26
	References	26
	Problems	28
2.	**First-Order Partial Differential Equations**	**35**
	2.1 The Method of Characteristics	36
	2.2 Boundary Conditions of First-Order Equations	45
	2.3 Nonlinear First-Order Equations	52
	2.4 The Complete Integral	60
	Further Reading	63
	References	64
	Problems	65
3.	**Second-Order Partial Differential Equations**	**77**
	3.1 Equations with Constant Coefficients	79
	3.2 Boundary Conditions of Second-Order Equations	84
	3.3 Geometry of Cauchy Boundary Conditions	89
	3.4 Hyperbolic Equations	91

	3.5 Parabolic Equations	97
	3.6 Elliptic Equations	101
	Further Reading	107
	References	107
	Problems	108
4.	**Separation of Variables and the Sturm–Liouville Problem**	**119**
	4.1 The Method of Separation of Variables	120
	4.2 Orthogonality of Functions	124
	4.3 Fourier Series	126
	4.4 Fourier Series Solutions of Differential Equations	133
	4.5 The Sturm–Liouville Problem	143
	Further Reading	146
	References	147
	Problems	148
5.	**Series Solutions of Ordinary Differential Equations**	**167**
	5.1 Singular Points of Differential Equations	168
	5.2 The Method of Frobenius	171
	5.3 Constructing Complementary Solutions	180
	5.4 Standard Forms of Equations with Singular Points	191
	Further Reading	194
	References	195
	Problems	196
6.	**Special Functions and Orthogonal Polynomials**	**207**
	6.1 Hermite Polynomials	208
	6.2 Legendre Polynomials	215
	6.3 Legendre Functions and Spherical Harmonics	222
	6.4 Bessel Functions	226
	6.5 Generating Functions	235
	Further Reading	240
	References	241
	Problems	242
7.	**Transform Methods and Green's Functions**	**261**
	7.1 The Fourier Transform	262
	7.2 The Laplace Transform and the Bromwich Integral	269
	7.3 Fundamental Solution of the Diffusion Equation	274
	7.4 Fundamental Solution of Poisson's Equation	279
	7.5 Fundamental Solution of the Wave Equation	283
	7.6 Green's Functions in the Presence of Boundaries: The Method of Images	286
	Further Reading	293
	References	293
	Problems	294

8. Integral Representations — 317
8.1 The Laplace Transform — 318
8.2 The Euler Transform — 322
8.3 Hypergeometric Functions — 328
8.4 Bessel Functions — 331
8.5 Hankel Functions — 338
Further Reading — 343
References — 343
Problems — 344

9. Introduction to Nonlinear Partial Differential Equations — 355
9.1 Nonlinearity in Partial Differential Equations — 356
9.2 The Burgers Equation—An Exact Solution — 361
9.3 Elementary Soliton Solutions — 366
9.4 Bäcklund Transformations — 371
9.5 Nonlinear Superposition Principles — 377
Further Reading — 385
References — 386
Problems — 388

10. The Method of Inverse Scattering — 403
10.1 Evolution of Eigenfunctions and Eigenvalues — 405
10.2 The Inverse Scattering Transform — 412
10.3 Solution of the Scattering Problem — 422
10.4 Lax's Method — 434
Further Reading — 437
References — 438
Problems — 439

Index — 453

Chapter 1

Introduction

Many physical phenomena are described by functions whose value at a given point depends on its values at neighboring points. The equation determining this function thus contains derivatives of the function, such as a first derivative to indicate the slope at a point, a second derivative to indicate the curvature and so on. Such an equation is called a differential equation. Familiar physical situations that are described by differential equations are the flow of a fluid, the vibrations of a drum head and the dissipation of heat in a material.

There are two basic types of differential equation, which are distinguished by the number of independent variables that enter the equation. A differential equation for a function of a single independent variable contains only ordinary derivatives of the function and is called an **ordinary differential equation**. For a function of two or more independent variables a **partial differential equation** expresses a relation among the partial derivatives of that function. This book is concerned with methods that have been developed for finding particular solutions to partial differential equations that arise in physical applications.

One aspect of solving partial differential equations that exhibits a different level of richness and variety compared with ordinary differential equations is the role that initial and boundary conditions play in determining whether a physically sensible solution can be obtained for a given equation. A partial differential equation contains only part of the information needed to obtain a particular solution. The additional information required to solve a particular problem is provided in the form of auxiliary conditions that specify the initial behavior of a function or the values the solution takes on the boundaries in the region of interest. These are the initial conditions and the boundary conditions. Choosing the right boundary conditions for a partial differential equation is not as straightforward as for ordinary differential equations. In fact, neither the existence nor even the uniqueness of solutions to a partial differential equation can be guaranteed for every set of boundary conditions.

We begin this chapter with an introduction to some of the basic definitions and nomenclature used for differential equations. This is followed by an outline of the book and a brief introduction to *Mathematica*, which will be used to supplement the discussion in the text and to provide examples and problems at the end of each chapter. We then turn our attention to some of the mathematical background that will be needed in the following chapters. This discussion will make use of *Mathematica* both to provide an introduction to the use of this package and to help illustrate the mathematical concepts themselves. Problems at the end of the chapter provide some additional background material both with and without *Mathematica*.

1.1 Introduction to Partial Differential Equations

The general form of an ordinary differential equation for a function y of a single independent variable x can be written in terms of a function F of the arguments x and y together with the derivatives of y as:

$$F(x, y, y', y'', \ldots) = 0 \tag{1.1}$$

The primes in this expression signify derivatives of y with respect to x:

$$y' \equiv \frac{dy}{dx}, \quad y'' \equiv \frac{d^2 y}{dx^2} \tag{1.2}$$

The following notation is also used frequently to signify the nth derivative of y:

$$y^{(n)} \equiv \frac{d^n y}{dx^n} \tag{1.3}$$

The **order** of a differential equation is the order of the highest derivative appearing as an argument of F in (1.1). For example, the most general form of a first-order ordinary differential equation is

$$F(x, y, y') = 0 \tag{1.4}$$

The general form of an nth-order ordinary differential equation is therefore given by the expression

$$F(x, y, y', \ldots, y^{(n)}) = 0 \tag{1.5}$$

If the function F in (1.5) is a polynomial in the highest-order derivative of y appearing in the argument list, then the **degree** of the differential equation is the power to which the highest derivative is raised, i.e. the degree of the polynomial. An equation is said to be **linear** if F is of first degree in y and in each of the derivatives appearing as arguments of F. Thus, the general form of a linear nth-order ordinary differential equation is

$$a_0(x)y(x) + a_1(x)\frac{dy}{dx} + \cdots + a_n(x)\frac{d^n y}{dx^n} = f(x) \qquad (1.6)$$

where $f(x)$ and the coefficients $a_1(x), \ldots, a_n(x)$ are known functions.

Example 1.1. Consider the following ordinary differential equations:

$$\frac{d^2 y}{dt^2} + y - \epsilon\frac{dy}{dt}(1 - y^2) = 0 \qquad (1.7a)$$

$$\frac{d^2 y}{dx^2} + \frac{1}{x}\frac{dy}{dx} + y = 0 \qquad (1.7b)$$

$$\left(\frac{dy}{dx}\right)^2 + x\frac{dy}{dx} + y = 0 \qquad (1.7c)$$

In Equation (1.7a), ϵ is a constant. Equation (1.7a), which is the **van der Pol equation**, is of *second* order and of *first* degree. This is a nonlinear equation because of the term $\epsilon y^2 y'$. Equation (1.7b) is a particular case of **Bessel's equation**, which will be solved in Chapter 6, and is seen to be a *linear second*-order equation. The equation in (1.7c) is of *first* order and *second* degree and is therefore a nonlinear equation. ∎

For one unknown function u of two independent variables, x and y, the most general partial differential equation can be written in a form analogous to that in (1.1):

$$F(x, y, u, u_x, u_y, u_{xx}, u_{xy}, u_{yy}, \ldots) = 0 \qquad (1.8)$$

The subscripts in (1.8) are a standard notation for partial differentiation with respect to the indicated variables:

$$u_x = \frac{\partial u}{\partial x}, \quad u_y = \frac{\partial u}{\partial y}, \quad u_{xy} = \frac{\partial^2 u}{\partial x \partial y}, \quad u_{yx} = \frac{\partial^2 u}{\partial y \partial x}, \ldots \qquad (1.9)$$

The **order** of a partial differential equation is defined in analogy with that for an ordinary differential equation as the highest-order derivative appearing in (1.8). Thus, the most general first-order partial differential equation may be written as a function F of the arguments x, y, u, u_x and u_y

$$F(x, y, u, u_x, u_y) = 0 \tag{1.10}$$

Similarly, the most general second-order partial differential equation may be written in terms of another function F as

$$F(x, y, u, u_x, u_y, u_{xx}, u_{xy}, u_{yy}) = 0 \tag{1.11}$$

and so on for higher-order equations.

If the functions F in Equations (1.8), (1.10), and (1.11) are polynomials of degree r in the highest-order partial derivative, then the partial differential equation is said to be of **degree** r. In particular, if F is linear in u and in all partial derivatives of u, the partial differential equation is **linear**. For example, the general linear first-order partial differential equation can be written as

$$a(x,y)\frac{\partial u}{\partial x} + b(x,y)\frac{\partial u}{\partial y} + c(x,y)u = d(x,y) \tag{1.12}$$

where $a(x,y)$, $b(x,y)$, $c(x,y)$ and $d(x,y)$ are known functions.

Example 1.2. Consider the following partial differential equations:

$$\frac{\partial u}{\partial t} + u\frac{\partial u}{\partial x} = 0 \tag{1.13a}$$

$$\left(\frac{\partial u}{\partial x}\right)^2 + \left(\frac{\partial u}{\partial y}\right)^2 = 1 \tag{1.13b}$$

$$\frac{\partial u}{\partial t} - u\frac{\partial u}{\partial x} + \frac{\partial^3 u}{\partial x^3} = 0 \tag{1.13c}$$

Equation (1.13a) is a partial differential equation of *first* order and *first* degree, while (1.13b) is of *first* order and *second* degree. Equations such as (1.13a), which are of first degree in the highest-order derivatives but contain other nonlinear terms are termed **quasi-linear**. Equations such as (1.13b) which are of degree greater than one in the highest-order derivatives are termed **nonlinear**. Quite often, however, the term 'nonlinear' is used to designate an equation that contains nonlinear terms of any type. The significance of the distinction between nonlinear and quasi-linear equations lies in the methodology used for solving the two types of equations. This will be most apparent in the solutions of quasi-linear and nonlinear first-order equations in Chapter 2. Equation (1.13c) is a quasi-linear *third*-order equation of *first* degree, called the **Korteweg–de Vries equation**, which will be solved in Chapter 10. ∎

An important factor underlying the versatility of linear ordinary and partial differential equations is the superposition principle. This principle, which is a direct consequence of the linearity of the equations, states that any linear combination of solutions to a differential equation is also a solution of that differential equation. Almost all of the techniques to be discussed in the following chapters rely upon this property to construct a linear combination of elementary solutions to form a flexible solution for satisfying initial and boundary conditions. The various methods differ only in the functions used to form the linear combinations. Examples of methods that will be developed include the separation of variables and eigenfunction expansions, special functions and orthogonal polynomials, transform methods and Green's functions.

Existence and uniqueness theorems have also played an important role in establishing the widespread utility of differential equations. These theorems identify the conditions under which solutions to differential equations exist and are unique. Proving existence and uniqueness theorems for solutions of partial differential equations (Garabedian, 1964) is a far more complicated task than that for ordinary differential equations (Coddington and Levinson, 1955; Ince, 1956). The **general solution** of an ordinary differential equation is an expression that involves arbitrary *constants*. A particular solution is then obtained by specifying the required number of boundary conditions. For partial differential equations, the general solution is expressed in terms of arbitrary *functions*. The boundary conditions that are needed to specialize the general solution for a particular problem must be chosen carefully in order to guarantee not only both the existence and uniqueness of the solution, but also to avoid spurious behavior associated with an inappropriate choice. The proofs of these statements, which are particular cases of the Cauchy–Kowalewski theorem (Garabedian, 1964) can be carried out, but only under very restrictive conditions, namely that the functions F appearing in Equations (1.10) and (1.11) are analytic in each of the arguments, and that the boundary conditions are expressed in terms of analytic functions. This procedure will be carried out for second-order equations in Chapter 3.

The situation for nonlinear partial differential equations is even more complex. There is no simple linear superposition principle, so methods of solution have often relied on inspired guesswork and families of ad hoc techniques (Ames 1965, 1972). Nevertheless, certain equations have been found to have 'nonlinear superposition principles', from which known solutions can be used to generate new solutions purely by algebraic operations. Furthermore, recent developments have led to a technique called the inverse scattering method, which shows some promise of providing a systematic way of solving certain types of nonlinear equation for particular types of initial condition. However, there is yet no systematic way of determining whether a given equation is amenable to this type of solution without having first to solve the equation explicitly. A comprehensive review of the current state of affairs for nonlinear partial differential equations is provided by Ablowitz and Clarkson (1991).

1.2 Outline of Book

This book is divided into two parts. The first part, which comprises Chapters 2 through 8, covers linear partial differential equations. Chapters 9 and 10 form the second part and are concerned with nonlinear equations. Although these two parts can be read independently, there are several places in the discussion of nonlinear equations that rely strongly on results obtained in the earlier chapters either as a basis of comparison or for a particular result.

First-order equations are the subject of Chapter 2. Quasi-linear equations are considered first to illustrate the geometrical content of Lagrange's method of characteristics before the method is generalized to nonlinear equations. Even though most equations encountered in applications are of higher order, first-order equations are useful for illustrating concepts that are also seen in higher-order equations but with the advantage that solutions are more readily obtained. Although existence and uniqueness theorems will not be discussed at length in this book, the techniques developed in Chapter 2 are in effect existence theorems for first-order equations, since the solutions are constructed explicitly in terms of auxiliary *ordinary* differential equations. The question of uniqueness is addressed by specifying boundary conditions that the solution must satisfy. For quasi-linear equations, we will see that not every boundary condition leads to a solution or even necessarily to a unique solution of the differential equation. This provides an important forewarning of the role of boundary conditions in higher-order equations.

In Chapter 3, solutions of second-order equations will be developed by building on the conceptual foundation established in Chapter 2. The theory of second-order equations will be seen to be more diverse than that of first-order equations. This will be most clearly evident in the classification of second-order equations according to the coefficients of the second-order derivatives and the importance of identifying the right type of boundary condition to insure a physically sensible solution. As in Chapter 2, the discussion of boundary conditions identifies the conditions for finding unique solutions for each type of second-order equation. Examples will be given to show the consequences of inappropriately-chosen boundary conditions.

The first appearance of the superposition principle as a way of solving partial differential equations is made in Chapter 4 with the method of separation of variables. For the equations considered, the method of separation of variables reduces the problem of solving a partial differential equation in n independent variables to solving n ordinary differential equations for each of the independent variables. A particular solution to the partial differential equation is then obtained by forming a product of the solutions of the differential equations. By forming a linear combination of these product solutions a more useful solution results in that boundary conditions can be satisfied with an appropriate choice of the expansion coefficients. In all but a few cases, this provides a much more practical way of finding solutions to second-order and higher-order equations with boundary conditions than by first finding the general solution. The simplest solution of this type, and one that is generally used as a prototype, is the Fourier series, which is composed of a series of trigonometric functions. The properties that make the trigonometric functions useful for this purpose will be examined and shown to be the result of the form of the governing differential equation. This motivates the discussion of the Sturm–Liouville problem and more general eigenfunction expansions.

The generalization of the Fourier series expansion to other types of equation is taken up in Chapters 5 and 6. The solutions to these equations are not always given in terms of elementary functions, i.e. expressions that can be constructed from simple algebraic operations involving exponential functions, trigonometric functions and their inverses. Thus, an approach is taken whereby the solution is written as a power series with expansion coefficients that are to be determined. This series is then substituted into the differential equation and the expansion coefficients are obtained by requiring the power series to be a solution of the equation. The details of this procedure, which is known as the method of Frobenius, are described in Chapter 5.

In Chapter 6, this method is applied to obtain several well-known special functions and orthogonal polynomials. Chapter 6 also introduces quantities known as generating functions, which allow the eigenfunctions of a differential equation to be obtained as the coefficients of a power series expansion of a relatively simple function. Generating functions can be used to derive a number of useful properties of eigenfunctions, such as recursion relations, which are interrelationships among different eigenfunctions (possibly including their derivatives) corresponding to neighboring eigenvalues. A general treatment of generating functions is given in Chapter 8.

An alternative implementation of the superposition principle is discussed in Chapter 7. The idea here is to work with the partial differential equation itself and to represent the solution for given initial and boundary conditions in terms of solutions for elementary impulses. The full solution is then written as an integral over these impulse solutions, called Green's functions, weighted by the initial conditions and the boundary conditions. This integral representation is more flexible than the Fourier series representation in that solutions with slowly-convergent Fourier series can be easily represented in terms of Green's functions. The integral representation of solutions also has the virtue of being a closed-form expression that incorporates explicitly the initial and boundary conditions. The methodology for obtaining Green's functions is closely related to integral transform methods, and the most common of these, Fourier and Laplace transforms, are discussed at the beginning of the chapter.

Chapter 8 is concerned with integral representations of solutions to ordinary differential equations. These representations provide a complementary way of writing solutions and have a similar relationship to the series representations developed in Chapter 5 that the Green's function representation has to the orthogonal function representation of solutions to partial differential equations. The advantages of integral representations are both aesthetic and practical. Integral representations provide a closed-form solution whose analytic properties are more clearly displayed than those of a series. One particularly useful result that emerges from this analysis is that the generating function for the eigenfunctions of a differential equation can be derived systematically from the integral representation of the solutions to that differential equation.

Chapters 9 and 10 provide an introduction to the modern theory of nonlinear partial differential equations. In Chapter 9, the origin of nonlinear equations is discussed from a heuristic point of view and some solutions of particular equations are obtained. Included among these are 'solitons', which are nondispersive propagating solutions of certain types of nonlinear equations. Solitons have the interesting property that their form is retained after interactions with other solitons. Multi-soliton solutions of several equations are constructed by using 'nonlinear superposition principles' and the dynamics of the solitons within these solutions are explored.

Although methods for solving nonlinear partial differential equations are not nearly as abundant as those available for linear equations, one method has shown itself to be applicable to solving initial-value problems for a number of nonlinear equations. This is the method of inverse scattering and is the subject of Chapter 10. The method was initially devised for a particular equation, the Korteweg–de Vries equation in (1.13c), but was later shown to be applicable to several other equations as well. For the Korteweg–de Vries equation this method provides a way of identifying features in the solution by regarding the initial value of the solution as the potential of a quantum mechanical scattering problem. For example, the number of bound states corresponds to the number of solitons that emerge from the initial value. This and other intriguing aspects of this method will be discussed at length.

1.3 Introduction to *Mathematica*

Mathematica was developed by Wolfram Research, Inc. for performing mathematics on a computer. There are three basic types of functions that can be performed with *Mathematica*: symbolic manipulations, such as algebra and calculus, numerical computations, such as the numerical evaluation of various expressions and numerical integration of ordinary and partial differential equations, and graphical representations either of functions generated in *Mathematica* or from data produced elsewhere and read into *Mathematica*. Any of these facilities can be combined within a program. Included in the graphical capability of *Mathematica* is the animation of a sequence of frames. This is useful not only for representing the development of a function with time, but also the changes of a function or an expansion of a function as a parameter is varied.

What makes *Mathematica* particularly useful for partial differential equations is the complete integration of the symbolic, numerical and graphical features into a structured programming environment. Thus, symbolic manipulations can be performed to obtain a formula for the solution to a differential equation which can then be evaluated numerically and visualized in a number of ways. This allows solutions to be explored in far greater depth than would otherwise be possible because many of the algebraic manipulations required to solve particular partial differential equations can be quite cumbersome and repetitive. Summaries of the symbolic, numerical and graphical capabilities of *Mathematica* are as follows.

Symbolic. The ability to perform symbolic manipulations on the computer removes much of the drudgery of lengthy algebraic manipulations. A specific example where this has useful consequences for differential equations is showing that a series representation of a function does in fact solve a particular differential equation. Even for instances where closed-form solutions are available, the operations required by the substitution into the differential equation can result in lengthy expressions. Examples of this abound in Chapters 9 and 10, where the complexity of many of the manipulations would require hours with pencil and paper, but is reduced to a few seconds with *Mathematica*.

Numerical and symbolic programming. The programming capability of *Mathematica* allows both symbolic and numerical manipulations to be combined with graphical representations. This mimics many computations that scientists and engineers carry out routinely, wherein algebraic steps are performed until a numerical solution must be found. *Mathematica* also includes many of the standard mathematical functions such as trigonometric and hyperbolic functions and most of the special functions encountered in applications. Any functions not included can be defined by the user. Thus, investigating the detailed properties of specific solutions to differential equations becomes a straightforward and often unexpectedly rewarding task. Indeed, many of the techniques to be developed in this book are sufficiently systematic in their execution that a symbolic program can be written to carry out the required steps and then the solution evaluated numerically and displayed. Examples include the solution of first-order quasi-linear equations (Chapter 2), the solution of differential equations with a series expansion method (Chapter 5), and the generation of hierarchies of solutions to nonlinear equations using the nonlinear superposition principles (Chapter 9).

Graphical. The visualization of data is one of the most important tools required by scientists and engineers for assessing various features of a solution and for building an intuition about the properties of solutions to differential equations. *Mathematica* provides a variety of ways to visualize data. One of the most useful of these is certainly through animation. With animation we can watch a solution change not only with time, but also as a parameter is varied. Applications include the bound-states and scattering properties of a square-well potential as the potential depth changes, the poles of a complex function as one of the parameters changes and, of course, the time development of a solution to a differential equation.

There are several ways that the output of *Mathematica* is used in this book. Graphical output generated by *Mathematica* is displayed as a figure. Symbolic manipulation is shown as a *Mathematica* session including both the input commands (in `slanted Courier font`) and the associated *Mathematica* output (in `upright Courier font`). The results of numerical computations are displayed either in tabular form or as an x-y plot, a contour plot, or a surface plot. The parts of the text that either illustrate the use of *Mathematica* or use *Mathematica* to derive a particular result are enclosed within a box.

1.4 Mathematical Preliminaries

In this section we will briefly review some of the required mathematical background that forms the basis of many of the techniques developed in this book. More comprehensive treatments of individual topics can be found in the references listed at the end of the chapter. In carrying out this brief review, we will also show how *Mathematica* can be used to illustrate and clarify some of the mathematical machinery.

We first introduce the idea of linear independence of functions. Two functions $f_1(x)$ and $f_2(x)$ are said to be independent over an interval if there is no linear combination of the two functions that vanishes over the interval, i.e. if the only choice of constants c_1 and c_2 that satisfies

$$c_1 f_1(x) + c_2 f_2(x) = 0 \tag{1.14}$$

over the interval is $c_1 = 0$ and $c_2 = 0$. Linear independence can be extended to any number of functions. Thus, the n functions $f_1, f_2, \ldots, f_n$ are linearly independent over an interval if the only choice of constants c_k that satisfies

$$c_1 f_1(x) + c_2 f_2(x) + \cdots + c_n f_n(x) = 0 \tag{1.15}$$

over the interval is $c_k = 0$ for all k.

To find a way of identifying linearly independent sets of functions, suppose that there are n functions that form a linearly *dependent* set over some interval. Thus, there is a set of constants c_k such that Equation (1.15) is satisfied. We suppose that these functions can each be differentiated $n-1$ times with respect to x. Since Equation (1.15) is satisfied over the entire interval in question and the functions have the required number of derivatives, up to $n-1$ derivatives of this equation may be taken to obtain the $n-1$ equations:

$$c_1 \frac{d^k f_1}{dx^k} + \cdots + c_n \frac{d^k f_n}{dx^k} = 0 \tag{1.16}$$

for $k = 1, 2, \ldots, n-1$. Thus, Equations (1.15) and (1.16) form a system of n equations for the n constants c_k. For these equations to have a solution with not all of the c_k being equal to zero, the determinant of the matrix of coefficients must vanish:

$$\begin{vmatrix} f_1 & f_2 & \cdots & f_n \\ f_1^{(2)} & f_2^{(2)} & \cdots & f_n^{(2)} \\ \vdots & \vdots & & \vdots \\ f_1^{(n-1)} & f_2^{(n-1)} & \cdots & f_n^{(n-1)} \end{vmatrix} = 0 \tag{1.17}$$

In writing this determinant, which is called the **Wronskian** of the functions f_k, the notation in (1.3) has been used for the entries of the determinant. We will use the notation

$$\mathcal{W}(f_1, f_2, \ldots, f_n) \tag{1.18}$$

for the Wronskian. Since the vanishing of $\mathcal{W}$ is a *necessary* condition for the functions f_k to be linearly dependent, then a *sufficient* condition for the functions to be linearly independent is that the Wronskian is *nonvanishing*. Thus, the sufficient condition for linear independence of two (nonzero) functions f_1 and f_2 takes the form

$$\mathcal{W}(f_1, f_2) = f_1 f_2' - f_1' f_2 \neq 0 \tag{1.19}$$

If $\mathcal{W}(f_1, f_2) = 0$, then f_1 and f_2 are easily seen to be proportional to one another.

Example 1.3. The evaluation of the Wronskian is straightforward if only two functions are involved, but can become quite cumbersome for a larger number of functions. *Mathematica* can be used to carry out the required operations. The array of functions f_k is first defined as the vector **F**. In the example below, this is carried out for three functions, though the procedure can be extended to any number of functions.

```
F={f1[x],f2[x],f3[x]}

{f1[x], f2[x], f3[x]}
```

The functions `f1[x]`, `f2[x]`, and `f3[x]` are generic; specific functions will be used below. The Wronskian is then assembled as a `Table`:

```
W[F_]:=Table[D[F,{x,i-1}],{i,3}]
```

This table of terms can be displayed as a matrix using `MatrixForm`,

```
MatrixForm[W[F]]
```

```
f1[x]     f2[x]     f3[x]

f1'[x]    f2'[x]    f3'[x]

f1''[x]   f2''[x]   f3''[x]
```

and the determinant can be evaluated with `Det`.

To carry out a specific evaluation, we take as an example the three functions $f_1(x) = \sin x$, $f_2(x) = \cos x$ and $f_3(x) = e^x$. Then,

```
Simplify[Det[W[{Sin[x],Cos[x],Exp[x]}]]]
```

```
       x        2         2
   -2 E  (Cos[x]  + Sin[x] )
```

which clearly does not vanish over any finite part of the real line. As a second example, suppose the trigonometric functions in the preceding example are replaced with their hyperbolic counterparts: $f_1(x) = \sinh x$, $f_2(x) = \cosh x$ and $f_3(x) = e^x$. The corresponding evaluation yields:

```
Simplify[Det[W[{Sinh[x],Cosh[x],Exp[x]}]]]
```

```
0
```

The Wronskian vanishes because $e^x = \cosh x + \sinh x$.

■

Power series, Taylor series and algebraic operations involving power series are used in several places in this book to represent solutions of ordinary and partial differential equations. In fact, the representation of solutions to differential equations as series expansions is one of the recurring themes in Chapters 3–8, both in terms of making general statements about the existence and uniqueness of solutions and in the actual solution of particular equations.

A **series** is an expression of the form

$$A_1 + A_2 + \cdots A_n = \sum_{k=0}^{n} A_k \tag{1.20}$$

where n is an integer. An **infinite series** is obtained by letting n approach infinity in (1.20). In other words the number of terms in the summation is infinite. A **partial sum** of a series, which is signified by S_n, is the sum of the first n terms of the series. If the sequence of partial sums converges to a value S as $n \to \infty$, the series is said to *converge*. The quantity S is then the sum of the series and we write

$$S = \sum_{k=0}^{\infty} A_k \tag{1.21}$$

An infinite series *diverges* if the sequence of partial sums diverges.

A **power series** is a particular type of series obtained when the terms in (1.20) are given by $A_k = a_k x^k$, where x is a variable. The general form of a power series is obtained by replacing x by $x - x_0$, where x_0 is a reference point:

$$\sum_{k=0}^{\infty} a_k (x - x_0)^k = a_0 + a_1(x - x_0) + a_2(x - x_0)^2 + \cdots \tag{1.22}$$

The numbers a_k are termed **coefficients** of the power series. For the range of values of x where the limit of the partial sums

$$\lim_{N \to \infty} \sum_{k=0}^{N} a_k (x - x_0)^k \tag{1.23}$$

exists, the power series is said to converge. The series (1.22) will either converge or diverge, depending upon the choice of x_0 and x. For a fixed x_0, there is an interval centered around x_0 of length $2R \geq 0$ such that the series converges if x lies within the interval and diverges if x lies outside the interval. If x lies on the boundary of the interval the series may or may not converge. R is called the **radius of convergence** of the series. These considerations also apply if x and x_0 are complex; hence the term *radius* of convergence, since the loci points satisfying $|x - x_0| < R$ is a line segment on the real axis, but a *circle* in the complex plane.

An important type of power series for determining series solutions of differential equations is the expansion of a function about a point x_0:

$$f(x) = \sum_{k=0}^{\infty} a_k(x - x_0)^k \tag{1.24}$$

Differentiating (1.24) n times with respect to x and evaluating the resulting expression at $x = x_0$ yields $k! a_k = f^{(k)}(x_0)$, where $f^{(k)}(x_0)$ is the kth derivative of f at the point x_0. Solving this expression for a_k and substituting into (1.24) yields the familiar Taylor series expansion of f about x_0:

$$f(x) = \sum_{k=0}^{\infty} \frac{1}{k!} f^{(k)}(x_0)(x - x_0)^k \tag{1.25}$$

A function f which has a convergent power series representation about a point x_0 is said to be **analytic** about x_0.

Of the various tests for convergence of series and power series (Knopp, 1956), the ratio test is one of the easiest to apply. For the series in (1.22) this test states that if the absolute value of the ratio of successive terms is bounded by some number ρ as $n \to \infty$,

$$\lim_{k \to \infty} \left| \frac{a_{k+1}(x - x_0)^{k+1}}{a_k(x - x_0)^k} \right| = |x - x_0| \lim_{k \to \infty} \left| \frac{a_{k+1}}{a_k} \right| = \rho \tag{1.26}$$

then the series converges if $\rho < 1$ and diverges if $\rho > 1$. If $\rho = 1$, then the test fails and some other method must be used. If in (1.26) we define

$$\lim_{k \to \infty} \left| \frac{a_{k+1}}{a_k} \right| = \frac{1}{R} \tag{1.27}$$

(where the limit exists), then the criterion $|x - x_0| < R$ identifies R as the radius of convergence for the series (1.22). The series converges if x lies within the *open* interval $(x_0 - R, x_0 + R)$, diverges if x lies outside of this interval, and the test fails if x lies on the end-points of the interval.

Several remarks are in order concerning the ratio test. The ratio test is a test for *absolute* convergence. A series $\sum A_k$ is absolutely convergent if the series $\sum |A_k|$ converges. In particular, if a series converges absolutely, then the series $\sum A_k$ converges (Knopp, 1956). The ratio test gives no information about a series that is not absolutely convergent. The series $\sum (-1)^k / k$ is an example of a series that is convergent but not absolutely convergent. (Such a series is said to be **conditionally convergent**.) The ratio test is thus not a very subtle indication of the convergence of a series, since the divergence of a series is deduced only from the terms of a series not tending to zero as $k \to \infty$.

When the ratio test fails, another test must be applied to determine whether or not a series converges. One useful test that is simple to apply is the Gauss test (Knopp, 1956). Suppose that the series $\sum_k A_k$ is composed only of *positive* terms and that for values of k greater than some integer N, the ratio A_{k+1}/A_k can be written as

$$\lim_{k \to \infty} \frac{a_{k+1}}{a_k} = 1 - \frac{\alpha}{k} + \frac{\beta(k)}{k^{1+\delta}} \tag{1.28}$$

where α is a constant, $\beta(k)$ is a quantity that remains bounded as $k \to \infty$ and $\delta > 0$. Then, according to the Gauss test, the series converges if $\alpha > 1$ and diverges if $\alpha \leq 1$.

Example 1.4. To illustrate the application of the ratio and Gauss tests to assess the convergence of a series, consider the function with the Taylor series expansion

$$F(a, b, c; x) = \sum_{k=0}^{\infty} \frac{(a + k - 1)!(b + k - 1)!}{(c + k - 1)!k!} x^k \tag{1.29}$$

This series, which is proportional to the hypergeometric function that will be discussed in Chapters 6 and 8, is seen to be of the form (1.24) with $x_0 = 0$ and with the expansion coefficients given by

$$a_k = \frac{(a + k - 1)!(b + k - 1)!}{(c + k - 1)!k!} \tag{1.30}$$

The radius of convergence of this series is determined by applying the ratio test in Equation (1.26). The limit of the ratio a_{k+1}/a_k is straightforward to calculate from (1.30) and we obtain

$$\lim_{k \to \infty} \frac{a_{k+1}}{a_k} = \lim_{k \to \infty} \frac{(a + k)(b + k)}{(c + k)(k + 1)} = 1 \tag{1.31}$$

Thus, the series converges for $|x| < 1$ and diverges for $|x| > 1$. To examine the behavior of the series for $x = 1$, we apply the Gauss test.

This Gauss test is most easily applied by making the transformation $k \to n^{-1}$, expanding the resulting expression to second order in n and then comparing with (1.28). The first step is carried out as follows:

```
Simplify[((n+a)(n+b))/((n+c)(n+1))/.n->1/m]

(1 + a m) (1 + b m)
-------------------
(1 + m) (1 + c m)
```

The expansion in powers of m can now carried out with `Series` and the substitution $m \to n^{-1}$ made afterwards:

```
Series[%,{m,0,2}]/.m->1/n
```

```
      -1 + a + b - c
1 + ---------------- +
            n

                                            2
    1 - a + (-1 + a) b - (-1 + a + b) c + c          1  3
    ----------------------------------------- + O[-]
                         2                            n
                        n
```

This expression is readily seen to be of the form (1.28), with the identifications

$$\alpha = -(a+b-c-1)$$
$$\beta = 1-a+(a-1)b-(a+b-1)c+c^2 \tag{1.32}$$

and with $\delta = 1$. Thus, we conclude that the series (1.29) converges for $x=1$ if $c-a-b>0$ and diverges otherwise.

■

The addition of two series is straightforward to define. Consider two series with terms A_n and B_n. The sums of the two series will be signified by A and B:

$$A = \sum_n A_n, \qquad B = \sum_n B_n \tag{1.33}$$

Then, the sum and difference of two series is given by

$$\sum_{n=0}^{\infty} A_n \pm \sum_{n=0}^{\infty} B_n = \sum_{n=0}^{\infty} (A_n \pm B_n) \tag{1.34}$$

with the sum converging to $A+B$ and the difference converging to $A-B$. If the two series are power series, i.e. if $A_n = a_n x^n$ and $B_n = b_n x^n$, then the corresponding expressions for the sum and difference are

$$\sum_{n=0}^{\infty} a_n x^n \pm \sum_{n=0}^{\infty} b_n x^n = \sum_{n=0}^{\infty} (a_n \pm b_n) x^n \tag{1.35}$$

The radius of convergence of the sum or difference is the *smaller* of the radii of convergence of the two power series in (1.33).

The series of the product AB is

$$\sum_{n=0}^{\infty} A_n \sum_{n=0}^{\infty} B_n = \sum_{n=0}^{\infty}\left[\sum_{k=0}^{n} A_k B_{n-k}\right] \quad (1.36)$$

This product converges to the product of the limits of the two series, i.e. the 'correct' value, if at least one of the series is absolutely convergent. If the two series are power series then this definition yields a natural way of grouping the product of two power series in x according to powers of x:

$$\sum_{n=0}^{\infty} a_n x^n \sum_{n=0}^{\infty} b_n x^n = \sum_{n=0}^{\infty}\left[\sum_{k=0}^{n} a_k b_{n-k}\right] x^n \quad (1.37)$$

If the two series are absolutely convergent series, then according to a theorem due to Cauchy (Knopp, 1956), the product of the series as defined in (1.36) is also absolutely convergent, with the value of the sum equal to AB. In general, absolutely convergent series can be manipulated in much the same way as finite sums.

The structure of the product of two series can be determined symbolically in *Mathematica*. We first define the partial sums A[n] and B[n]:

A[n_]:=Sum[a[k]x^k,{k,0,n}]

B[n_]:=Sum[b[k]x^k,{k,0,n}]

The product in (1.34) is now constructed using Expand and the coefficients of powers of x are assembled with Coefficient. The results for the first four terms are shown below:

Do[Print[Coefficient[Expand[A[4]B[4]],x^j]],{j,1,4}]

a[1] b[0] + a[0] b[1]
a[2] b[0] + a[1] b[1] + a[0] b[2]
a[3] b[0] + a[2] b[1] + a[1] b[2] + a[0] b[3]
a[4] b[0] + a[3] b[1] + a[2] b[2] + a[1] b[3] + a[0] b[4]

These coefficients are precisely of the form shown on the right-hand side of Equation (1.37).

The method of separation of variables (Chapter 4) is connected closely with orthogonal curvilinear coordinate systems. We will review briefly some general properties of coordinate transformations and then obtain expressions for the gradient and the Laplacian in a general orthogonal curvilinear coordinate system. The corresponding expressions for the divergence and the curl are taken up in the problems at the end of the chapter.

A transformation from rectangular coordinates (x, y, z) to new coordinates (u_1, u_2, u_3) is obtained by writing each of the original coordinates as functions of the new coordinates:

$$x = x(u_1, u_2, u_3), \quad y = y(u_1, u_2, u_3), \quad z = z(u_1, u_2, u_3) \tag{1.38}$$

The **Jacobian** of the transformation is defined by

$$J = \begin{vmatrix} x_{u_1} & y_{u_1} & z_{u_1} \\ x_{u_2} & y_{u_2} & z_{u_2} \\ x_{u_3} & y_{u_3} & z_{u_3} \end{vmatrix} \tag{1.39}$$

where we have used the notation in (1.9) for the entries of this determinant:

$$x_{u_k} = \frac{\partial x}{\partial u_k}, \quad y_{u_k} = \frac{\partial y}{\partial u_k}, \quad z_{u_k} = \frac{\partial z}{\partial u_k} \tag{1.40}$$

In regions where the Jacobian of the transformation is nonvanishing, the expressions in Equation (1.38) may be inverted to express each of the u_i in terms of x, y and z. Within these regions, the correspondence between the two sets of coordinate systems is unique. If in (1.38) a particular u_i is varied while the other u_j $(j \neq i)$ are held fixed, the resulting curve is termed a u_i coordinate curve. If the quantities u_i are chosen such that the coordinate curves are mutually perpendicular at each point in space, the coordinates are called **orthogonal curvilinear coordinates**.

Example 1.5. The transformation from rectangular coordinates (x, y, z) to spherical polar coordinates (r, θ, ϕ) is defined by

$$x = r \sin\theta \cos\phi, \quad y = r \sin\theta \sin\phi, \quad z = r \cos\theta \tag{1.41}$$

where the ranges of r, θ and ϕ are

$$0 \leq r < \infty, \quad 0 \leq \theta \leq \pi, \quad 0 \leq \phi < 2\pi \tag{1.42}$$

The Jacobian of the transformation is given by

$$J = \begin{vmatrix} \sin\theta \cos\phi & \sin\theta \sin\phi & \cos\theta \\ r\cos\theta \cos\phi & r\cos\theta \sin\phi & -r\sin\theta \\ -r\sin\theta \sin\phi & r\sin\theta \cos\phi & 0 \end{vmatrix} = r^2 \sin\theta \tag{1.43}$$

which is nonvanishing except if $r = 0$, where the r coordinate curves intersect, or if $\sin\theta = 0$, i.e. $\theta = 0$ or $\theta = \pi$, where the θ coordinate curves intersect. Away from these regions, the correspondence between rectangular and spherical polar coordinates is unique. ∎

Let **r** be the position vector of a point. In rectangular coordinates, the coordinates of this point are (x, y, z) and **r** is written as

$$\mathbf{r} = x\,\mathbf{i} + y\,\mathbf{j} + z\,\mathbf{k} \tag{1.44}$$

The vectors **i**, **j** and **k** are the usual unit vectors along the x, y and z directions, respectively. According to the coordinate transformation (1.38), we may also regard **r** as being a function of the coordinates u_1, u_2 and u_3. If **r** traces out a space curve parametrized by the quantity s (see Problem 19), then an application of the chain rule shows that the tangent, or 'velocity', vector to this curve, $d\mathbf{r}/ds$, is given in terms of u_1, u_2 and u_3 by

$$\begin{aligned}\frac{d\mathbf{r}}{ds} &= \frac{\partial \mathbf{r}}{\partial u_1}\frac{du_1}{ds} + \frac{\partial \mathbf{r}}{\partial u_2}\frac{du_2}{ds} + \frac{\partial \mathbf{r}}{\partial u_3}\frac{du_3}{ds} \\ &\equiv \mathbf{U}_1\frac{du_1}{ds} + \mathbf{U}_2\frac{du_2}{ds} + \mathbf{U}_3\frac{du_3}{ds}\end{aligned} \tag{1.45}$$

The partial derivatives $\mathbf{U}_i \equiv \partial \mathbf{r}/\partial u_i$ are the changes of **r** with respect to u_i, with the remaining u_j ($j \neq i$) held fixed. In other words, $\mathbf{U}_i$ is the tangent vector to the coordinate curve of the variable u_i.

Example 1.6. In rectangular coordinates, $u_1 = x$, $u_2 = y$ and $u_3 = z$, and the position vector **r** as a function of the parameter s is

$$\mathbf{r}(s) = x(s)\,\mathbf{i} + y(s)\,\mathbf{j} + z(s)\,\mathbf{k} \tag{1.46}$$

The tangent vector in Equation (1.45) is then given by

$$\frac{d\mathbf{r}}{ds} = \frac{\partial \mathbf{r}}{\partial x}\frac{dx}{ds} + \frac{\partial \mathbf{r}}{\partial y}\frac{dy}{ds} + \frac{\partial \mathbf{r}}{\partial z}\frac{dz}{ds} \tag{1.47}$$

By comparing Equations (1.45) with (1.47) the $\mathbf{U}_i$ are identified as

$$\mathbf{U}_1 = \frac{\partial \mathbf{r}}{\partial x}, \quad \mathbf{U}_2 = \frac{\partial \mathbf{r}}{\partial y}, \quad \mathbf{U}_3 = \frac{\partial \mathbf{r}}{\partial z} \tag{1.48}$$

which, from (1.44), allows these vectors to be equated to the unit vectors along the x, y and z directions,

$$\mathbf{U}_1 = \mathbf{i}, \quad \mathbf{U}_2 = \mathbf{j}, \quad \mathbf{U}_3 = \mathbf{k} \tag{1.49}$$

Thus, Equation (1.47) reduces to

$$\frac{d\mathbf{r}}{ds} = \frac{dx}{ds}\mathbf{i} + \frac{dy}{ds}\mathbf{j} + \frac{dz}{ds}\mathbf{k} \tag{1.50}$$

as expected from Equation (1.44). The coordinate curves of a rectangular coordinate system are simply straight lines parallel to the three coordinate axes.

In spherical polar coordinates, the position vector $\mathbf{r}$ is obtained by using the expressions in (1.41):

$$\mathbf{r} = r\sin\theta\cos\phi\,\mathbf{i} + r\sin\theta\sin\phi\,\mathbf{j} + r\cos\theta\,\mathbf{k} \tag{1.51}$$

With the identifications $u_1 = r$, $u_2 = \theta$ and $u_3 = \phi$, we obtain

$$\begin{aligned}
\mathbf{U}_r &= \frac{\partial \mathbf{r}}{\partial r} = \sin\theta\cos\phi\,\mathbf{i} + \sin\theta\sin\phi\,\mathbf{j} + \cos\theta\,\mathbf{k} \\
\mathbf{U}_\theta &= \frac{\partial \mathbf{r}}{\partial \theta} = r\cos\theta\cos\phi\,\mathbf{i} + r\cos\theta\sin\phi\,\mathbf{j} - r\sin\theta\,\mathbf{k} \\
\mathbf{U}_\phi &= \frac{\partial \mathbf{r}}{\partial \phi} = -r\sin\theta\sin\phi\,\mathbf{i} + r\sin\theta\cos\phi\,\mathbf{j}
\end{aligned} \tag{1.52}$$

Taking pair-wise inner products shows that the $\mathbf{U}_i$ are mutually orthogonal:

$$\mathbf{U}_r \cdot \mathbf{U}_\theta = 0, \quad \mathbf{U}_r \cdot \mathbf{U}_\phi = 0, \quad \mathbf{U}_\theta \cdot \mathbf{U}_\phi = 0 \tag{1.53}$$

We also have

$$|\mathbf{U}_r|^2 = 1, \quad |\mathbf{U}_\theta|^2 = r^2, \quad |\mathbf{U}_\phi|^2 = r^2\sin^2\theta \tag{1.54}$$

which shows that the length of each of the $\mathbf{U}_i$ is a strong function of position. In particular, the $\mathbf{U}_i$ in this coordinate system are not unit vectors. ∎

If, as in the preceding example, the coordinate curves of a particular coordinate system are orthogonal, then the tangent vectors to these curves are also orthogonal:

$$\mathbf{U}_i \cdot \mathbf{U}_j = 0, \quad \text{if } i \neq j \tag{1.55}$$

Thus, if we introduce the quantities h_i as the squared magnitude of the $\mathbf{U}_i$:

$$h_i^2 = \mathbf{U}_i \cdot \mathbf{U}_i \tag{1.56}$$

for $i = 1, 2$ and 3, then the corresponding unit tangent vectors $\hat{\mathbf{u}}_i$ to the coordinate curves are given by

$$\mathbf{U}_i = h_i \hat{\mathbf{u}}_i \tag{1.57}$$

Since the $\hat{u}_i$ are mutually orthogonal and are normalized to unity, they are termed **orthonormal**. The quantities h_i are called **scale factors** and are the central quantities used for transforming scalar and vector relations between two orthogonal coordinate systems.

Consider, as an example, the differential volume element in rectangular coordinates, $dV = dx\, dy\, dz$. In the coordinate system (u_1, u_2, u_3) this can be expressed in terms of the Jacobian in (1.39) as $dV = J\, du_1\, du_2\, du_3$. Comparing (1.39) with the definition of the $\mathbf{U}_i$ in (1.45) shows that the Jacobian can be written as $J = |\mathbf{U}_1 \cdot \mathbf{U}_2 \times \mathbf{U}_3|$. Thus, in terms of the unit vectors $\hat{u}_i$, we obtain

$$\begin{aligned} J &= |\mathbf{U}_1 \cdot \mathbf{U}_2 \times \mathbf{U}_3| \\ &= |\hat{u}_1 \cdot \hat{u}_2 \times \hat{u}_3| h_1 h_2 h_3 \\ &= h_1 h_2 h_3 \end{aligned} \qquad (1.58)$$

since the $\hat{u}_i$ are an orthonormal set of vectors. The relation $J = h_1 h_2 h_3$ can be verified immediately for spherical polar coordinates by comparing (1.43) and (1.52).

To determine derivative operations in the u_i coordinate system, we begin with the differential, df, of a scalar function f. In rectangular coordinates, this quantity can be written as the inner product of the gradient of f and the differential position vector:

$$df = \frac{\partial f}{\partial x}\, dx + \frac{\partial f}{\partial y}\, dy + \frac{\partial f}{\partial z}\, dz = \nabla f \cdot d\mathbf{r} \qquad (1.59)$$

Since the differentials of f and of the position vector $\mathbf{r}$ can be calculated in a given coordinate system, an expression for the gradient of f in this coordinate system can be obtained from (1.59). Thus, in the (u_1, u_2, u_3) coordinate system, with f viewed as a function of the u_i, the chain rule yields

$$df = \frac{\partial f}{\partial u_1}\, du_1 + \frac{\partial f}{\partial u_2}\, du_2 + \frac{\partial f}{\partial u_3}\, du_3 \qquad (1.60)$$

Equation (1.45) can be used to obtain an expression for the differential of $\mathbf{r}$:

$$\begin{aligned} d\mathbf{r} &= \mathbf{U}_1 du_1 + \mathbf{U}_2 du_2 + \mathbf{U}_3 du_3 \\ &= h_1 \hat{u}_1 du_1 + h_2 \hat{u}_2 du_2 + h_3 \hat{u}_3 du_3 \end{aligned} \qquad (1.61)$$

We now write the gradient of f in the u_i coordinate system as

$$\nabla f = F_1 \hat{u}_1 + F_2 \hat{u}_2 + F_3 \hat{u}_3 \qquad (1.62)$$

where the F_i are functions of the u_i and are to be determined. Taking the inner product of (1.61) and (1.62) yields $\nabla f \cdot \mathrm{d}\mathbf{r}$ in the u_i coordinate system:

$$\nabla f \cdot \mathrm{d}\mathbf{r} = h_1 F_1 du_1 + h_2 F_2 du_2 + h_3 F_3 du_3 \tag{1.63}$$

We now use the relation $\mathrm{d}f = \nabla f \cdot \mathrm{d}\mathbf{r}$ to equate (1.60) and (1.63):

$$\frac{\partial f}{\partial u_1} du_1 + \frac{\partial f}{\partial u_2} du_2 + \frac{\partial f}{\partial u_3} du_3 = h_1 F_1 \, du_1 + h_2 F_2 \, du_2 + h_3 F_3 \, du_3 \tag{1.64}$$

Since the u_i are independent variables, we may equate the coefficients of each of the quantities du_i on the two sides of the equation to obtain the following expressions for the F_i:

$$F_i = \frac{1}{h_i} \frac{\partial f}{\partial u_i} \tag{1.65}$$

for $i = 1, 2, 3$. Substituting (1.65) into (1.62), the gradient of a scalar function f in the u_i coordinate system is obtained as

$$\nabla f = \frac{1}{h_1} \frac{\partial f}{\partial u_1} \hat{\mathbf{u}}_1 + \frac{1}{h_2} \frac{\partial f}{\partial u_2} \hat{\mathbf{u}}_2 + \frac{1}{h_3} \frac{\partial f}{\partial u_3} \hat{\mathbf{u}}_3 \tag{1.66}$$

Another vector operation frequently encountered is $\nabla^2 f$, known as the **Laplacian** of a function f. The behavior of this quantity under coordinate transformations is especially important for expressing many second-order equations in different coordinate systems. Included are Laplace's equation, the diffusion equation, the wave equation and Schrödinger's equation. To determine the form of this operator acting on a scalar function in a general set of orthogonal curvilinear coordinates, we begin with the relation $\nabla^2 f = \nabla \cdot \nabla f$ and use (1.66) to write

$$\begin{aligned}\nabla^2 f &= \nabla \cdot \left[\frac{1}{h_1} \frac{\partial f}{\partial u_1} \hat{\mathbf{u}}_1 + \frac{1}{h_2} \frac{\partial f}{\partial u_2} \hat{\mathbf{u}}_2 + \frac{1}{h_3} \frac{\partial f}{\partial u_3} \hat{\mathbf{u}}_3 \right] \\ &= \frac{\hat{\mathbf{u}}_1}{h_1} \cdot \frac{\partial}{\partial u_1}\left(\frac{\hat{\mathbf{u}}_1}{h_1} \frac{\partial f}{\partial u_1}\right) + \frac{\hat{\mathbf{u}}_2}{h_2} \cdot \frac{\partial}{\partial u_2}\left(\frac{\hat{\mathbf{u}}_2}{h_2} \frac{\partial f}{\partial u_2}\right) + \frac{\hat{\mathbf{u}}_3}{h_3} \cdot \frac{\partial}{\partial u_3}\left(\frac{\hat{\mathbf{u}}_3}{h_3} \frac{\partial f}{\partial u_3}\right)\end{aligned} \tag{1.67}$$

To proceed further, we observe that Equation (1.66) implies that the gradient of each of the coordinates is given by $\nabla u_i = \hat{\mathbf{u}}_i / h_i$. Since the curl of a gradient must vanish, we have

$$\nabla \times \frac{\hat{\mathbf{u}}_i}{h_i} = \nabla \times \nabla u_i = 0 \tag{1.68}$$

Furthermore, since $\hat{u}_1$, $\hat{u}_2$ and $\hat{u}_3$ are orthonormal vectors, we can label the coordinates such that $\hat{u}_i = \hat{u}_j \times \hat{u}_k$ for any cyclic permutation of i, j and k. Thus, we can write

$$\frac{\hat{u}_1}{h_2 h_3} = \frac{\hat{u}_2}{h_2} \times \frac{\hat{u}_3}{h_3}, \quad \frac{\hat{u}_2}{h_3 h_1} = \frac{\hat{u}_3}{h_3} \times \frac{\hat{u}_1}{h_1}, \quad \frac{\hat{u}_3}{h_1 h_2} = \frac{\hat{u}_1}{h_1} \times \frac{\hat{u}_2}{h_2} \quad (1.69)$$

We can then use Equation (1.68) to show that the divergence of each of these quantities vanishes:

$$\nabla \cdot \frac{\hat{u}_i}{h_j h_k} = \nabla \cdot \frac{\hat{u}_j}{h_j} \times \frac{\hat{u}_k}{h_k}$$

$$= \left(\frac{\hat{u}_k}{h_k} \cdot \nabla \times \frac{\hat{u}_j}{h_j} \right) - \left(\frac{\hat{u}_j}{h_j} \cdot \nabla \times \frac{\hat{u}_k}{h_k} \right) = 0 \quad (1.70)$$

The subscripts i, j and k are chosen to be any cyclic permutation of 1, 2 and 3, as in Equation (1.69). Using this result, the Laplacian of a scalar function f in orthogonal curvilinear coordinates can be written as

$$\nabla^2 f = \frac{\hat{u}_1}{h_1} \cdot \frac{\partial}{\partial u_1} \left(\frac{\hat{u}_1}{h_2 h_3} \frac{h_2 h_3}{h_1} \frac{\partial f}{\partial u_1} \right) + \frac{\hat{u}_2}{h_2} \cdot \frac{\partial}{\partial u_2} \left(\frac{\hat{u}_2}{h_1 h_3} \frac{h_1 h_3}{h_2} \frac{\partial f}{\partial u_2} \right)$$

$$+ \frac{\hat{u}_3}{h_3} \cdot \frac{\partial}{\partial u_3} \left(\frac{\hat{u}_3}{h_1 h_2} \frac{h_1 h_2}{h_3} \frac{\partial f}{\partial u_3} \right)$$

$$= \frac{1}{h_1 h_2 h_3} \left[\frac{\partial}{\partial u_1} \left(\frac{h_2 h_3}{h_1} \frac{\partial f}{\partial u_1} \right) + \frac{\partial}{\partial u_2} \left(\frac{h_1 h_3}{h_2} \frac{\partial f}{\partial u_2} \right) \right.$$

$$\left. + \frac{\partial}{\partial u_3} \left(\frac{h_1 h_2}{h_3} \frac{\partial f}{\partial u_3} \right) \right] \quad (1.71)$$

This general expression, and those given in Equation (1.66) and in Problems 11 and 12, reduce the determination of vector operations in any orthogonal curvilinear coordinate system simply to a matter of calculating the scale factors in that coordinate system.

Example 1.7. In spherical polar coordinates, the scale factors h_i are determined from (1.56) to be

$$h_r = 1, \quad h_\theta = r, \quad h_\phi = r \sin \theta \quad (1.72)$$

The unit vectors corresponding to (1.52) are therefore given by

$$\hat{u}_r = \sin \theta \cos \phi \, \mathbf{i} + \sin \theta \sin \phi \, \mathbf{j} + \cos \theta \, \mathbf{k}$$
$$\hat{u}_\theta = \cos \theta \cos \phi \, \mathbf{i} + \cos \theta \sin \phi \, \mathbf{j} - \sin \theta \, \mathbf{k} \quad (1.73)$$
$$\hat{u}_\phi = -\sin \phi \, \mathbf{i} + \cos \phi \, \mathbf{j}$$

From (1.53), these vectors are readily seen to form an orthonormal set. The gradient of a scalar function $f(r,\theta,\phi)$ is obtained by substituting the scale factors in (1.72) into the general expression in (1.66):

$$\nabla f = \frac{\partial f}{\partial r}\hat{u}_r + \frac{1}{r}\frac{\partial f}{\partial \theta}\hat{u}_\theta + \frac{1}{r\sin\theta}\frac{\partial f}{\partial \phi}\hat{u}_\phi \qquad (1.74)$$

The Laplacian of $f(r,\theta,\phi)$ is similarly obtained from Equation (1.71) as

$$\nabla^2 f = \frac{1}{r^2}\frac{\partial}{\partial r}\left(r^2\frac{\partial f}{\partial r}\right) + \frac{1}{r^2\sin\theta}\frac{\partial}{\partial \theta}\left(\sin\theta\frac{\partial f}{\partial \theta}\right) + \frac{1}{r^2\sin^2\theta}\frac{\partial^2 f}{\partial \phi^2} \qquad (1.75)$$

The results obtained in this section can be used to write a few simple commands in *Mathematica* to determine various vector operations for any orthogonal curvilinear coordinate system. The package `VectorAnalysis.m` could also be used to obtain these quantities directly, but the explicit constructions carried out below are a useful illustration of the interplay between scalar and vector quantities in *Mathematica*.

We consider the spherical polar coordinates as an example. We first define the vectors u and h, whose entries `u[[i]]` and `h[[i]]` are the coordinates, u_i, and the scale factors, h_i, respectively:

`u:= {r,theta,phi}`

`h:= {1,r,r Sin[theta]}`

The entries of h must be in the same order as those of u, i.e. `h[[1]]`=h_r, `h[[2]]`=h_θ and `h[[3]]`=h_ϕ. We then calculate the Jacobian using Equation (1.58) and construct the expressions for, say, the gradient of a scalar function in Equation (1.66) and the divergence of a vector function obtained in Problem 12:

`J=h[[1]]h[[2]]h[[3]];`

`Grad[f_]:=Table[D[f,u[[i]]]/h[[i]],{i,1,3}]`

`Div[f_]:=(1/J)Sum[D[(J f[[i]])/h[[i]],u[[i]]],{i,1,3}]`

Notice that `Grad` returns a vector function, as indicated by the construction of a `Table`, and that `Div` operates on vector functions with components given by `f[[i]]`. The expression for the gradient of a function in spherical polar coordinates is then obtained as

```
Grad[f[r,theta,phi]]
```

$$\left\{ f^{(1,0,0)}[r, theta, phi], \frac{f^{(0,1,0)}[r, theta, phi]}{r}, \frac{f^{(0,0,1)}[r, theta, phi]}{r\,\mathrm{Sin}[theta]} \right\}$$

which is the expression obtained in (1.74).

We can also combine Div and Grad to obtain the Laplacian of a scalar function f by taking the divergence of the gradient of f:

```
Expand[Div[Grad[f[r,theta,phi]]]]
```

$$\frac{f^{(0,0,2)}[r, theta, phi]}{r^2\,\mathrm{Sin}[theta]^2} + \frac{\mathrm{Cos}[theta]\, f^{(0,1,0)}[r, theta, phi]}{r^2\,\mathrm{Sin}[theta]} + \frac{f^{(0,2,0)}[r, theta, phi]}{r^2} + \frac{2\, f^{(1,0,0)}[r, theta, phi]}{r} + f^{(2,0,0)}[r, theta, phi]$$

which agrees with (1.75) when the derivatives are carried out.

Further Reading

An introduction to the theory of ordinary differential equations is given by Boyce and DiPrima (1977), and a more advanced treatment may be found in the book by Ince (1956). Morse and Feshbach (1953) provide a comprehensive treatment of many aspects of ordinary and partial differential equations, including detailed discussions of the mathematical machinery needed to develop the methods of solution. Similarly, Hildebrand (1962) and Whittaker and Watson (1963) provide general texts in which many of the mathematical topics required for solving ordinary and partial differential equations are developed.

The most complete reference to *Mathematica* is the book by Wolfram (1988), which serves as both a guide and a users' manual. Other sources that are useful in gaining familiarity with various aspects of *Mathematica* are the book by Maeder (1990), written by one of the original developers of the system, which is a guide to programming in *Mathematica* and the book by Blachman (1992), which provides an introduction to *Mathematica* that assumes little prior knowledge. The book by Crandall (1991) provides a broad range of applications of *Mathematica* in physics, mathematics and related areas. Crandall (1991) also includes an introduction to many of the graphical, symbolic and programming features of *Mathematica*.

References

Ablowitz M. J. and Clarkson P. A. (1991). *Solitons, Nonlinear Evolution Equations and Inverse Scattering*. Cambridge: Cambridge University Press.

Ames W. F. (1965). *Nonlinear Partial Differential Equations in Engineering* Vol. 1. New York: Academic.

Ames W. F. (1972). *Nonlinear Partial Differential Equations in Engineering* Vol. 2. New York: Academic.

Blachman N. (1992). *Mathematica: A Practical Approach*. Englewood Cliffs NJ: Prentice-Hall.

Boyce W. E. and DiPrima R. C. (1977). *Elementary Differential Equations and Boundary Value Problems* 3rd edn. New York: Wiley.

Coddington E. A. and Levinson N. (1955). *Theory of Ordinary Differential Equations*. New York: McGraw-Hill.

Crandall R. E. (1991). *Mathematica for the Sciences*. Redwood City, CA: Addison-Wesley.

Garabedian P. R. (1964). *Partial Differential Equations*. New York: Wiley.

Hildebrand F. B. (1962). *Advanced Calculus for Applications*. Englewood Cliffs NJ: Prentice-Hall.

Ince E. L. (1956). *Ordinary Differential Equations*. New York: Dover.

Knopp K. (1956). *Infinite Sequences and Series*. New York: Dover.
Maeder R. (1990). *Programming in Mathematica*. Redwood City, CA: Addison-Wesley.
Morse P. M. and Feshbach H. (1953). *Methods of Theoretical Physics* Vol. 1. New York: McGraw-Hill.
Whittaker E. T. and Watson G. N. (1963). *A Course of Modern Analysis* 4th edn. Cambridge: Cambridge University Press.
Wolfram S. (1988). *Mathematica: A System for Doing Mathematics by Computer*. Redwood City, CA: Addison-Wesley.

Problems

1. For each set of functions below, calculate the Wronskian to determine whether or not the functions are linearly independent:

(a) $1, x, x^2$

(b) $\sin x, \sin 2x, \sin 3x$

(c) $\sin x, \cos x, e^{ix}$

(d) $\sin x, \cos x, e^{2ix}$

(e) $\sin x, \sinh x, e^x$

(f) $\sin x, \sinh x, e^{ix}$

2. Extend the *Mathematica* commands for the evaluation of the Wronskian described in Example 1.3 to the case where there are n functions, where n is assigned as an argument.

3. Using the construction of the Wronskian in Problem 2, test for the linear independence of the set of functions

$$1, x, x^2, \ldots, x^n$$

for different (integer) values of n. Is there any point where these functions are not linearly independent? Suppose that the function

$$\frac{1-x^m}{1-x}$$

where m is an integer, is included in this set. Calculate the Wronskian for $m \leq n$ and $m > n$.

4. Determine the radius of convergence of each of the following series. If the radius of convergence is finite, determine the behavior of the series at the end-points.

(a) $y(x) = \sum_{k=0}^{\infty} \frac{x^k}{k!}$

(b) $y(x) = \sum_{k=0}^{\infty} \frac{x^k}{(k!)^2}$

(c) $y(x) = \sum_{k=0}^{\infty} k! x^k$

(d) $y(x) = \sum_{k=0}^{\infty} \dfrac{x^{2k}}{2k-1}$

5. Use *Mathematica* to define a function that calculates the partial sums for each of the series in Problem 4. Then use this function to examine the convergence of these series by listing sequences with increasing values of n. For any of the series that have a finite radius of convergence, examine the convergence for values of x on either side of the radius of convergence.

6. Use **Series** to generate the first several terms of the Taylor series about $x=0$ for each of the following pairs of functions. Then multiply these series together and compare the terms obtained with those calculated directly from the product function.

(a) $\sin x$ and $\cos x$

(b) $\sin x$ and $\log(1+x)$

(c) e^x and $\tanh x$

(d) $\log(1+x)$ and $\log(1+x)$

(e) $\sin x$ and $1/\cos x$

(f) $\sinh x$ and $1/\cosh x$

7. If a function $f(x)$ has a power series, then the power series for the inverse of that function, $f^{-1}(x)$, is called the **inverse** of the power series for $f(x)$. By using **Series** and **InverseSeries** show that the power series of the following pairs of functions are inverses of one another.

(a) $\log x$ and e^x

(b) $\sin x$ and $\arcsin x$

(c) $\sinh x$ and $\text{arcsinh } x$

8. Determine the scale factors for each of the coordinate systems listed below:

(a) **Cylindrical polar** coordinates r, ϕ, z:

$$x = r\cos\phi, \quad y = r\sin\phi, \quad z = z$$

(b) **Parabolic** coordinates ξ, η, ϕ:

$$x = \xi\eta\cos\phi, \quad y = \xi\eta\sin\phi, \quad z = \tfrac{1}{2}(\xi^2 - \eta^2)$$

(c) **Parabolic cylinder** coordinates ξ, η, z:
$$x = \tfrac{1}{2}(\xi^2 - \eta^2), \quad y = \xi\eta, \quad z = z$$

(d) **Elliptic cylinder** coordinates ξ, η, z:
$$x = a\cosh\xi\cos\eta, \quad y = a\sinh\xi\sin\eta, \quad z = z$$

(e) **Prolate spheroidal** coordinates ξ, η, ϕ:
$$x = a\sqrt{\xi^2 - 1}\sqrt{1 - \eta^2}\cos\phi, \quad y = a\sqrt{\xi^2 - 1}\sqrt{1 - \eta^2}\sin\phi, \quad z = a\xi\eta$$

(f) **Oblate spheroidal** coordinates ξ, η, ϕ:
$$x = a\sqrt{\xi^2 + 1}\sqrt{1 - \eta^2}\cos\phi, \quad y = a\sqrt{\xi^2 + 1}\sqrt{1 - \eta^2}\sin\phi, \quad z = a\xi\eta$$

9. Use the *Mathematica* command `ParametricPlot3D` to display the coordinate systems in Problem 8 in terms of surfaces where one of the coordinates is held fixed while the other two are varied. Two examples are shown in Figure 1.1 for cylindrical polar coordinates (Problem 8(a)) and parabolic cylinder coordinates (Problem 8(c)). The individual coordinate curves can be constructed with `SpaceCurve`.

10. Verify that the transformation matrix A between the unit vectors $\mathbf{i}$, $\mathbf{j}$ and $\mathbf{k}$ of the rectangular coordinate system and unit vectors $\mathbf{u}_r$, $\mathbf{u}_\theta$ and $\mathbf{u}_\phi$ of the spherical polar coordinate system is an **orthogonal** matrix, i.e. $A^{-1} = A^{\mathrm{T}}$, where 'T' indicates the transpose of the matrix. Show that the determinant of A is unity and that the rows and columns form mutually orthogonal unit vectors.

Consider two vectors $\mathbf{a}$ and $\mathbf{b}$ in the rectangular coordinates and the same vectors $\mathbf{a}'$ and $\mathbf{b}'$ in the spherical coordinate system. Show explicitly and from the properties of orthogonal matrices that the inner product of these vectors is the same in the two coordinate systems:
$$\mathbf{a} \cdot \mathbf{b} = \mathbf{a}' \cdot \mathbf{b}'$$

11. Consider a vector function $\mathbf{F}$ in the coordinate system (u_1, u_2, u_3) introduced in Section 1.4:
$$\mathbf{F} = f_1\mathbf{u}_1 + f_2\mathbf{u}_2 + f_3\mathbf{u}_3$$
where each of the components f_i are functions of u_1, u_2 and u_3. The curl of $\mathbf{F}$ in this coordinate system is defined as
$$\nabla \times \mathbf{F} = \nabla \times (f_1\mathbf{u}_1 + f_2\mathbf{u}_2 + f_3\mathbf{u}_3)$$
$$= \nabla \times (f_1\mathbf{u}_1) + \nabla \times (f_2\mathbf{u}_2) + \nabla \times (f_3\mathbf{u}_3)$$

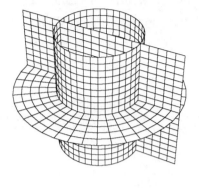

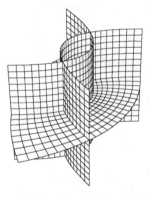

Figure 1.1 (Top) Representation of cylindrical polar coordinates using the command `ParamatricPlot3D` of *Mathematica*, showing the surfaces of constant r (circular cylinder), constant ϕ (rectangular plane) and constant z (circular plane). (Bottom) The corresponding representation showing the surfaces of constant ξ and η (parabolic cylinders) and constant z (plane).

Use Equation (1.68) to obtain the following expression for the curl of F in a curvilinear coordinate system:

$$\nabla \times \mathbf{F} = \frac{1}{h_1 h_2 h_3} \begin{vmatrix} h_1 \mathbf{u}_1 & h_2 \mathbf{u}_2 & h_3 \mathbf{u}_3 \\ \dfrac{\partial}{\partial u_1} & \dfrac{\partial}{\partial u_2} & \dfrac{\partial}{\partial u_3} \\ h_1 f_1 & h_2 f_2 & h_3 f_3 \end{vmatrix}$$

12. In the notation of Problem 11, the divergence of $\mathbf{F}$ in the coordinate system (u_1, u_2, u_3) is given by

$$\nabla \cdot \mathbf{F} = \nabla \cdot (f_1 \mathbf{u}_1 + f_2 \mathbf{u}_2 + f_3 \mathbf{u}_3)$$
$$= \nabla \cdot (f_1 \mathbf{u}_1) + \nabla \cdot (f_2 \mathbf{u}_2) + \nabla \cdot (f_3 \mathbf{u}_3)$$

Use Equation (1.68) to show that this expression can be written as

$$\nabla \cdot \mathbf{F} = \frac{1}{h_1 h_2 h_3}\left[\frac{\partial}{\partial u_1}(h_2 h_3 f_1) + \frac{\partial}{\partial u_2}(h_3 h_1 f_2) + \frac{\partial}{\partial u_3}(h_1 h_2 f_3)\right]$$

13. Using the *Mathematica* commands described in Example 1.7, obtain expressions for the gradient, the curl, the divergence and the Laplacian for each of the coordinate systems in Problem 8.

14. The behavior of equations under the transformation of variables is used in several places in this book. As an example consider the following equation

$$\frac{\partial^2 u}{\partial x^2} - \frac{\partial^2 u}{\partial t^2} = 0$$

which is the wave equation. The following sequence of *Mathematica* commands changes the variables in this equation from x and t to x1[x,t] and x2[x,t]:

```
Eqn=D[z[x1[x,t],x2[x,t]],{x,2}]
    -D[z[x1[x,t],x2[x,t]],{t,2}]==0;
Simplify[Eqn/.
    x->x/.Solve[{x1==x1[x,t],x2==x2[x,t]},{x,t}][[1]]/.
    t->t/.Solve[{x1==x1[x,t],x2==x2[x,t]},{x,t}][[1]]/.
    x1[x,t]->x1/.x2[x,t]->x2]
```

Express the wave equation in the following coordinates and check your results with *Mathematica*.

(a) $x_1 = x e^t$, $x_2 = x e^{-t}$

(b) $x_1 = x + t$, $x_2 = x - t$

(c) $x_1 = t \cosh x$, $x_2 = t \sinh x$

(d) $x_1 = xy$, $x_2 = x/y$

15. For each coordinate transformation in Problem 14, calculate the Jacobian and identify any places where the transformation breaks down. How are the points where the transformation becomes ill-defined reflected in the transformed equation?

16. One of the origins of *ordinary* differential equations is through the elimination of arbitrary constants in an expression involving the dependent and independent variables. In general, such an expression involving n arbitrary constants will lead to an nth-order differential equation. By taking the required number of derivatives, eliminate the arbitrary constants (indicated by

A, B, and C) in each of the expressions below to obtain the ordinary differential equation for which the expression is a solution.

(a) $y = Ax^2 + Bx + C$

(b) $y = A\cos x + B\sin x$

(c) $y = A e^x + B e^{2x} + C e^{3x}$

(d) $x^2 y + y^2 x = A$

17. The elimination of arbitrary constants from expressions can be carried out in *Mathematica* using the steps described below. For example, the following sequence of commands eliminates the quantity A from the implicit expression for y given by $x^2 y - xy^2 = A$:

```
F[x,y]=x^2y[x]+(y[x])^2x-A;
Eqn1=      F[x,y]==0/.y[x]->y;
Eqn2= D[F[x,y],x]==0/.y[x]->y/.y'[x]->yx;
Eliminate[{Eqn1,Eqn2},{A}]
```

Mathematica returns the differential satisfied by $y(x)$:

```
yx x (2 y + x) == y (-y - 2 x)
```

Two points should be noted in these commands. The first is that the x-dependence of y is declared explicitly in the defining expression. The second is that once the derivatives of this expression have been taken, the functions y(x) and y'(x) are replaced by ordinary variables, signified by y and yx, respectively, since their role as functions is no longer required. This replacement considerably widens the scope of expressions to which this procedure can be applied.

Eliminate the constants in the following expressions to obtain the differential equations satisfied by y. The constants are signified by A, B and C, as above.

(a) $x^2 y - xy^4 = A$

(b) $A + Bxy^2 = xy^{-1}$

(c) $y^4 = Ax + By + Cxy$

18. Some elementary properties of curves in two-dimensional and three-dimensional spaces will be required in Chapters 2 and 3. A two-dimensional curve maps part of the real line into a two-dimensional space. Such a curve is an example of a **plane curve**, since it lies within a plane. The general form of this curve is

$$C(t) = (f_1(t), f_2(t))$$

where the coordinate functions f_1 and f_2 are differentiable. Use **Parametric-Plot** to plot each of the following curves for a range of values of t.

(a) $(\cos t, \sin t)$

(b) $(t \cos t, t \sin t)$

(c) $(t^{-1} \cos t, t^{-1} \sin t)$

(d) $(\log t, \sin t)$

19. A three-dimensional **space curve** is defined as a differentiable function that maps a part of the real line into a three-dimensional space. The general form of a space curve can then be written as

$$C(t) = (f_1(t), f_2(t), f_3(t))$$

where the coordinate functions f_1, f_2 and f_3 are differentiable functions. Use `SpaceCurve` to plot each of the following curves over a range of values of t.

(a) (t^n, t, t), $\quad n = 1, 2, 3$

(b) (t^2, t, t^3)

(c) $(\sin t, \cos t, t)$

(d) $(\cosh t, \sinh t, t)$

20. A **surface** can be represented either as a function of two independent variables x and y or, by generalizing the representation in Problem 19, as

$$S(s,t) = (f_1(s,t), f_2(s,t), f_3(s,t))$$

Use both `Plot3D` and `ParametricPlot3D` to visualize the following surfaces.

(a) $(s, t, s - t)$

(b) $(t + 1, s, s - t)$

(c) $(s + t, s - t, s^2 + t^2)$

(d) $(s^2 - t^2, 2st, s^2 + t^2)$

(e) $(s + t, s - t, st)$

(f) $(s, t, t^3 - 3s^2 t)$

Chapter 2

First-Order Partial Differential Equations

There are many applications in physics, engineering and applied mathematics that involve the formulation and solution of first-order partial differential equations. From a mathematical point of view, first-order equations have the advantage of providing a conceptual basis that can be used for higher-order equations, but with the attractive feature that there are methods of solution which are particularly suited to first-order equations. In this chapter, we will obtain the solution of first-order partial differential equations using Lagrange's method of characteristics and its generalizations. We will then turn our attention to finding solutions for specified boundary conditions and to the wider issue of identifying appropriate boundary conditions for a particular differential equation.

We will first consider partial differential equations of the form

$$P(x,y,u)\frac{\partial u}{\partial x} + Q(x,y,u)\frac{\partial u}{\partial y} = R(x,y,u) \qquad (2.1)$$

for a function u of two independent variables x and y. Lagrange's method can be applied to an equation of this form involving any number of independent variables, but the use of only two independent variables allows the geometric features of the method to be readily visualized in three spatial dimensions. The coefficients P, Q and R in general are functions of x, y and u, but not of any derivatives of u. Equations of the type in (2.1) are called **quasi-linear**. *Linear* equations are a special type of quasi-linear equation for which P and Q are functions only of x and y and R contains terms only to linear order in u. Thus, the general linear first-order equation is

$$P(x,y)\frac{\partial u}{\partial x} + Q(x,y)\frac{\partial u}{\partial y} = R_0(x,y) + R_1(x,y)u \qquad (2.2)$$

The form (2.1) allows nonlinear terms involving only products of u with single factors of derivatives, e.g. $u^2 u_x$, $u^2 u_y$ and u^2. An equation containing nonlinear terms in the partial derivatives, e.g. u_x^2 or $u_x u_y$, is called **nonlinear**. The presence of these latter type of nonlinearity introduces qualitative differences into the Lagrange method as compared to the quasi-linear case, which is the reason for the distinction between nonlinear and quasi-linear equations.

Once Lagrange's method has been illustrated for quasi-linear equations, we will apply the same philosophy to solving equations of the more general form

$$F(x,y,u,u_x,u_y) = 0 \qquad (2.3)$$

There is now no restriction placed on the type of nonlinear terms that are contained in this expression. It is a remarkable feature of first-order equations that a procedure can be developed for solving this equation, regardless of its complexity. Of course, putting the method into practice to obtain an explicit expression for the solution may not always be a straightforward task. However, the methodology does provide a systematic way of simplifying the problem by reducing the nonlinear equation to a system of coupled first-order *ordinary* differential equations.

2.1 The Method of Characteristics

The general quasi-linear equation in (2.1) can be written in an equivalent, but more useful, form in terms of the **solution surface** of (2.1):

$$f(x,y,u) = 0 \qquad (2.4)$$

Given the values of any two of the arguments of f, the value of the third argument is fixed. This yields the equation of a surface. Equation (2.4) is a solution surface of (2.1) if, when solved for u as a function of x and y and substituted into (2.1), an identity results. We will now obtain the differential equation satisfied by f by using the fact that u is a solution of (2.1).

We first take the differential of (2.4):

$$df = \frac{\partial f}{\partial x}dx + \frac{\partial f}{\partial y}dy + \frac{\partial f}{\partial u}du = 0 \qquad (2.5)$$

By regarding u as a function of the independent variables x and y, we obtain

$$\begin{aligned} df &= \frac{\partial f}{\partial x}\,dx + \frac{\partial f}{\partial y}\,dy + \frac{\partial f}{\partial u}\left[\frac{\partial u}{\partial x}\,dx + \frac{\partial u}{\partial y}\,dy\right] \\ &= \left[\frac{\partial f}{\partial x} + \frac{\partial f}{\partial u}\frac{\partial u}{\partial x}\right]dx + \left[\frac{\partial f}{\partial y} + \frac{\partial f}{\partial u}\frac{\partial u}{\partial y}\right]dy = 0 \end{aligned} \quad (2.6)$$

Equation (2.6) could also have been obtained directly by applying the chain rule to calculate the differential of the expression $f[x,y,u(x,y)]=0$. Since x and y are independent variables, the coefficients of dx and dy must separately vanish if (2.6) is to be satisfied. This yields the two equations

$$\frac{\partial f}{\partial x} + \frac{\partial f}{\partial u}\frac{\partial u}{\partial x} = 0, \qquad \frac{\partial f}{\partial y} + \frac{\partial f}{\partial u}\frac{\partial u}{\partial y} = 0 \qquad (2.7)$$

Since, by hypothesis, $\partial f/\partial u \neq 0$, the two equations in (2.7) may be solved for the two partial derivatives of u:

$$\frac{\partial u}{\partial x} = -\frac{\partial f/\partial x}{\partial f/\partial u}, \qquad \frac{\partial u}{\partial y} = -\frac{\partial f/\partial y}{\partial f/\partial u} \qquad (2.8)$$

Substituting these expressions for u_x and u_y into (2.1), we obtain after some simple rearrangement the differential equation for f:

$$P(x,y,u)\frac{\partial f}{\partial x} + Q(x,y,u)\frac{\partial f}{\partial y} + R(x,y,u)\frac{\partial f}{\partial u} = 0 \qquad (2.9)$$

This is nothing other than the original differential equation in (2.1) but written in terms of the solution surface f rather than the function u.

The advantage of rewriting Equation (2.1) in the form (2.9) stems from the fact that the latter equation admits a simple and useful geometric interpretation. Recall that the gradient of a function f is given by the expression

$$\nabla f = \frac{\partial f}{\partial x}\mathbf{i} + \frac{\partial f}{\partial y}\mathbf{j} + \frac{\partial f}{\partial u}\mathbf{k} \qquad (2.10)$$

where $\mathbf{i}$, $\mathbf{j}$ and $\mathbf{k}$ are the usual unit vectors along the x, y and u directions, respectively. Consider now the vector $\mathbf{V}$ defined by

$$\mathbf{V}(x,y,u) \equiv P(x,y,u)\mathbf{i} + Q(x,y,u)\mathbf{j} + R(x,y,u)\mathbf{k} \qquad (2.11)$$

Since P, Q and R are functions of x, y and u, Equation (2.11) associates a vector $V(x,y,u)$ with every point (x,y,u). For this reason, $\mathbf{V}$ is called a **vector field**. Thus, the gradient of f is also a vector field; every point is assigned to a vector which is the normal to the surface of constant f that passes through that point. By combining the definitions in Equations (2.10)

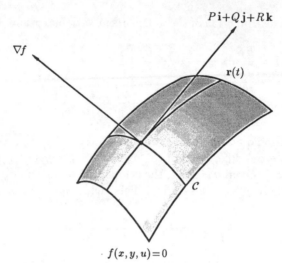

Figure 2.1 Schematic illustration of the geometric interpretation of Equation (2.12), showing a section of the solution surface $f(x, y, u) = 0$, the vector field $V = P\mathbf{i} + Q\mathbf{j} + R\mathbf{k}$, the gradient ∇f, and the field lines, or characteristics, $\mathbf{r}(t)$. Also shown for later reference is the curve $\mathcal{C}$ which the solution surface is required to contain. This is the boundary condition for the differential equation.

and (2.11), Equation (2.9) can be written in a more suggestive form as the scalar product between the vectors $\mathbf{V}$ and ∇f:

$$\mathbf{V} \cdot \nabla f = 0 \qquad (2.12)$$

Since ∇f is a vector field normal to surfaces of constant f, (2.12) states that $\mathbf{V}$ is everywhere perpendicular to ∇f. Therefore, $\mathbf{V}$ is a vector field that is tangent to the surfaces of constant f at every point and, in particular, is a tangent vector field to the solution surface $f = 0$ (Figure 2.1). This suggests a representation of the solution surface of (2.1) in terms of the **integral curves** of $\mathbf{V}$, i.e. the curves whose tangents are parallel to $\mathbf{V}$. These integral curves are the **characteristics** of (2.1).

The geometric interpretation of (2.12) depicted in Figure 2.1 can be used to obtain a prescription for finding the general solution of (2.1) by constructing the characteristics. Let $\mathbf{r}$ denote the position vector of a point in the three-dimensional space (x, y, u). In terms of Cartesian components x, y and u the position vector is given by

$$\mathbf{r} = x\mathbf{i} + y\mathbf{j} + u\mathbf{k} \qquad (2.13)$$

Suppose now that x, y and u are regarded as functions of a parameter t. Then $\mathbf{r}$ also becomes a function of t,

$$\mathbf{r}(t) = x(t)\mathbf{i} + y(t)\mathbf{j} + u(t)\mathbf{k} \qquad (2.14)$$

As t is varied continuously, $\mathbf{r}(t)$ sweeps out a curve in space. The tangent vectors to this curve are obtained by taking the derivative of (2.14) with respect to t:

$$\frac{d\mathbf{r}}{dt} = \frac{dx}{dt}\mathbf{i} + \frac{dy}{dt}\mathbf{j} + \frac{du}{dt}\mathbf{k} \tag{2.15}$$

If the parameter t corresponds to the time, then this tangent vector corresponds to the instantaneous velocity along the curve at the point $\mathbf{r}(t)$.

Suppose now that we require $\mathbf{r}(s)$ to correspond to an integral curve of the vector field $\mathbf{V}$. This can be insured by setting the tangent vectors in (2.15) equal to (2.11):

$$\frac{d\mathbf{r}}{dt} = \frac{dx}{dt}\mathbf{i} + \frac{dy}{dt}\mathbf{j} + \frac{du}{dt}\mathbf{k} = P\mathbf{i} + Q\mathbf{j} + R\mathbf{k} \tag{2.16}$$

This equation may be regarded as the *definition* of the integral curves of $\mathbf{V}$; clearly, the equality need not hold for *every* curve. By equating the individual vector components in (2.16), the equations for the Cartesian components of the characteristics are obtained in terms of the functions P, Q and R of the partial differential equation:

$$\frac{dx}{dt} = P, \qquad \frac{dy}{dt} = Q, \qquad \frac{du}{dt} = R \tag{2.17}$$

We can use these equations to write (2.9) as

$$\frac{\partial f}{\partial x}\frac{dx}{dt} + \frac{\partial f}{\partial y}\frac{dy}{dt} + \frac{\partial f}{\partial u}\frac{du}{dt} = 0 \tag{2.18}$$

Then, recognizing the left-hand side of this equation as the total derivative of u with respect to t, the differential equation (2.9) for f can be expressed in characteristic coordinates simply as

$$\frac{df}{dt} = 0 \tag{2.19}$$

This provides the first illustration that the characteristics provide the most 'natural' coordinate system for quasi-linear first-order partial differential equations, in the sense that the equation is reduced to the simplest form.

There remains only the determination of explicit expressions for the characteristics. By eliminating the quantity dt in (2.17), the three equations can be combined into a single expression as

$$\frac{dx}{P} = \frac{dy}{Q} = \frac{du}{R} \tag{2.20}$$

The solutions to (2.20) are the characteristics, and are obtained by solving any two independent ordinary differential equations chosen from the three equations

$$\frac{dx}{P} = \frac{dy}{Q}, \qquad \frac{dx}{P} = \frac{du}{R}, \qquad \frac{dy}{Q} = \frac{du}{R} \tag{2.21}$$

or combinations of these equations. For example, the pair-wise equalities in (2.21) can be written in the form of three ordinary differential equations

$$\frac{dy}{dx} = \frac{Q}{P}, \quad \frac{du}{dx} = \frac{R}{P}, \quad \frac{du}{dy} = \frac{R}{Q} \tag{2.22}$$

Thus, we see that by reformulating (2.1) in terms of the integral curves of the vector field **V**, solving a *partial* differential equation has been reduced to solving two independent *ordinary* differential equations.

Each of the terms in (2.20) and (2.22) is well-defined provided that the respective denominator is nonvanishing. If one of the coefficient functions does vanish, the appropriate starting point for obtaining solutions for the characteristics is Equation (2.17), which is the equation that defines the characteristics. Suppose, for example, that for a particular equation $R=0$. Referring to (2.17), we see that this implies $du/dt=0$, which immediately yields the solution $u = $ constant as one of the solutions to (2.22). The remaining solution is determined from $dx/P = dy/Q$, where any explicit dependence on u is treated as a constant.

In general, the two independent solutions of (2.20) are functions of x, y and u. We will signify these solutions by

$$\xi(x, y, u) = c_1, \quad \eta(x, y, u) = c_2 \tag{2.23}$$

The quantities c_1 and c_2 are constants of integration. The functions ξ and η are required to be independent in that at least one of the Jacobians with respect to all distinct pairs of independent variables is nonzero at each point where the solution is required. For example,

$$\frac{\partial(\xi, \eta)}{\partial(x, y)} \equiv \begin{vmatrix} \xi_x & \eta_x \\ \xi_y & \eta_y \end{vmatrix} \neq 0 \tag{2.24}$$

or, similarly, $\partial(\xi, \eta)/\partial(x, u) \neq 0$ or $\partial(\xi, \eta)/\partial(y, u) \neq 0$.

For fixed values of c_1 and c_2, ξ and η are equations of two specific surfaces whose intersection is a characteristic. Therefore, as c_1 and c_2 are varied, ξ and η represent two families of surfaces. The intersections of these surfaces thus generate a three-dimensional space of characteristic curves as c_1 and c_2 are varied. Since the solution surface for a particular boundary condition must include a subset of these characteristics, the equation describing *any* solution surface in this space is obtained by writing the general expression for the intersection of the two families of surfaces ξ and η. This can be constructed as a function that associates values of c_1 with values of c_2, in other words, a function that specifies which pairs of the families of the two surfaces in (2.23) intersect. The form of the relationship between c_1 and c_2 can be written in several ways. For example, by writing $\mathcal{F}(c_1, c_2) = 0$, the general solution to (2.1) becomes

$$\mathcal{F}(\xi, \eta) = 0 \tag{2.25}$$

Provided only that $\mathcal{F}$ is once-differentiable in each argument, (2.25) is a solution of (2.1) (Problem 6). Two other ways of representing the general solution (2.25) are obtained by writing $c_1 = \mathcal{G}_1(c_2)$ and $c_2 = \mathcal{G}_2(c_1)$:

$$\xi = \mathcal{G}_1(\eta), \qquad \eta = \mathcal{G}_2(\xi) \tag{2.26}$$

where $\mathcal{G}_1$ and $\mathcal{G}_2$ are also once-differentiable, but otherwise arbitrary, functions. The solution may sometimes be expressed as an expression for u in terms of x and y, though an implicit expression, where u appears on both sides of the solutions in (2.26), often occurs even for simple equations, and is an equally acceptable form of solution.

The specification of the arbitrary functions in (2.25) and (2.26) for a particular problem amounts to identifying the pairs of planes in the ξ and η families that intersect to form the solution surface. This will be described in the next section.

Example 2.1. To illustrate the geometrical relationships among the various quantities introduced in this section, we will solve the linear equation

$$x\frac{\partial u}{\partial x} + y\frac{\partial u}{\partial y} = u \tag{2.27}$$

with the method of characteristics. This equation has the form of (2.1) with the identifications

$$P = x, \qquad Q = y, \qquad R = u \tag{2.28}$$

The vector field $\mathbf{V}$ is given by

$$\mathbf{V} = x\mathbf{i} + y\mathbf{j} + u\mathbf{k} \tag{2.29}$$

This vector field is seen to be a radial vector field, i.e. the vector at each point is directed along the straight line from the origin to that point, and the magnitude of the vector depends only on the distance from the origin (Figure 2.2).

The equation determining f is

$$x\frac{\partial f}{\partial x} + y\frac{\partial f}{\partial y} + u\frac{\partial f}{\partial u} = 0 \tag{2.30}$$

and the Equations (2.20) that determine the characteristics are given by

$$\frac{dx}{x} = \frac{dy}{y} = \frac{du}{u} \tag{2.31}$$

For this particular example, the variables in (2.31) are already separated, so the differential equations obtained from (2.31) may be integrated directly. Choosing the two differential equations

$$\frac{dx}{x} = \frac{dy}{y}, \qquad \frac{dx}{x} = \frac{du}{u} \tag{2.32}$$

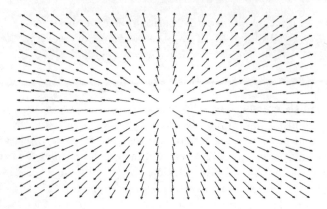

Figure 2.2 The vector field $\mathbf{V} = x\mathbf{i} + y\mathbf{j} + u\mathbf{k}$ in the x-y coordinate plane. This diagram was generated with `VectorField`. The length of each vector has been normalized to unity.

we obtain the solutions

$$\ln x + a_1 = \ln y + a_2, \qquad \ln x + a_3 = \ln u + a_4 \tag{2.33}$$

where the a_i are constants of integration. These solutions may be rewritten in the form of (2.23) as

$$\xi(x,y) = \frac{x}{y} = c_1, \qquad \eta(x,u) = \frac{x}{u} = c_2 \tag{2.34}$$

To check the linear independence of the two solutions in (2.34), we compute the Jacobians

$$\frac{\partial(\xi,\eta)}{\partial(x,y)} = \frac{x}{y^2 u}, \qquad \frac{\partial(\xi,\eta)}{\partial(x,u)} = -\frac{x}{yu^2}, \qquad \frac{\partial(\xi,\eta)}{\partial(y,u)} = \frac{x^2}{y^2 u^2} \tag{2.35}$$

These quantities are nonvanishing provided $x=0$. The two families of planes $\xi = c_1$ and $\eta = c_2$ are shown in Figure 2.3. The plane $x=0$, which is common to both families and corresponds to $\xi(x,y)=0$ and $\eta(x,u)=0$, is indicated by shading. Thus, the quantities in (2.34) form a linearly independent pair for any region that does not contain this plane.

The characteristics are the intersections of the two families of planes in (2.34). Parametric representations of the characteristics in the form (2.14) can be obtained by solving for x in each of the two solutions and parametrizing the x coordinate with t:

$$\mathbf{r}(t) = t\mathbf{i} + t c_1^{-1}\mathbf{j} + t c_2^{-1}\mathbf{k} \tag{2.36}$$

The characteristics are seen to comprise all straight lines that pass through the origin as c_1 and c_2 are varied. This is to be expected since both surfaces

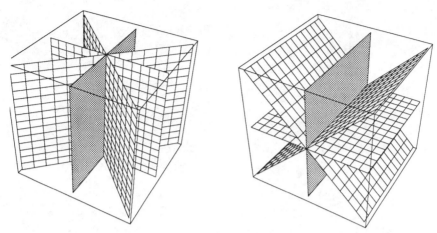

Figure 2.3 The families of surfaces in (2.34) generated with `Parametric-Plot3D`, with $\xi(x,y) = x/y = c_1$ shown at left and $\eta(x,u) = x/u = c_2$ shown at right. The two families of surfaces are linearly independent everywhere except for the shaded region, which corresponds to the plane $x = 0$ and is common to both families ($c_1 = c_2 = 0$).

in (2.34) are planes that contain the origin. Two examples of characteristics formed from the intersections of the surfaces in (2.34) are shown in Figure 2.4. The fact that all of the characteristics pass through the origin is indicative of the differential equation vanishing there. This is an example of a **singular point** of a differential equation and its consequences will be seen below.

Using (2.34), the solution surface of (2.27) can be written as

$$f(x,y,u) = \mathcal{F}\left(\frac{x}{y}, \frac{x}{u}\right) = 0 \tag{2.37}$$

The function $\mathcal{F}$ is required only to be differentiable with respect to each of its arguments; otherwise there is no restriction. For this particular example, we see that the variable u enters only one of the terms in the argument of $\mathcal{F}$. Therefore, we may solve (2.37) for η in terms of ξ to obtain the solution as an expression for u in terms of x and y:

$$u = x\mathcal{G}\left(\frac{x}{y}\right) \tag{2.38}$$

where $\mathcal{G}$ is also required only to be differentiable once.

Another way to see that (2.38) is indeed the general solution of (2.27) is to introduce new variables $\bar{x}$ and $\bar{u}$ defined by $\bar{u} = u/x$ and $\bar{x} = x/y$. The variable y is left unchanged, i.e. $\bar{y} = y$. We leave as an exercise the steps showing that under this change of variables the differential equation in (2.27) is transformed to the simpler form $\partial \bar{u}/\partial \bar{y} = 0$. The general solution

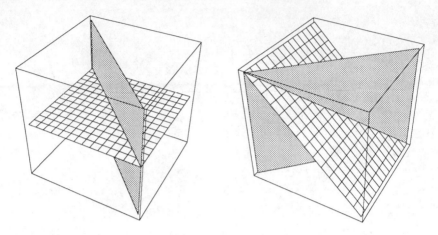

Figure 2.4 Two examples of intersections of the surfaces in (2.34) obtained from `ParametricPlot3D`. The planes $x/y = c_1$ are indicated by shading while the planes $x/u = c_2$ are marked with grid lines. The intersections are seen to be straight lines through the origin.

of this latter equation is clearly $\bar{u} = \mathcal{G}(\bar{x})$, or $\mathcal{F}(\bar{x}, \bar{u}) = 0$, which is equivalent to (2.37) in the original variables.

Mathematica can be used to show directly that the expression in Equation (2.37) solves the differential equation in (2.30) for any differentiable function $\mathcal{F}$. We first define the function f according to the general solution in (2.37):

`f=F[x/y,x/u];`

The general solution is now substituted into the differential equation in (2.30) and `Simplify` is used to show that an identity results for any function $\mathcal{F}$:

`Simplify[x D[f,x] + y D[f,y] + u D[f,u] == 0]`

True

Analogous steps can be carried out to verify that the expression $u = xf(x/y)$ solves Equation (2.27) for any function f that can be differentiated.

The transformation of (2.27) carried out after Equation (2.38) can also be performed within *Mathematica*. The transformed function u is defined as $u(x,y) = xf(x/y,y)$ and substituted into the original differential equation in (2.27). `Simplify` is then used to combine terms to obtain the transformed differential equation.

```
u=x f[x/y,y];

Simplify[x D[u,x] + y D[u,y] - u == 0]

          (0,1)  x
x y f          [-, y] == 0
            y
```

which implies $\partial f/\partial \bar{y}=0$.

∎

The generalization of the solution (2.25) to a quasi-linear first-order partial differential equation for a function u of k independent variables is now straightforward. If the independent variables are signified by $x_1, x_2, \ldots, x_k$, then the general form of the differential equation is

$$P_1 \frac{\partial u}{\partial x_1} + P_2 \frac{\partial u}{\partial x_2} + \cdots + P_k \frac{\partial u}{\partial x_k} = R \tag{2.39}$$

Each of the functions P_i and R are functions of the x_i and u. Then writing the solution 'surface' as $f(x_1, x_2, \ldots, x_k, u) = 0$ an equation of the form (2.9) can be derived from which the following characteristic equations are obtained:

$$\frac{\mathrm{d}x_1}{P_1} = \frac{\mathrm{d}x_2}{P_2} = \cdots = \frac{\mathrm{d}x_k}{P_k} = \frac{\mathrm{d}u}{R} \tag{2.40}$$

The solution to these equations, which are the characteristics, can be expressed in terms of the intersections of k 'surfaces' obtained from k independent solutions of the equations in (2.40). The general solution therefore can be written as

$$\mathcal{F}(\xi_1, \xi_2, \ldots, \xi_k) = 0 \tag{2.41}$$

where $\xi_i = \xi_i(x_1, x_2, \ldots, x_k, u)$.

2.2 Boundary Conditions of First-Order Equations

As they stand, (2.25) and (2.26) are general expressions for the intersections between two families of surfaces. The solution surface that is required to contain a specified curve may then be viewed as being constructed from characteristics that intersect that curve (Figure 2.1). Consider, therefore, a curve $\mathcal{C}$ which is required to lie in the solution surface. We will represent this curve as in Equation (2.14) in terms of a parameter s with the three coordinate functions $x_0(s)$, $y_0(s)$ and $u_0(s)$:

$$\mathcal{C}(s) = x_0(s)\,\mathbf{i} + y_0(s)\,\mathbf{j} + u_0(s)\,\mathbf{k} \tag{2.42}$$

Along $\mathcal{C}$ (2.25) takes the form

$$\mathcal{F}[\xi_0(s), \eta_0(s)] = 0 \qquad (2.43)$$

where we have used the abbreviations

$$\begin{aligned}\xi_0(s) &\equiv \xi[x_0(s), y_0(s), u_0(s)] \\ \eta_0(s) &\equiv \eta[x_0(s), y_0(s), u_0(s)]\end{aligned} \qquad (2.44)$$

Suppose first that $\mathcal{C}$ is a characteristic. Then there are two constants c_1 and c_2 such that $\mathcal{C}$ is the intersection of the two surfaces given by the equations $\xi_0(s) = c_1$ and $\eta_0(s) = c_2$, where the constants c_1 and c_2 are the same for *all* values of s. Thus, along $\mathcal{C}$, (2.43) reduces to

$$\mathcal{F}(c_1, c_2) = 0 \qquad (2.45)$$

which means that *any* function of two variables that vanishes for the particular values c_1 and c_2 of the arguments satisfies the boundary conditions. Since any function of the variables ξ and η satisfies the differential equation, the requirement that $\mathcal{C}$ be a characteristic does *not* lead to a unique solution of (2.1) (Problem 15).

As a second special case, suppose that the boundary curve is specified as the intersection of two surfaces $\mathcal{S}_1 = \mathcal{S}_1(x, y, u)$ and $\mathcal{S}_2 = \mathcal{S}_2(x, y, u)$. If one of the surfaces, say $\mathcal{S}_1$, corresponds to one of the solutions (2.23) for a particular value of c_1, then the intersection of $\mathcal{S}_1$ and $\mathcal{S}_2$ is not a characteristic, but the curve is contained entirely within one of the two surfaces in (2.23) for a particular value of c_1. Then along $\mathcal{C}$, the general solution can be written in the form $\mathcal{F}[c_1, \eta_0(s)] = 0$. Solving for c_1,

$$c_1 = \mathcal{G}[c_2(s)] \qquad (2.46)$$

which has the solution $\mathcal{G} \equiv c_1$. If, instead of solving for c_1, the equation $\mathcal{F}[c_1, \eta_0(s)] = 0$ is solved for $\eta_0(s)$, $\eta_0(s) = g(c_1)$ then the problem has no solution. In Equation (2.46) $\mathcal{G}$ maps $c_2(s)$ onto c_1 for *every* value of s. Thus, the inverse of $\mathcal{G}$, which maps c_1 to $c_2(s)$, is not defined. The solution of $\mathcal{F}[\xi_0(s), \eta_0(s)] = 0$ along $\mathcal{C}$ either for ξ in terms of η, or η in terms of ξ, relies upon *every* value of $c_1(s)$ being assigned a unique value of $c_2(s)$. Thus, the solution to the equation in this case is the characteristic surface $\xi(x, y, u) = c_1$. However, the problem is still underspecified, since the specification of *any* surface of intersection $\mathcal{S}_2$ yields the same solution.

Finally, if $\mathcal{C}$ is not a characteristic and does not lie in one of the surfaces (2.23), then $\mathcal{C}$ intersects characteristics corresponding to a range of values of c_1 and c_2. For these values of c_1 and c_2, $\mathcal{F}$ is completely determined and a unique solution is obtained. Thus, to completely determine $\mathcal{F}$ in a particular region of space, the boundary curve contained by the solution surface must intersect each of the characteristics.

Example 2.2. To obtain the particular solution surface of (2.27) through the line $x = y = u$ we begin with the general solution (2.38) in the form $u = x\mathcal{G}(x/y)$. Since the parametric form of the line is

$$C(s) = (s, s, s) \tag{2.47}$$

the solution along this curve,

$$u_0(s) = x_0(s)\mathcal{G}[x_0(s)/y_0(s)] \tag{2.48}$$

must satisfy $s = s\mathcal{G}(1)$. Requiring the solution surface to pass through the line $x = y = u$ thus restricts $\mathcal{G}$ only to the condition that $\mathcal{G}(1) = 1$ on this line. The solution of (2.27) passing through this line is therefore

$$u = x\mathcal{G}(x/y), \quad \mathcal{G}(1) = 1 \tag{2.49}$$

There is clearly an infinity of solutions of this form, since any function with the property $\mathcal{G}(1) = 1$ satisfies the differential equation. This situation arises because the boundary condition is itself a characteristic corresponding to $c_1 = c_2 = 1$ in (2.36).

Now consider the problem of finding a solution of (2.27) with the requirement that $u = y^3$ when $x = y^2$. The line of intersection of these two surfaces can be represented in parametric form as

$$C(s) = (s^2, s, s^3) \tag{2.50}$$

and is shown in Figure 2.5. Along this curve, the solution (2.48) reduces to $t = \mathcal{G}(t)$. This determines the function $\mathcal{G}$, and we obtain the particular solution by substituting the form of this function into (2.38):

$$u = x \cdot \frac{x}{y} = \frac{x^2}{y} \tag{2.51}$$

A plot of the solution (2.51) is shown in Figure 2.5. The reason a unique solution is obtained in this case can be seen by examining the parametric form of the curve C. Comparing the parametric form of the curve in (2.50) with that in (2.36) for the characteristics shows that for a given value of s, the boundary curve intersects the characteristic corresponding to the constants c_1 and c_2 given by $c_1 = s$ and $c_2 = s^{-1}$ (Problem 16). Thus, each point along C intersects a unique characteristic.

This can be used to provide an alternative representation of the solution surface that more clearly shows the characteristics as the constituents.

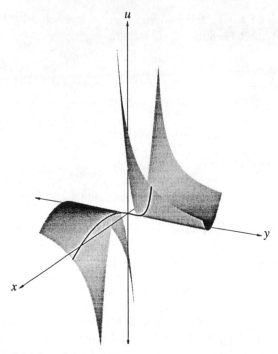

Figure 2.5 The line of intersection of the two surfaces $u=y^3$ and $x=y^2$ and the solution surface (2.51).

Substituting the expressions deduced above for $c_1(s)$ and $c_2(s)$ into equation (2.36) for the characteristics yields

$$\mathbf{r}(s,t) = t\mathbf{i} + ts^{-1}\mathbf{j} + ts\,\mathbf{k} \qquad (2.52)$$

This equation is a representation of the solution surface (2.51) in terms of the coordinates $\mathbf{r}(s,t)$ of the points on the surface as a function of the two parameters s and t (Figure 2.6). If s is held constant, $\mathbf{r}(s,t)$ represents a straight line—a characteristic—that passes through the origin as t is varied. The role of the characteristics is shown directly in this representation of the solution surface. ∎

In many physical problems involving first-order equations, one of the independent variables corresponds to the time and the solution is expressed as an evolution from some initial condition. The ability to generate closed-form solutions using the method of characteristics enables these solutions to be examined in detail and provides a physical realization of the characteristics themselves. In the next example, a first-order equation from fluid mechanics is solved to illustrate the application of Lagrange's method to a simple model for a physical problem. A detailed discussion of the origin of the equation will not be given, but may be found in the references given at the end of the chapter.

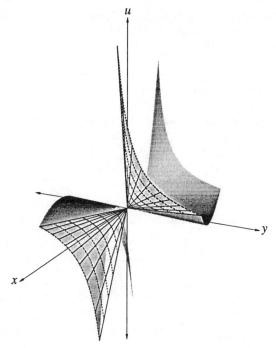

Figure 2.6 A section of the solution surface (2.51) based upon the parametric representation (2.52).

Example 2.3. An equation that is sometimes used to illustrate how shock waves can develop in fluid systems is

$$\frac{\partial u}{\partial t} + u \frac{\partial u}{\partial x} = 0 \qquad (2.53)$$

with the initial condition

$$u(x, 0) = f(x) \qquad (2.54)$$

The characteristic equations are

$$dt = \frac{dx}{u}, \qquad du = 0 \qquad (2.55)$$

where, as discussed in Section 2.1, the characteristic equation $du = 0$ is obtained because, in the notation of Equation (2.1), $R = 0$. The solution to this equation is clearly

$$u = c_1 \qquad (2.56)$$

Thus, regarding u as a constant, the solution of $dt = dx/u$ is

$$x - ut = c_2 \qquad (2.57)$$

The general solution can therefore be written as $c_1 = f(c_2)$, which has the implicit form

$$u = \mathcal{F}(x - ut) \qquad (2.58)$$

since u appears on both sides of the equation.

To determine the particular solution for the initial function given by (2.54), we observe that at $t=0$, (2.58) becomes

$$u(x, 0) = \mathcal{F}(x) \qquad (2.59)$$

Thus, comparing this equation with (2.54), we deduce that $\mathcal{F}(x) = f(x)$. The solution to the initial value problem in (2.53) and (2.54) is therefore given by

$$u(x, t) = f[x - u(x, t)t] \qquad (2.60)$$

where the dependence of u upon x and t has been displayed on both sides of the equation to emphasize the implicit nature of the solution.

The characteristics of (2.53), which are the intersections of the surfaces in (2.56) and (2.57), are straight lines that lie in planes of constant u with slopes that are determined by the value of u in the particular plane. To see the implications of this for the evolution of solutions to (2.53), we consider a particular initial value problem.

The time development of the solution in (2.60) from a particular initial condition can be constructed with remarkable ease using *Mathematica*. Suppose the initial condition in (2.54) is given by

$$f(x) = \begin{cases} 1 - x^2 & \text{if } |x| \leq 1 \\ 0 & \text{if } |x| > 1 \end{cases} \qquad (2.61)$$

This initial function has been chosen because the solution in (2.60) can be solved for u as a function of x and t (Problem 18). However, the procedure described below for constructing the solution using *Mathematica* can be applied to initial functions that do not permit such an explicit solution.

The evolution of the point $x = x_0$, $u = f(x_0)$ on the initial function at $t=0$ can be determined from (2.56) and (2.57). At $t=0$, we have $c_2 = x_0$, so at time t the point initially at x_0 has moved to

$$x = x_0 + f(x_0)t \qquad (2.62)$$

These are the characteristics of Equation (2.53).

To construct the characteristics and the evolution of the solution in *Mathematica*, the initial profile (2.61) and the characteristics (2.62) are first entered as

```
f[x_]:=If[Abs[x]<1,1-x^2,0]

x[t_,x0_]:=x0+f[x0]t
```

Both the characteristics and the evolution of the solution can now be displayed with `ParametricPlot`. The trajectory of the characteristic emanating from the point $x_0=0$ up to time $t=1$ is displayed with

```
ParametricPlot[{x[t,0],t},{t,0,1}]
```

Similarly, the solution at time t over the region $-2 \leq x_0 \leq 2$ is displayed with the command

```
u[t_]:=ParametricPlot[{x[t,x0],f[x0]},{x0,-2,2}]
```

Notice that in generating the solution with `ParametricPlot` the quantity being plotted is the pair of points $[x(x_0,t), f(x_0)]$, which corresponds to the position at time t of the point on the profile $u(x,t)$ that was initially located at x_0. Thus, this construction represents a point-by-point assembly of the solution at time t in terms of the initial profile and the characteristics.

The trajectories of several characteristics projected on to the x-t plane and the early stages of evolution of the solution from the initial profile (2.61) are shown in Figure 2.7. For short times the initial solution is deformed because the characteristics move with different speeds for different values of u. At the point where two characteristics first cross, the solution becomes double-valued, and the slope in the u-x plane becomes infinite. From this point onward, the solution is no longer a continuous function of position, and corresponds to the onset of a 'shock' wave. This type of behavior is due to the form of the characteristics and not to a specific choice of initial condition (Problem 19).

In Figure 2.8 this solution is shown in the three-dimensional space (x,t,u). The projection of the characteristics on to the x-t plane (Figure 2.7) is also shown for comparison. Intersections of several of the characteristics with individual 'time slices' of the solution are included to illustrate the evolution of the profile of the solution in terms of the characteristics emanating from the initial function. In particular, the crossing at $t=1$ of the two characteristics initially at $x=0$ and $x=1$ is indicative of the double-valuedness of the solution at the point $x=1$ at time $t=1$. ∎

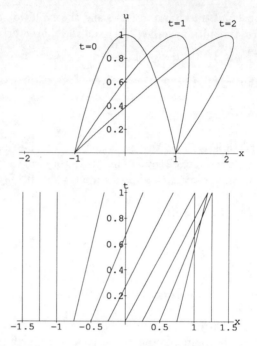

Figure 2.7 (Top) The early stages of the evolution of the solution (2.60) obtained by using `ParametricPlot`. The profiles are shown for $t=0$, $t=1$ and $t = 2$. (Bottom) The characteristics with the initial profile (2.61) projected onto the x-t plane, also obtained with `ParametricPlot`. The crossing of characteristics, which first occurs at $t=\frac{1}{2}$, corresponds to double-valuedness of the solution (see Figure 2.8).

2.3 Nonlinear First-Order Equations

In this section we return to the general form of a first-order partial differential equation in (2.3):

$$F(x, y, u, u_x, u_y) = 0 \qquad (2.63)$$

We have seen in the preceding sections that the general solution of a quasi-linear equation can be written as

$$\mathcal{F}[\eta(x,y,u), \xi(x,y,u)] = 0 \qquad (2.64)$$

The quantities η and ξ are determined by the particular equation, but the function $\mathcal{F}$ is arbitrary, apart from the requirement that it be once-differentiable. The geometric interpretation of (2.64) is that ξ and η are surfaces in the space of x, y and u, and their intersections, which are represented in (2.64), are the characteristics. The boundary conditions are used to determine which pairs of surfaces intersect to form the solution surface.

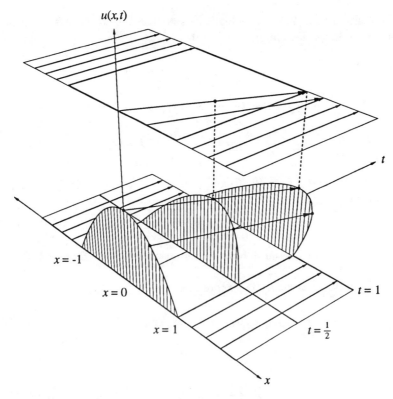

Figure 2.8 The solution in Figure 2.7 shown in the three-dimensional (x, t, u) space for the times $t=0$, $t=\frac{1}{2}$ and $t=1$ with the projection of the characteristics on to the x-t plane. The crossing at $t=1$ of the characteristic emanating from x_0 with that emanating from $x_0=1$ corresponds to the solution being double-valued at that point, as indicated.

As we have seen in the preceding section (and in Problems 6 and 14), there is a correspondence between quasi-linear equations and general solutions involving arbitrary functions. Thus, to apply the ideas developed in the preceding sections to nonlinear equations, we must look to other representations of solutions. Consider therefore the expression

$$F(x, y, u, \alpha, \beta) = 0 \tag{2.65}$$

where F is a *known* function. The quantities α and β are parameters, so Equation (2.65) represents a two-parameter family of surfaces. The equation for which this expression is a solution may be obtained by differentiating (2.65) in turn with respect to x and y and eliminating α and β. The result is a first-order partial differential equation for $u(x, y)$, though not necessarily one that is quasi-linear. The expression in (2.65) is called the **complete integral** or the **complete solution** of the differential equation.

Example 2.5. To illustrate some of the ideas just mentioned, suppose we have the complete integral

$$F(x,y,u,\alpha,\beta) = u - \alpha x - \beta y = 0 \tag{2.66}$$

Differentiating in turn with respect to x and y, we obtain

$$\frac{\partial u}{\partial x} - \alpha = 0, \qquad \frac{\partial u}{\partial y} - \beta = 0 \tag{2.67}$$

Eliminating α and β from (2.66) and (2.67), we see that (2.66) is a complete integral for

$$x\frac{\partial u}{\partial x} + y\frac{\partial u}{\partial y} = u \tag{2.68}$$

which is the linear partial differential equation solved in Examples 2.1 and 2.2.

Similarly, consider the expression

$$F(x,y,u,\alpha,\beta) = u - \alpha x - \beta y - \alpha\beta = 0 \tag{2.69}$$

Again differentiating with respect to x and y,

$$\frac{\partial u}{\partial x} - \alpha = 0, \qquad \frac{\partial u}{\partial y} - \beta = 0 \tag{2.70}$$

and eliminating α and β, we now obtain

$$x\frac{\partial u}{\partial x} + y\frac{\partial u}{\partial y} + \frac{\partial u}{\partial x}\frac{\partial u}{\partial y} = u \tag{2.71}$$

which is a *nonlinear* equation. ∎

The complete integral represents a two-parameter family of surfaces, since for any choice of α and β, the expression in (2.65) corresponds to a particular surface. If the solution surface is required to contain the space curve $\mathcal{C}$ parametrized by $x_0(t)$, $y_0(t)$ and $u_0(t)$, then (2.65) becomes

$$F[x_0(t), y_0(t), u_0(t), \alpha, \beta] = 0 \tag{2.72}$$

Since F, x_0, y_0 and u_0 are known functions, this equation shows that α and β cannot be independent. Thus, suppose that (2.72) is solved for β as a function of α, $\beta = \phi(\alpha)$, which we substitute into the complete integral (2.65),

$$F[x, y, u, \alpha, \phi(\alpha)] = 0 \tag{2.73}$$

If we now differentiate this expression with respect to α, we obtain

$$\frac{\partial F}{\partial \alpha} + \frac{\partial F}{\partial \beta}\frac{\partial \phi}{\partial \alpha} = 0 \tag{2.74}$$

The quantities F_α and F_β are both functions only of x, y, u and α and ϕ is a function only of α. Thus this equation provides an expression for α as a function of x, y and u, which we write as $\alpha(x,y,u)$. Substituting this functional dependence into Equation (2.73), we obtain

$$F\{x,y,u,\alpha(x,y,u),\phi[\alpha(x,y,u)]\} = 0 \tag{2.75}$$

By solving this equation formally for ϕ, the complete integral can be written as

$$G[x,y,u,\alpha(x,y,u)] = \phi[\alpha(x,y,u)] \tag{2.76}$$

Since G is a *known* function, while ϕ is in effect an arbitrary function since the boundary conditions have not yet been specified, the expression in (2.76) is of the form of a general solution. Thus, in principle, the general solution of a partial differential equation can be obtained from the complete integral.

Example 2.6. Consider the complete integral in Example 2.5, with $\beta = \phi(\alpha)$:

$$F[x,y,u,\alpha,\phi(\alpha)] = u - \alpha x - y\phi(\alpha) = 0 \tag{2.77}$$

Differentiating with respect to α yields

$$-x - y\phi'(\alpha) = 0 \tag{2.78}$$

This equation can be solved for $\phi'(\alpha)$ to obtain

$$\phi'(\alpha) = -\frac{x}{y} \tag{2.79}$$

The left-hand side of this equation is some function of α. Thus, solving for α as a function of x/y,

$$\alpha = \alpha(x/y) \tag{2.80}$$

and substituting this expression into (2.77), we obtain after some manipulation the solution (2.77) in the form

$$\begin{aligned} u &= \alpha x - y\phi(\alpha) \\ &= x[\alpha - y/x\phi(\alpha)] \\ &= x\{\alpha(x/y) - (y/x)\phi[\alpha(x/y)]\} \\ &= x\mathcal{G}(x/y) \end{aligned} \tag{2.81}$$

where $\mathcal{G}$ is an arbitrary function (since ϕ is arbitrary). Therefore, this solution is the same as the general solution in (2.38). It must be stressed that obtaining the general solution from the complete integral in practice is not always a feasible task. The example here is meant only to provide a simple illustration of this procedure when it can be carried out. ∎

With these preliminary considerations as background, we turn our attention to constructing the solution to a general first-order differential equation of the form (2.63) by generalizing Lagrange's method of characteristics. Although the basic strategy is the same as for solving quasi-linear equations there are some important geometric differences encountered for nonlinear equations that complicate the simple picture developed in Section 2.2. These differences are best appreciated by comparing the results at intermediate stages of the derivation to the corresponding constructions for quasi-linear equations.

We first introduce some standard notation. The two first-order partial derivatives of u are abbreviated as $p \equiv u_x$ and $q \equiv u_y$. With this notation, the general first-order partial differential equation is written as

$$F(x, y, u, p, q) = 0 \qquad (2.82)$$

In this and subsequent equations, u is regarded as a function of the independent variables x and y. The complete integral of this equation is then an expression of the form $u = u(x, y, \alpha, \beta)$, where α and β are constants as defined above. Specifying a space curve through which the complete integral must pass generates a solution surface.

The tangent plane at each point (x, y, u) of the solution surface has the equation

$$p(x - x') + q(y - y') - (u - u') = 0 \qquad (2.83)$$

where x', y' and u' are the points on the tangent plane (the running coordinates of the plane). Thus, Equation (2.82) can be interpreted as providing a relation between the points on the solution surface and the orientation of the tangent planes at each point on the solution surface. In fact, since (2.82) provides a relation between p and q for *any* solution surface there is a family of tangent planes corresponding to different values of p and q. We can obtain an equation that describes the envelope swept out by these planes by considering the intersection between (2.83) and the tangent plane (at the same point) corresponding to neighboring values of p and q:

$$(p + dp)(x - x') + (q + dq)(y - y') - (u - u') = 0 \qquad (2.84)$$

The intersection of (2.83) and (2.84) is obtained by setting the two equations equal to one another, with the result

$$dp(x - x') + dq(y - y') = 0 \qquad (2.85)$$

This equation can be rewritten in terms of quantities specific to the differential equation at hand by taking the differential of (2.82) at the point (x, y, u) while holding x, y and u fixed:

$$\frac{\partial F}{\partial p}dp + \frac{\partial F}{\partial q}dq = 0 \tag{2.86}$$

Eliminating dp and dq from (2.85) and (2.86) yields

$$(x - x')\frac{\partial F}{\partial q} - (y - y')\frac{\partial F}{\partial p} = 0 \tag{2.87}$$

The equation of the surface for which the tangent planes are the envelope is thus obtained by eliminating p and q from (2.82), (2.83) and (2.87). This surface, which we signify by T is a cone for nonlinear equations and is called the **Monge cone** (Figure 2.9). Recognizing that the normal to the tangent plane (2.83) has the representation

$$\frac{x - x'}{p} = \frac{y - y'}{q} = -(u - u') \tag{2.88}$$

we can solve in turn for p and q and substitute the resulting expressions into (2.82) to obtain

$$F\left(x, y, u, -\frac{x - x'}{u - u'}, -\frac{y - y'}{u - u'}\right) = 0 \tag{2.89}$$

Since the envelope of the tangent planes is a cone, the surface swept out by the associated tangents is also a cone. This cone, which we signify by N, is described by Equation (2.89). For quasi-linear equations, where F is linear in p and q, the cone T enveloped by the tangent planes becomes the *straight line* (Problem 24)

$$\frac{x - x'}{P} = \frac{y - y'}{Q} = \frac{u - u'}{R} \tag{2.90}$$

and the cone N swept out by the normals to these planes becomes the *plane*

$$P(x - x') + Q(y - y') + R(u - u') = 0 \tag{2.91}$$

For the purposes of generalizing the method of characteristics in Section 2.2, we write Equation (2.87) as

$$\frac{x - x'}{F_p} = \frac{y - y'}{F_q} \tag{2.92}$$

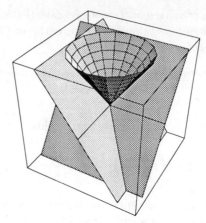

Figure 2.9 Two tangent planes from the family of planes whose envelope is the Monge cone, which is also shown. For quasi-linear equations, the equation itself specifies the orientation of the tangent planes to the solution surface. For the general nonlinear equation in (2.63), the Monge cone represents the family of possible orientations of the tangent plane to the solution surface at the point corresponding to the vertex of the cone.

where $F_p = \partial F/\partial p$ and $F_q = \partial F/\partial q$. Combining this equation with (2.84), we obtain

$$\frac{x-x'}{F_p} = \frac{y-y'}{F_q} = \frac{u-u'}{pF_p + qF_q} \tag{2.93}$$

This equation describes a line on the tangent plane to a solution surface at the point (x, y, u). To obtain differential equations for the characteristics as this point moves along a solution surface, we let u' approach u, so that we can write $u' = u + du$, and we obtain

$$\frac{dx}{F_p} = \frac{dy}{F_q} = \frac{du}{pF_p + qF_q} \tag{2.94}$$

These equations are analogous to the Equations (2.20) for the characteristics of a quasi-linear equation. Notice an important difference, however. Although both sets of equations depend upon the unknown solution u, the characteristic equations for the nonlinear equation also depend upon p and q. This means that the characteristics also depend on the *orientation* of the tangent planes on the Monge cone at each point. As our geometrical analysis indicates this situation arises because for nonlinear equations there is a cone of tangent planes to the solution surface and thus a cone of characteristic directions. For quasi-linear equations, the direction of the characteristics is determined solely by the vector field in Equation (2.11) at each point of the solution surface. Thus, for a nonlinear equation, Equations (2.94) are not a closed set, and so need to be supplemented by equations for the quantities p and q. To determine these equations, we first write Equations (2.94) as differential equations for the Cartesian components of the characteristics as in (2.17) by introducing the running coordinate s along the characteristics:

$$\frac{dx}{ds} = F_p, \qquad \frac{dy}{ds} = F_q, \qquad \frac{du}{ds} = pF_p + qF_q, \qquad (2.95)$$

To close the equations in (2.94) and (2.95), a natural choice is to examine the equations for dp/ds and dq/ds. Using the chain rule and Equation (2.95), we obtain

$$\frac{dp}{ds} = p_x \frac{dx}{ds} + p_y \frac{dy}{ds} = p_x F_p + p_y F_q \qquad (2.96a)$$

$$\frac{dq}{ds} = q_x \frac{dx}{ds} + q_y \frac{dy}{ds} = q_x F_p + q_y F_q \qquad (2.96b)$$

These equations involve the partial derivatives of p and q with respect to x and y. To eliminate these quantities, we first differentiate Equation (2.82) in turn with respect to x and y,

$$F_x + F_u u_x + F_p p_x + F_q q_x = 0 \qquad (2.97a)$$

$$F_y + F_u u_y + F_p p_y + F_q q_y = 0 \qquad (2.97b)$$

Using the fact that $p_y = q_x$, which may be regarded as an integrability condition for the solution surface, and then substituting (2.97a) into (2.96a) and (2.97b) into (2.96b), produces

$$\frac{dp}{ds} = -F_x - pF_u, \qquad \frac{dq}{ds} = -F_y - qF_u \qquad (2.98)$$

These equations, combined with (2.95), form a closed set of equations for x, y, u, p and q. Eliminating the quantity ds from these equations yields the **Charpit equations** for the characteristics:

$$\frac{dx}{F_p} = \frac{dy}{F_q} = \frac{du}{pF_p + qF_q} = -\frac{dp}{F_x + pF_u} = -\frac{dq}{F_y + qF_u} \qquad (2.99)$$

The Charpit equations consist of five first-order ordinary differential equations for x, y, u, p and q. These equations are therefore solved by specifying not only the initial values $x_0(s)$, $y_0(s)$ and $u_0(s)$ of a space curve through which a solution surface must pass, as in the case of a quasi-linear equation, but also the corresponding values $p_0(s)$ and $q_0(s)$ indicating the orientation of the surface through the curve. These five quantities cannot all be chosen independently, since we must observe the relation

$$\frac{du}{ds} = \frac{du}{dx}\frac{dx}{ds} + \frac{du}{dy}\frac{dy}{ds} = p\frac{dx}{ds} + q\frac{dy}{ds} \qquad (2.100)$$

The quantities consisting of the spatial coordinates $x_0(s)$, $y_0(s)$ and $u_0(s)$ and the orientations of the tangent planes $p_0(s)$ and $q_0(s)$ are referred to as the **initial strip**, since the initial condition can be viewed as the initial curve together with initial tangent planes attached. Similarly, the Charpit equations are referred to as the **characteristic strip equations**, and (2.100) is the **strip condition**.

2.4 The Complete Integral

The differential equation in (2.63) provides a relation among the variables x, y, u, p and q. If a second relation among these variables and an arbitrary constant α can be found that is independent of (2.63), then p and q can be solved in terms of these quantities,

$$p = p(x, y, u, \alpha), \qquad q = q(x, y, u, \alpha) \qquad (2.101)$$

and substituted into the equation for the differential of u:

$$du = p(x, y, u, \alpha)dx + q(x, y, u, \alpha)dy \qquad (2.102)$$

Integrating this equation produces a second constant, β, so the solution, $u = u(x, y, \alpha, \beta)$ is the complete integral of the differential equation (2.63). The Charpit equations derived in the preceding section provide a way of obtaining an expression that is independent of (2.63). We consider a few special cases where these constructions can be carried out explicitly.

Case I. $F = F(p, q)$. If the differential equation involves only the derivatives of u, then Equations (2.98) reduce to

$$\frac{dp}{ds} = 0, \qquad \frac{dq}{ds} = 0 \qquad (2.103)$$

which states that p and q are constants along the characteristics. The differential equation provides a relation among these constants. Thus, let $p = \alpha$. Then $F(\alpha, q) = 0$ which, when solved for q, yields $q = f(\alpha)$. The differential for u is therefore given by

$$du = \alpha dx + f(\alpha)dy \qquad (2.104)$$

This equation may be integrated readily to obtain the complete integral in the form

$$u = \alpha x + f(\alpha)y + \beta \qquad (2.105)$$

Case II. $F = F(u, p, q)$. If the differential equation does not involve the variables x and y explicitly, then the Charpit equations in (2.99) yield

$$-\frac{dp}{pF_u} = -\frac{dq}{qF_u} \qquad (2.106)$$

Cancelling the factor F_u from both sides of this equation and integrating shows that p and q are proportional. Accordingly, let $q = \alpha p$. Then the differential equation provides the relation $F(u, p, \alpha p) = 0$ which can be solved for p: $p = f(u, \alpha)$. Thus, $q = \alpha f(u, \alpha)$, so the differential of u is

$$du = f(u, \alpha) + \alpha f(u, \alpha) dy \tag{2.107}$$

Upon integration, we obtain the complete integral

$$\int^u \frac{1}{f(s, \alpha)} ds = x + \alpha y + \beta \tag{2.108}$$

Case III. $F = F(x, p, q)$. If the only variable that explicitly enters the differential equation is x, then the equations in (2.98) become

$$\frac{dp}{ds} = -F_x, \qquad \frac{dq}{ds} = 0 \tag{2.109}$$

The second of these equations states that q is a constant along the characteristics, $q = \alpha$. The differential equation then yields an expression for p in terms of x and α, $p = f(x, \alpha)$, so the differential of u reads

$$du = f(x, \alpha) dx + \alpha dy \tag{2.110}$$

Upon integration this equation yields the complete integral

$$u = \int^x f(s, \alpha) ds + \alpha y + \beta \tag{2.111}$$

A similar sequence of steps yields the complete integral for differential equations of the form $F(y, p, q) = 0$.

Case IV. $F = F_1(x, p) - F_2(y, q)$. In this case, the equations in (2.95) and (2.98) become

$$\frac{dx}{ds} = F_{1,p}, \quad \frac{dy}{ds} = -F_{2,q}, \quad \frac{dp}{ds} = -F_{1,x}, \quad \frac{dq}{ds} = F_{2,y} \tag{2.112}$$

Combining the first and third of these equations yields a differential equation for F_1:

$$F_{1,x} dx + F_{1,p} dp = 0 \tag{2.113}$$

which yields

$$F_1(x, p) = F_2(y, q) = \alpha \tag{2.114}$$

Thus, p can be solved in terms of x and α, and q can be solved in terms of y and α, $p=f_1(x,\alpha)$ and $q=f_2(y,\alpha)$. The differential of u is thus given by

$$du = f_1(x,\alpha)dx + f_2(y,\alpha)dy \tag{2.115}$$

A straightforward integration yields the complete integral:

$$u = \int^x f_1(s,\alpha)\,ds + \int^y f_2(s,\alpha)\,ds + \beta \tag{2.116}$$

Sometimes an equation that is not in one of these special forms can be made so by a transformation of variables. Some examples are given in the problems at the end of the chapter.

Example 2.7. Consider the nonlinear partial differential equation

$$\frac{\partial u}{\partial x} + \frac{\partial u}{\partial y} = \frac{\partial u}{\partial x}\frac{\partial u}{\partial y} \tag{2.117}$$

We require the solution that contains the curve obtained by the intersection of the surfaces $u=2x$ and $y=0$. This equation is seen to be of the type in Case I with $F=p+q-pq$. The steps used to obtain (2.105) can be carried out easily for this case and we obtain the complete integral

$$u = \alpha x + \frac{\alpha}{\alpha-1}y + \beta \tag{2.118}$$

Alternatively, we can obtain a complete integral directly from the Charpit equations:

$$\frac{dx}{ds} = 1-q, \qquad \frac{dy}{ds} = 1-p, \qquad \frac{du}{ds} = p(1-q)+q(1-p) \tag{2.119a}$$

$$\frac{dp}{ds} = 0, \qquad \frac{dq}{ds} = 0 \tag{2.119b}$$

Equations (2.119b) say that p and q do not change from their values on the boundary curve. Thus, we set $p=p_0$ and $q=q_0$. Equations (2.119a) can now be integrated to obtain

$$x(s) = (1-q_0)s + x_0, \qquad y(s) = (1-p_0)s + y_0$$
$$u(s) = p_0(1-q_0)s + q_0(1-p_0)s + u_0 \tag{2.120}$$

where x_0, y_0 and u_0 are the values of x, y and u on the boundary curve, which we have taken to correspond to $s=0$. By using the solutions for $x(s)$

and $y(s)$ to eliminate the explicit appearance of s in the solution for $u(s)$, we obtain

$$u(s) = p_0 x(s) + \frac{p_0}{p_0 - 1} y(s) + u_0 - p_0 x_0 - q_0 y_0 \qquad (2.121)$$

This expression is seen to be of the same form as that in (2.118) with the identifications

$$\alpha = p_0, \qquad \beta = u_0 - p_0 x_0 - q_0 y_0 \qquad (2.122)$$

Turning our attention to finding the solution that contains the required curve, we set $y = 0$ and $u = 2x$ in Equation (2.118). This immediately yields $\alpha = 2$ and $\beta = 0$, so the required solution is

$$u = 2x + 2y \qquad (2.123)$$

A parametric representation of this solution can be obtained by first writing the boundary curve in parametric form as $(t, 0, 2t)$. This shows that

$$x_0 = t, \qquad y_0 = 0, \qquad u_0 = 2t \qquad (2.124)$$

To complete the solution, we require the values of p_0 and q_0. These can be obtained by combining the expression for the differential of u along the initial curve and the differential equation itself. The parametric form of the initial curve yields $dx = dt$, $dy = 0$ and $du = 2dt$. Thus, the differential of u along the initial curve is

$$2dt = p_0 dt \qquad (2.125)$$

which yields $p_0 = 2$. Then, the differential equation requires

$$p_0 + q_0 - p_0 q_0 = 0 \qquad (2.126)$$

With $p_0 = 2$, this equation produces $q_0 = 2$. By substituting these values and those in (2.124) into the equations in (2.120), we obtain the solution in parametric form:

$$x(s) = t - s, \qquad y(s) = -s, \qquad u(s) = 2t - 4s \qquad (2.127)$$

By eliminating s and t from these expressions, we obtain the solution in the form shown in Equation (2.123). ∎

Further Reading

There are many books that develop the theory of quasi-linear first-order partial differential equations. The book by Hildebrand (1962) is an excellent general reference for the topics discussed in this chapter as well as for other related topics. Discussions similar to those in this chapter but with a somewhat more mathematical emphasis are given by Garabedian (1964) and Sneddon (1957). Whitham (1974) covers first-order quasi-linear equations in the context of fluids, including shock-wave formation of the type discussed in Section 2.2.

References

Crandall R. E. (1991). *Mathematica for the Sciences.* Redwood City CA: Addison-Wesley.
Garabedian P. R. (1964). *Partial Differential Equations.* New York: Wiley.
Goldstein H. (1950). *Classical Mechanics.* Reading MA: Addison-Wesley.
Hildebrand F. B. (1962). *Advanced Calculus for Applications.* Englewood Cliffs NJ: Prentice-Hall.
Sneddon I. N. (1957). *Elements of Partial Differential Equations.* New York: McGraw-Hill.
Sommerfeld A. (1954). *Optics.* New York: Academic.
Whitham G. B. (1974). *Linear and Nonlinear Waves.* New York: Wiley.

Problems

1. Find the general solutions to each of the differential equations below (a, b and c are constants).

(a) $a\dfrac{\partial u}{\partial x} + b\dfrac{\partial u}{\partial y} = c$

(b) $ax\dfrac{\partial u}{\partial x} + by\dfrac{\partial u}{\partial y} = cu$

(c) $a\dfrac{\partial u}{\partial x} + b\dfrac{\partial u}{\partial y} = 0$

(d) $ax\dfrac{\partial u}{\partial x} + by\dfrac{\partial u}{\partial y} = 0$

(e) $ax^2\dfrac{\partial u}{\partial x} + by^2\dfrac{\partial u}{\partial y} = cu$

2. Quasi-linear equations can be obtained by eliminating the arbitrary function from expressions of the form $\mathcal{F}[\xi(x,y,u), \eta(x,y,u)] = 0$, which is the general solution for such equations. Determine the quasi-linear equations for which the following expressions are the general solutions.

(a) $\mathcal{F}(x+u, y+u) = 0$

(b) $\mathcal{F}(x+u, yu) = 0$

(c) $\mathcal{F}(xu, yu) = 0$

(d) $\mathcal{F}(x^2+y^2, xu) = 0$

(e) $\mathcal{F}(xy, x/u) = 0$

(f) $\mathcal{F}(x\,e^y, x/u) = 0$

(g) $\mathcal{F}(xy, x^2+y^2+u^2) = 0$

(h) $\mathcal{F}(x^{-1}+y^{-1}, y^{-1}+u^{-1}) = 0$

3. The *Mathematica* command Eliminate can be used to carry out the steps in Problem 2 symbolically. The procedure for setting up the input to Eliminate is given below for eliminating the arbitrary function $\mathcal{F}$ from the expression $\mathcal{F}[f(x,y,z), g(x,y,z)] = 0$, where f and g are specified functions and z is regarded as a function of x and y. The following command sequence illustrates the procedure for the functions

$$f(x,y,z) = x+z, \qquad g(x,y,z) = y+z$$

```
f[x,y,z] = x + z[x,y]
g[x,y,z] = y + z[x,y]
Eqn   = F[f[x,y,z],g[x,y,z]] == 0
Eqn1 = D[Eqn,x]
Eqn2 = D[Eqn,y]
Eqn3 = D[Eqn,z]
{Eqn1,Eqn2,Eqn3} = {Eqn1,Eqn2,Eqn3}/.{z[x,y]->u,
  Derivative[1,0][z][x,y]->p,
  Derivative[0,1][z][x,y]->q,
  Derivative[1,0][F][f[x,y,z],g[x,y,z]]->F1,
  Derivative[0,1][F][f[x,y,z],g[x,y,z]]->F2}
Eliminate[{Eqn1,Eqn2,Eqn3},{F1,F2}][[1]]
```

Apply this procedure to verify the results obtained in Problem 2.

4. For a quasi-linear first-order partial differential equation in k independent variables

$$P_1 \frac{\partial u}{\partial x_1} + P_2 \frac{\partial u}{\partial x_2} + \cdots + P_k \frac{\partial u}{\partial x_k} = R$$

where the coefficient functions are functions of the independent variables x_i and the dependent variable u:

$$P_i = P_i(x_1, x_2, \ldots, x_k, u), \qquad R = R(x_1, x_2, \ldots, x_k, u)$$

for $i = 1, 2, \ldots, k$. Show that the general solution can be written in the form

$$\mathcal{F}(u_1, u_2, \ldots, u_k) = 0$$

where $u_i = u_i(x_1, x_2, \ldots, x_k, u)$ where $i = 1, 2, \ldots, k$.

5. The *Mathematica* commands used in Problem 3 can be generalized to eliminate the arbitrary function $\mathcal{F}$ from

$$\mathcal{F}[f(x, y, z, w), g(x, y, z, w), h(x, y, z, w)] = 0$$

to obtain the quasi-linear equation for $w(x, y, z)$ for which this expression is the general solution. Thus, eliminate the arbitrary function in each of the following expressions to obtain the quasi-linear equation satisfied by w.

(a) $\mathcal{F}(xy, yz, x + w) = 0$

(b) $\mathcal{F}(x + y, y + z, z + w) = 0$

(c) $\mathcal{F}(xy, y/z, xw) = 0$

(d) $\mathcal{F}(x^2 + y^2, y^2 + z^2, x + w) = 0$

(e) $\mathcal{F}(x^2 + y^2 + z^2, xyz, xz + w) = 0$

6. Suppose we have an expression of the form

$$\mathcal{F}[\xi(x,y,u), \eta(x,y,u)] = 0$$

where $\xi(x,y,u)$ and $\eta(x,y,u)$ are known functions. By differentiating this expression in turn with respect to x and y, show that if

$$\begin{vmatrix} \xi_x + \xi_u u_x & \eta_x + \eta_u u_x \\ \xi_y + \xi_u u_y & \eta_y + \eta_u u_y \end{vmatrix} = 0$$

then $\mathcal{F}_\xi$ and $\mathcal{F}_\eta$ cannot be determined uniquely. Expand this determinant and obtain a first-order partial differential equation for u:

$$P\frac{\partial u}{\partial x} + Q\frac{\partial u}{\partial y} = R$$

where the functions P, Q, and R are given by

$$P = \xi_y \eta_u - \xi_u \eta_y, \quad Q = \xi_u \eta_x - \xi_x \eta_u, \quad R = \xi_x \eta_y - \xi_y \eta_x$$

The general solution to this equation is the expression given above where $\mathcal{F}$ is any once-differentiable function.

7. Use *Mathematica* to show directly that the expression

$$\mathcal{F}[\xi(x,y,u), \eta(x,y,u)] = 0$$

is a solution of the equation derived in Problem 6. To do this, first solve for the partial derivatives u_x and u_y:

```
Fx = D[F[xi[x,y,u[x,y]],eta[x,y,u[x,y]]],x];
Fy = D[F[xi[x,y,u[x,y]],eta[x,y,u[x,y]]],y];
ux = Derivative[1,0][u][x,y]/.
    Solve[Fx==0,Derivative[1,0][z][x,y]][[1]]/.u[x,y]->u;
uy = Derivative[0,1][u][x,y]/.
    Solve[Fy==0,Derivative[0,1][u][x,y]][[1]]/.u[x,y]->u;
```

Then define a generic function for P, Q and R:

```
J[f_,g_,x_,y_] := D[f,x] D[g,y] - D[f,y] D[g,x]
```

Thus, construct the equation derived in Problem 6 and show that an identity results.

8. Use the expression derived in Problem 6 to verify the results obtained in Problem 2.

9. Generalize the procedure in Problem 6 to obtain the quasi-linear equation satisfied by the following expression:

$$\mathcal{F}[f(x,y,u), g(x,y,u), h(x,y,u)] = 0$$

10. Obtain the general solutions of each of the following quasi-linear equations.

(a) $x\dfrac{\partial u}{\partial x} - y\dfrac{\partial u}{\partial y} = u$

(b) $(y^2 - u^2)\dfrac{\partial u}{\partial x} - xy\dfrac{\partial u}{\partial y} = ux$

(c) $x\dfrac{\partial u}{\partial x} + y\dfrac{\partial u}{\partial y} = 2u$

(d) $x^2\dfrac{\partial u}{\partial x} - y^2\dfrac{\partial u}{\partial y} = y^2 - x^2$

(e) $x^2 u\dfrac{\partial u}{\partial x} + (yu^2 - x^2)\dfrac{\partial u}{\partial y} = u^3$

In equations such as that in (b), the general solution can be obtained by first obtaining one of the independent solutions of the characteristic equations, in this case $yu = A$, where A is a constant. To obtain the second independent solution, solve for y in terms of u and A, carry out the integration and then re-express c_1 in terms of y and u.

11. For each of the equations in Problem 10, obtain the particular solutions that contain the line

$$C(t) = t\mathbf{i} + t\mathbf{j} + t\mathbf{k}$$

Express your solutions as an expression involving x, y and u.

12. Express the solutions obtained in Problem 11 in parametric form, as in Equation (2.52) and compare with the solutions obtained in Problem 11. For example, the solution for (a) can be expressed in terms of coordinates as

$$u(x,y) = x$$

or in parametric form as

$$r(s,t) = t^2 s^{-1}\mathbf{i} + s\mathbf{j} + t^2 s^{-1}\mathbf{k}$$

Use `ParametricPlot3D` to plot the parametric forms of the solution surfaces and, where possible, compare these surfaces with those expressed in terms of coordinates using `Plot3D`.

13. With the quantities calculated in Example 2.3, use `ParametricPlot3D` to display the solution surface in the three-dimensional space (x, t, u).

14. Show that (2.25) is the general solution of (2.1) by first observing that if f and g are chosen to be constants in (2.26), then ξ and η are themselves solution surfaces:

$$P\frac{\partial \xi}{\partial x} + Q\frac{\partial \xi}{\partial y} + R\frac{\partial \xi}{\partial u} = 0$$

$$P\frac{\partial \eta}{\partial x} + Q\frac{\partial \eta}{\partial y} + R\frac{\partial \eta}{\partial u} = 0$$

Use this observation to calculate expressions for $\mathcal{F}_x$ and $\mathcal{F}_y$ and show that

$$P\mathcal{F}_x + Q\mathcal{F}_y = \mathcal{F}_u(Pu_x + Qu_y - R) = 0$$

Hence, since $\mathcal{F}$ is an arbitrary function, deduce that

$$Pu_x + Qu_y = R$$

15. Suppose that a solution of Equation (2.1) is required along a characteristic, which leads to the condition in Equation (2.45):

$$\mathcal{F}(c_1, c_2) = 0$$

Thus, deduce that for *any* function $\mathcal{F}$ of two variables, the expression

$$\mathcal{G}(\xi, \eta) = \mathcal{F}(\xi, \eta) - \mathcal{F}(c_1, c_2)$$

satisfies both the differential equation and the boundary condition.

16. Consider the point of intersection of the two curves in Equations (2.36) and (2.50):

$$\mathbf{r}(t) = t\mathbf{i} + t c_1^{-1}\mathbf{j} + t c_2^{-1}\mathbf{k}, \qquad \mathcal{C}(s) = s^2\mathbf{i} + s\mathbf{j} + s^3\mathbf{k}$$

By equating each vector component separately, deduce that $\mathcal{C}$ intersects the characteristic $\mathbf{r}(t)$ corresponding to

$$c_1 = s, \qquad c_2 = s^{-1}$$

17. Show that the parametric representation in Equation (2.52) of the solution of (2.27) implies the functional relationship in Equation (2.51).

18. Show that the solution of

$$\frac{\partial u}{\partial t} + u\frac{\partial u}{\partial x} = 0$$

with the initial condition

$$u(x, 0) = 1 - x^2$$

is given by

$$u_\pm(x, t) = \frac{1}{2t^2}\left[2xt - 1 \pm \sqrt{1 - 4xt + 4t^2}\right]$$

To make the connection with the solution in Example 2.3, show that in the limit of small t, only $u_+(x,t)$ provides an allowed solution if the initial condition is to be satisfied. Then show that for sufficiently large t both $u_+(x,t)$ and $u_-(x,t)$ contribute to the allowed solution.

By calculating the time at which characteristics first cross, discuss how the contributions of both signs of this solution is related to the onset of multi-valuedness.

Use `Plot` to display the solutions using the appropriate ranges of x for $u_+(x,t)$ and $u_-(x,t)$.

19. Consider the equation

$$\frac{\partial u}{\partial t} + 2u\frac{\partial u}{\partial x} = 0$$

with the initial condition

$$u(x, 0) = \begin{cases} 1 & \text{if } x \leq 0 \\ 1 - x & \text{if } 0 < x < 1 \\ 0 & \text{if } 1 \leq x \end{cases}$$

Show that the characteristics for this solution first intersect at the point where $x=1$ and $t=t^* = \frac{1}{2}$ in the x-t plane and sketch the characteristics emanating from the x axis at $t=0$. What happens to the value of u as a function of x at $t=t^*$? Sketch the evolution of this solution, indicating the form at $t=0$, $t<t^*$, $t=t^*$ and $t>t^*$.

20. Use `ParametricPlot` to display the characteristics and the evolution of the solution in Problem 19, and use `ParametricPlot3D` to display the solution surface, as in Problem 13.

21. An equation that arises in the statistical mechanics of phase transitions is the first-order quasi-linear equation

$$\frac{\partial t}{\partial \ell} = 2t + (2 - d)\left(x - 1 + \frac{t}{1+t}\right)\frac{\partial t}{\partial x}$$

This is the **renormalization-group equation** for the 'spherical' model. The point $x = 1$, $t = 0$ is a singular point of the equation which corresponds to a phase transition analogous to a ferromagnetic transition. Solve the characteristic equations to obtain the invariants

$$c_1 = te^{-2\ell}$$

$$c_2 = 2t^{(2-d)/2}\left[x - 1 - \tfrac{1}{2}t\int_0^t \frac{s^{(2-d)/2}}{1+s}\,ds\right]$$

and so obtain the general solution in the form of x solved as a function of t and ℓ.

22. The **Hamilton–Jacobi equation** arises in classical mechanics in constructing a canonical transformation from coordinates q and momenta p at time t to coordinates q_0 and p_0 that do not explicitly depend on time and which represent the initial values of the coordinates and momenta at $t = 0$ (Goldstein, 1950). In constructing such a transformation, the equations relating the original and transformed canonical variables yield the solution of the mechanical problem: $q = q(q_0, p_0, t)$, $p = p(q_0, p_0, t)$.

Consider the Hamilton–Jacobi equation for a harmonic oscillator with mass m attached to a spring with stiffness k:

$$\frac{1}{2m}\left(\frac{\partial S}{\partial q}\right)^2 + \frac{kq^2}{2} + \frac{\partial S}{\partial t} = 0$$

The quantity S is called Hamilton's principle function. Note that this is a *nonlinear* equation of first order and second degree. However, by differentiating this equation with respect to q, obtain an equation for the quantity $S' = \partial S/\partial q$:

$$\frac{S'}{m}\frac{\partial S'}{\partial q} + \frac{\partial S'}{\partial t} = -kq$$

which is a quasi-linear equation to which the method of characteristics as developed in Section 2.2 can be applied. Show that the quantity

$$\frac{1}{2m}S'^2 + \tfrac{1}{2}kq^2 = \frac{1}{2m}\left(\frac{\partial S}{\partial q}\right)^2 + \frac{kq^2}{2} = c_1$$

is constant along the characteristics. Then, by writing the characteristic equations in the parametric form shown in Equation (2.17) obtain solutions to these equations subject to the following initial conditions at $s = 0$:

$$t(0) = t_0, \qquad q(0) = q_0, \qquad S'(0) = S'_0$$

In particular, show that S' can be identified with the momentum p of the particle and so give the quantity c_1 a physical interpretation. The characteristic equations for q and S' can now be solved for particular initial conditions. Thus, obtain the following solutions for $q(t)$ and $p(t)$ subject to the initial conditions where the mass is initially displaced by a distance q_0 with no initial momentum:

$$q(t) = q_0 \cos \omega_0 t, \qquad p(t) = \sqrt{2mE - m^2\omega_0^2 q^2} \sin \omega_0 t$$

where $\omega_0 = \sqrt{k/m}$ is the natural frequency of the oscillator. Use SpaceCurve to show that these solutions represent spirals on the constant energy surfaces.

23. By using a *Mathematica* procedure analogous to that in Problem 3, or by direct calculation, eliminate the arbitrary constants a and b in the following expressions to obtain the partial differential equations for which the expressions are the complete integral.

(a) $u = ax + by + a^2 b^2$

(b) $u^2 = ax^2 + y - b$

(c) $(xu)^2 = ax + (y+b)^2$

(d) $(x+u)^2 = ax + (y+b)^2$

(e) $(ax+u)^2 = ax + (y+b)^2$

(f) $u = (ax+by)^2$

(g) $u^2 = (ax+y+b)^{1/2}$

24. For a general first-order partial differential equation,

$$F(x,y,x,p,q) = 0$$

the normals to all possible solution surfaces sweep out a cone $\mathcal{N}$ that is given by the equation

$$F\left(x,y,u,-\frac{x-x'}{u-u'},-\frac{y-y'}{u-u'}\right) = 0$$

Similarly, the tangent planes to the solution surfaces determine an envelope cone whose equation is obtained by eliminating p and q from

$$F(x,y,u,p,q) = 0$$

$$(x-x')\frac{\partial F}{\partial q} - (y-y')\frac{\partial F}{\partial p} = 0$$

$$p(x-x') + q(y-y') - (u-u') = 0$$

Show that for a quasi-linear equation, $Pp + Qq = R$, that $\mathcal{N}$ reduces to

$$P(x-x') + Q(y-y') + R(u-u') = 0$$

and, by eliminating p and q from the second set of equations, T becomes

$$\frac{x-x'}{P} = \frac{y-y'}{Q} = \frac{u-u'}{R}$$

25. Consider the general quasi-linear equation

$$F(x,y,u,p,q) = qQ(x,y,u) + pP(x,y,u) - R(x,y,u) = 0$$

where $p = u_x$ and $q = u_y$. Show that the characteristic equations (2.96) for this equation reduce to the corresponding equations (2.19) for the characteristics of a quasi-linear equation:

$$\frac{dx}{P} = \frac{dy}{Q} = \frac{du}{R}$$

so the quantities p and q do not appear in these equations.

26. Determine complete integrals for each of the quasi-linear equations in Problem 1. In (b) and (e) perform a change of variables to transform the equations into one of the standard forms discussed in Section 2.4.

27. Determine complete integrals for each of the following nonlinear partial differential equations.

(a) $\dfrac{\partial u}{\partial x}\dfrac{\partial u}{\partial y} = u^2$

(b) $\left(\dfrac{\partial u}{\partial x}\right)^2 + \left(\dfrac{\partial u}{\partial y}\right)^2 = u$

(c) $\left(\dfrac{\partial u}{\partial x}\right)^2 + \left(\dfrac{\partial u}{\partial y}\right)^2 = 1$

(d) $\left(\dfrac{\partial u}{\partial x}\right)^2 + \dfrac{\partial u}{\partial y} + x^2 = 0$

(e) $x^2\left(\dfrac{\partial u}{\partial x}\right)^2 + y^2\left(\dfrac{\partial u}{\partial x}\right)^2 = 1$

(f) $x^2\left(\dfrac{\partial u}{\partial x}\right)^2 + y^2\left(\dfrac{\partial u}{\partial x}\right)^2 = u^2$

(g) $x^4\left(\dfrac{\partial u}{\partial x}\right)^2 + y^2\left(\dfrac{\partial u}{\partial x}\right)^2 = u$

(h) $y^2 \left(\dfrac{\partial u}{\partial x}\right)^2 + x^2 \left(\dfrac{\partial u}{\partial y}\right)^2 = x^2 y^2 u^2$

In (e)–(h), perform a change of variables to transform the equations into one of the standard forms discussed in Section 2.4.

28. Consider the nonlinear partial differential equation

$$\frac{\partial u}{\partial x}\frac{\partial u}{\partial y} = xy$$

By solving the parametric equations (2.95) and (2.98), obtain the solutions in the form

$x(s) = x_0 \cosh s + q_0 \sinh s$

$y(s) = y_0 \cosh s + p_0 \sinh s$

$u(s) = u_0 - \frac{1}{2}(x_0 p_0 + y_0 q_0) + \frac{1}{2}(x_0 p_0 + y_0 q_0)\cosh 2s + (x_0 y_0 + p_0 q_0)\sinh 2s$

$p(s) = q_0 \cosh s + x_0 \sinh s$

$q(s) = p_0 \cosh s + y_0 \sinh s$

Then, use Equations (2.99) to obtain the complete integral

$$(u - \beta)^2 = (x^2 - \alpha) y^2$$

Consider the solution that is required to contain the line $u = -y$ in the y-z plane ($x=0$). Show that the parametric form of the solution is

$$x(s,t) = -\sinh s$$
$$y(s,t) = t \cosh s$$
$$u(s,t) = -\tfrac{1}{2}t - \tfrac{1}{2}t \cosh s$$

and that the complete integral reduces to

$$u(x,y) = -y\sqrt{x^2 + 1}$$

Show that these two forms of solution are equivalent. Then display the parametric form using `ParametricPlot3D` and the latter form with `Plot3D`.

29. In two spatial dimensions, the **eikonal equation** of geometrical optics (Sommerfeld, 1954) is the nonlinear first-order equation

$$\left(\frac{\partial u}{\partial x}\right)^2 + \left(\frac{\partial u}{\partial y}\right)^2 = n^2$$

where n is the index of refraction of the medium, which we assume to be a constant. Using the methods of Section 2.4, solve for $x(s)$, $y(s)$ and $u(s)$ and obtain the complete integral in the form

$$u(x,y) = (x - x_0)p_0 + (y - y_0)q_0 + u_0$$
$$= xp_0 + yq_0 + u_0 - x_0p_0 - y_0q_0$$

x_0, y_0, u_0, p_0 and q_0 are the initial values of x, y, u, p and q, respectively, with p_0 and q_0 related by $p_0^2 + q_0^2 = n^2$. Thus, with p_0 and q_0 written as $p_0 = n \cos \alpha$ and $q_0 = n \sin \alpha$, this condition is satisfied, and the complete integral for the eikonal equation can be written as

$$u(x,y) = n(x \cos \alpha + y \sin \alpha) + \beta$$

Suppose a solution of the eikonal equation is required to satisfy the boundary condition that $u(x,x) = nx$. Show that the strip condition (2.100) for p_0 and q_0 is

$$p_0 + q_0 = n$$

Hence, deduce that $\alpha = 0$ or $\alpha = \tfrac{1}{2}\pi$. Thus, obtain the solution to the eikonal equation:

$$u(x,y) = n(x \cos \alpha + y \sin \alpha) = \begin{cases} nx, & \text{if } \alpha = 0; \\ ny, & \text{if } \alpha = \tfrac{1}{2}\pi \end{cases}$$

Chapter 3

Second-Order Partial Differential Equations

Many physical phenomena are described by partial differential equations with higher-order derivatives. The presence of these derivatives has profound implications both for the general behavior of the solutions and for the types of boundary conditions that are appropriate for a particular equation. As an illustration of this, consider the three types of second-order equations commonly encountered in applications; the wave equation:

$$\frac{\partial^2 u}{\partial x^2} - \frac{\partial^2 u}{\partial t^2} = 0 \tag{3.1}$$

the diffusion, or heat, equation:

$$\frac{\partial^2 u}{\partial x^2} - \frac{\partial u}{\partial t} = 0 \tag{3.2}$$

and Laplace's equation:

$$\frac{\partial^2 u}{\partial x^2} + \frac{\partial^2 u}{\partial y^2} = 0 \tag{3.3}$$

We have used x and y to signify spatial variables and t to signify the time. The wave equation describes the propagation of a disturbance, such as the vibrations of a string. Since this equation contains a second-order time derivative, two initial conditions are required to determine how the solution changes with time. These can be taken as the initial displacement and the initial rate of the displacement. The diffusion equation describes the flow of a quantity, such as heat, or a concentration of particles. The total amount of the quantity does not change, i.e. both heat and mass are *conserved* quantities, but their distribution within a medium changes. The presence of a first-order time derivative means that the time development of the solution is completely determined once the initial temperature or mass distribution is specified, and once the distribution of any sources and sinks is known. The first-order time derivative also means that the equation is *irreversible*, in that changing the sign of the time changes the behavior of the solution. The wave equation, by contrast, is not affected by such a variable change, and so is reversible. Laplace's equation is used in two-dimensional electrostatics, incompressible fluid flow and for steady-state heat conduction problems. The solution within a region is completely determined once the potential or field has been specified around a boundary enclosing that region.

These three examples provide a strong indication that the theory of higher-order equations is much richer than that for first-order equations. In this chapter, we will develop the theory of second-order equations by focusing on quasi-linear equations of the form

$$a(x,y)\frac{\partial^2 u}{\partial x^2} + b(x,y)\frac{\partial^2 u}{\partial x \partial y} + c(x,y)\frac{\partial^2 u}{\partial y^2} = F(x,y,u,u_x,u_y) \qquad (3.4)$$

for a function u of the independent variables x and y. We will confine ourselves to the case where a, b and c are each functions only of x and y, and not functions of u. This form includes the three equations in (3.1)–(3.3) as particular cases. The different types of initial and boundary conditions used in solving the second-order equations mentioned above will be seen to be distinguished primarily by the quantities a, b and c, so this form is general enough for our purposes.

Much of our discussion will focus on Equations (3.1)–(3.3) because they are the simplest equation in each class. However, the concepts and techniques that will be developed in this chapter and the next few chapters can be applied with little or no modification to other equations of the general form (3.4) that are encountered in applications. Among those that we will solve in the following chapters are the telegraph equation,

$$\frac{\partial^2 u}{\partial x^2} - \frac{\partial^2 u}{\partial t^2} - \frac{\partial u}{\partial t} = 0 \qquad (3.5)$$

which is satisfied by the potential of a telegraph cable and is of the same type as the wave equation (3.1), the Fokker–Planck equation,

$$\frac{\partial u}{\partial t} = \frac{\partial^2 u}{\partial x^2} + x\frac{\partial u}{\partial x} + u \qquad (3.6)$$

which is used in non-equilibrium statistical mechanics and has several common features with the diffusion equation (3.2), and the two-dimensional diffusion equation,

$$\frac{\partial u}{\partial t} = \frac{\partial^2 u}{\partial x^2} + \frac{\partial^2 u}{\partial y^2} \qquad (3.7)$$

In the steady state, where the time derivative of u vanishes, this equation reduces to Laplace's equation (3.3). In fact, some of the qualitative features of solutions to Laplace's equation can be deduced directly when viewed in terms of steady-state solutions of (3.7).

3.1 Equations with Constant Coefficients

As an introduction to the classification of second-order equations, we consider a simplified form of (3.4) with $F=0$ and with *constant* coefficients a, b and c:

$$a\frac{\partial^2 u}{\partial x^2} + b\frac{\partial^2 u}{\partial x \partial y} + c\frac{\partial^2 u}{\partial y^2} = 0 \tag{3.8}$$

Equation (3.8) includes as special cases the wave equation in (3.1) and Laplace's equation in (3.3).

To examine the types of solutions that are obtained for different choices of a, b and c, we begin with a trial solution of the form:

$$u(x,y) = f(mx + y) \tag{3.9}$$

for a twice-differentiable function f, and for a constant m to be determined by the requirement that (3.9) be a solution to (3.8). This is a useful starting point for any nth-order partial differential equation with constant coefficients involving only nth-order derivatives (Problem 1). Using a prime to signify the derivative of f with respect to its argument, the required second partial derivatives of (3.9) are

$$\frac{\partial^2 u}{\partial x^2} = m^2 f'', \quad \frac{\partial^2 u}{\partial x \partial y} = m f'', \quad \frac{\partial^2 u}{\partial y^2} = f'' \tag{3.10}$$

Substitution of (3.10) into (3.8) produces

$$(am^2 + bm + c) f'' = 0 \tag{3.11}$$

from which we conclude that either $f'' = 0$ or the quantity enclosed in parentheses vanishes (or both). The case where $f'' = 0$ yields the solution $f = f_0 + mx + y$ which, being a linear function of x and y, is expressed in terms of two arbitrary *constants*, f_0 and m, rather than two arbitrary *functions* and is therefore not the most *general* solution. Thus, we disregard this possibility and instead require

$$am^2 + bm + c = 0 \tag{3.12}$$

In other words, by choosing a particular linear combination of the independent variables x and y, we obtain a solution of (3.8) for *any* function f that is twice differentiable. Solving the quadratic equation in (3.12) for m we obtain the two solutions

$$m = \frac{1}{2a}\left(-b + \sqrt{b^2 - 4ac}\right), \qquad m = \frac{1}{2a}\left(-b - \sqrt{b^2 - 4ac}\right) \qquad (3.13)$$

The importance of the coefficients in (3.8) is now evident, since the sign of the discriminant in (3.13) is crucial to determining the number and type of the solutions of (3.12). There are three cases to consider.

Case I. $b^2 - 4ac > 0$. There are two distinct real solutions of (3.12), m_1 and m_2. Any function of either of the arguments $m_1 x + y$ or $m_2 x + y$ solves (3.8). Therefore, the general solution to (3.8) is

$$u(x, y) = \mathcal{F}(m_1 x + y) + \mathcal{G}(m_2 x + y) \qquad (3.14)$$

where $\mathcal{F}$ and $\mathcal{G}$ are required only to be twice-differentiable functions; otherwise they may be chosen arbitrarily. There is another way to see that (3.14) is the most general solution of (3.8). In terms of the variables $\xi = m_1 x + y$ and $\eta = m_2 x + y$, Equation (3.8) is transformed into

$$\frac{\partial^2 u}{\partial \xi \partial \eta} = 0 \qquad (3.15)$$

By integrating this equation first with respect to ξ and then with respect to η, we obtain the solution to (3.14) as $u(\xi, \eta) = \mathcal{F}(\xi) + \mathcal{G}(\eta)$, where $\mathcal{F}$ and $\mathcal{G}$ are any twice-differentiable functions of integration. In terms of the original variables x and y, this solution is the same as that given in Equation (3.14).

Equations for which the coefficients in (3.8) take on values such that $b^2 - 4ac > 0$ are termed **hyperbolic**, and include the one-dimensional wave equation (3.1) as a particular case.

Case II. $b^2 - 4ac = 0$. We first assume that $b \neq 0$ and $a \neq 0$ (which implies that $c \neq 0$). Then there is a single degenerate root of (3.12) with the value $m_1 = -b/2a$. The simplest way to obtain the general solution for this case is to follow the path taken for the hyperbolic case and transform the original equation (3.8) with the change of variables $\xi = m_1 x + y$ and $\eta = y$:

$$a\frac{\partial^2 u}{\partial x^2} + b\frac{\partial^2 u}{\partial x \partial y} + c\frac{\partial^2 u}{\partial y^2}$$

$$= (am_1^2 + bm_1 + c)\frac{\partial^2 u}{\partial \xi^2} + (bm_1 + 2c)\frac{\partial^2 u}{\partial \xi \partial \eta} + c\frac{\partial^2 u}{\partial \eta^2} \qquad (3.16)$$

The first term on the right-hand side of (3.16) vanishes because m_1 is a root of (3.12). The second term vanishes because the coefficient can be written as $bm_1+c = -(b^2-4ac)/2a$, which vanishes by hypothesis. Therefore, since $c \neq 0$, Equation (3.8) is transformed to the simple form

$$\frac{\partial^2 u}{\partial \eta^2} = 0 \tag{3.17}$$

The general solution of this equation is $u(\xi,\eta) = \mathcal{F}(\xi) + x_2 \mathcal{G}(\eta)$, where $\mathcal{F}$ and $\mathcal{G}$ are again twice-differentiable, but otherwise arbitrary, functions of integration. In terms of the original variables, the solution is

$$u(x,y) = \mathcal{F}(m_1 x + y) + y\mathcal{G}(m_1 x + y) \tag{3.18}$$

If $b=0$ and $a=0$, then (3.8) becomes

$$\frac{\partial^2 u}{\partial y^2} = 0 \tag{3.19}$$

the general solution to which is clearly

$$u(x,y) = \mathcal{F}(x) + y\mathcal{G}(x) \tag{3.20}$$

with analogous expressions if $b=0$ and $c=0$.

Equations for which $b^2 - 4ac = 0$ are termed **parabolic**. The one-dimensional diffusion equation (3.2) is an example of a parabolic equation, but is not included as a special case of (3.8) because of the presence of the first-order derivative.

Case III. $b^2 - 4ac < 0$. There are two distinct *complex* roots m_1 and m_2 of (3.12) which are complex conjugates: $m_2 = m_1^*$. The general solution may be written in the form of Equation (3.14), but with complex arguments of real functions:

$$u(x,y) = \mathcal{F}(m_1 x + y) + \mathcal{G}(m_1^* x + y) \tag{3.21}$$

and (3.8) may be reduced to the form (3.15) with $\xi = \eta^*$, with a corresponding expression for the general solution. We can obtain an alternative simplified form of (3.8) by introducing new *real* variables v_1 and v_2 by $\xi = v_1 + iv_2$ and $\eta = v_1 - iv_2$. Carrying out the required operations, we find

$$\frac{\partial^2 u}{\partial \xi \partial \eta} = \frac{1}{4}\frac{\partial^2 u}{\partial v_1^2} + \frac{1}{4}\frac{\partial^2 u}{\partial v_2^2} \tag{3.22}$$

Thus, Equation (3.8) may be written as

$$\frac{\partial^2 u}{\partial v_1^2} + \frac{\partial^2 u}{\partial v_2^2} = 0 \tag{3.23}$$

Equations for which $b^2 - 4ac < 0$ are termed **elliptic**. Equation (3.23) shows that the two-dimensional Laplace equation is an elliptic equation.

Since the functions $\mathcal{F}$ and $\mathcal{G}$ appearing in (3.21) are real, the real and imaginary parts of this expression are separately solutions of Laplace's equation. This has important consequences for solving two-dimensional potential problems, such as those in electrostatics and fluid mechanics. The relationship between complex variables and two-dimensional electrostatics is discussed by Panofsky and Phillips (1962), and will be explored later in this chapter.

In order for the classification of Equation (3.8) as hyperbolic, parabolic or elliptic to be meaningful, a change of coordinates should leave the classification of an equation unaltered. To examine the behavior of our classification under a general transformation of coordinates, we introduce new variables $\xi = \xi(x, y)$ and $\eta = \eta(x, y)$ and determine the classification of the transformed equation by calculating the quantity $b^2 - 4ac$ for the transformed equation. The required quantities for transforming the original Equation (3.8) from the (x, y) to the (ξ, η) coordinate system are expressions for the second derivatives. A simple application of the chain yields

$$u_{xx} = u_{\xi\xi}\xi_x^2 + 2u_{\xi\eta}\xi_x\eta_x + u_{\eta\eta}\eta_x^2 + u_\xi\xi_{xx} + u_\eta\eta_{xx}$$
$$u_{xy} = u_{\xi\xi}\xi_x\xi_y + u_{\xi\eta}\xi_x\eta_y + u_{\xi\eta}\xi_y\eta_x + u_{\eta\eta}\eta_x\eta_y + u_\xi\xi_{xy} + u_\eta\eta_{xy} \quad (3.24)$$
$$u_{yy} = u_{\xi\xi}\xi_y^2 + 2u_{\xi\eta}\xi_y\eta_y + u_{\eta\eta}\eta_y^2 + u_\xi\xi_{yy} + u_\eta\eta_{yy}$$

Substituting (3.24) into (3.8) and grouping coefficients of the same second derivatives, we obtain

$$au_{xx} + bu_{xy} + cu_{yy}$$
$$= (a\xi_x^2 + b\xi_x\xi_y + c\xi_y^2)u_{\xi\xi} + (2a\xi_x\eta_x + b\xi_x\eta_y$$
$$+ b\xi_y\eta_x + 2c\xi_y\eta_y)u_{\xi\eta} + (a\eta_x^2 + b\eta_x\eta_y + c\eta_y^2)u_{\eta\eta} + \cdots$$
$$\equiv Au_{\xi\xi} + Bu_{\xi\eta} + Cu_{\eta\eta} + \cdots \quad (3.25)$$

In the last line of this equation, the quantities A, B and C are used as abbreviations for the coefficients of $u_{\xi\xi}$, $u_{\xi\eta}$ and $u_{\eta\eta}$, respectively. The terms involving first-order derivatives of u have not been explicitly written out in (3.25) since they are not required for our calculation. To calculate the quantity $B^2 - 4AC$, which is the discriminant in the new coordinate system, we have

$$B^2 = 4a^2\xi_x^2\eta_x^2 + b^2\xi_x^2\eta_y^2 + b^2\xi_y^2\eta_x^2 + 4c^2\xi_y^2\eta_y^2 + 4ab\xi_x^2\eta_x\eta_y$$
$$+ 4ab\xi_x\xi_y\eta_x^2 + 8ac\xi_x\xi_y\eta_x\eta_y + 2b^2\xi_x\xi_y\eta_x\eta_y$$
$$+ 4bc\xi_x\xi_y\eta_y^2 + 4bc\xi_y^2\eta_x\eta_y \quad (3.26)$$

and

$$4AC = 4a^2\xi_x^2\eta_x^2 + 4ab\xi_x^2\eta_x\eta_y + 4ac\xi_x^2\eta_y^2 + 4ab\eta_x^2\xi_x\xi_y$$
$$+ 4b^2\xi_x\xi_y\eta_x\eta_y + 4bc\xi_x\xi_y\eta_y^2 + 4ac\xi_y^2\eta_x^2$$
$$+ 4bc\xi_y^2\eta_x\eta_y + 4c^2\xi_y^2\eta_y^2 \tag{3.27}$$

Thus,

$$\begin{aligned}B^2 - 4AC &= (b^2 - 4ac)(\xi_x^2\eta_y^2 - 2\xi_x\xi_y\eta_x\eta_y + \xi_y^2\eta_x^2)\\ &= (b^2 - 4ac)(\xi_x\eta_y - \xi_y\eta_x)^2\\ &\equiv (b^2 - 4ac)J^2\end{aligned} \tag{3.28}$$

where J is the Jacobian of the transformation,

$$J = \begin{vmatrix} \xi_x & \xi_y \\ \eta_x & \eta_y \end{vmatrix} \tag{3.29}$$

which determines the change of the differential volume (area) element under change of coordinates: $dx\,dy = J\,d\xi\,d\eta$. Equation (3.28) shows that if $J \neq 0$, the classification of (3.8) as hyperbolic, parabolic or elliptic is independent of the coordinate system and is therefore a property of the equation. In the more general case where a, b and c are functions of x and y, the sign of the quantity $b^2(x,y) - 4a(x,y)c(x,y)$ may be different in different regions of the x-y plane, so the classification of second-order quasi-linear partial differential equations is not necessarily valid over the entire x-y plane. In the next section we will see the importance of this classification of second-order equations even for equations where $F \neq 0$ in (3.4).

The operations carried out in Equations (3.24)–(3.29), which involve a straightforward but somewhat lengthy sequence of algebraic steps, can be performed easily in *Mathematica*. We first define the function u as a function of $\xi(x,y)$ and $\eta(x,y)$:

```
f=u[xi[x,y],eta[x,y]];
```

We now carry out the operations required to obtain (3.29). The original differential equation is transformed into the new variables, with the derivatives carried out using the chain rule automatically.

```
Eqn=Expand[a D[f,{x,2}] + b D[f,x,y] +c D[f,{y,2}]];
```

The coefficients A, B and C of the second-order derivatives in the transformed variables are then obtained using the command `Coefficient`. For example,

```
A=Coefficient[Eqn,Derivative[2,0][u][xi[x,y],eta[x,y]]]

        (0,1)   2         (0,1)           (1,0)
   c xi      [x, y]  + b xi      [x, y] xi      [x, y] +

        (1,0)   2
   a xi      [x, y]
```

Similarly, for obtaining B and C we use the commands (for brevity, the output from these commands is suppressed)

```
B=Coefficient[Eqn,Derivative[1,1][u][xi[x,y],eta[x,y]]];

C=Coefficient[Eqn,Derivative[0,2][u][xi[x,y],eta[x,y]]];
```

Finally, the quantity $B^2 - 4AC$ is constructed, and the result in (3.28) is obtained:

```
Factor[B^2-4 A C]

   2              (0,1)        (1,0)
 (b  - 4 a c) (eta     [x, y] xi     [x, y] -

      (0,1)         (1,0)    2
    xi     [x, y] eta     [x, y])
```

3.2 Boundary Conditions of Second-Order Equations

The discussion in the preceding section showed how the general solutions of second-order equations are influenced by the coefficients of the second-order derivatives for a special class of second-order partial differential equations. To proceed further to address the question of boundary conditions, we adopt in this section a different approach. We will attempt to calculate the solution to (3.4) by performing a Taylor series expansion from a curve C_0 along which $u(x, y)$ is specified. This is given in terms of the coordinate functions $x_0(s)$ and $y_0(s)$. As s is varied a curve is traced out which, because it lies in the x-y plane, is called a **base curve**:

$$C_0(s) = [x_0(s), y_0(s)] \tag{3.30}$$

The values taken by u along C_0, together with $x_0(s)$ and $y_0(s)$, produce another space curve, C, that lies in the solution surface:

$$C(s) = \{x_0(s), y_0(s), u[x_0(s), y_0(s)]\} \tag{3.31}$$

Thus, specifying the values that u must take when x and y trace out a particular base curve C_0 is equivalent to specifying a curve through which the solution surface must pass. From the discussion of Section 2.2, we recognize that requiring the solution surface to contain a particular space curve is sufficient to determine only one arbitrary function. Since the general solution of a second-order quasi-linear equation is composed of two arbitrary functions, a second condition is required to uniquely determine the solution. This will emerge naturally in the following discussion.

In order to carry out a Taylor series expansion of u about C_0, the quantities F, u, a, b and c in (3.4) are required to have convergent power series expansions in a neighborhood of C_0. The functions that make up the boundary conditions are also assumed to satisfy this requirement. These assumptions are somewhat restrictive but enable the conditions determining the convergence of the Taylor series for u to be systematically calculated. The resulting existence and uniqueness of the solution u to (3.4) with appropriately specified boundary conditions is known as the **Cauchy–Kowalewski theorem** (Garabedian, 1964). We will not be concerned with the details of the proof of this theorem here, though we will show how this method can be used to construct solutions later in this chapter.

The expansion of u about C_0 takes the form:

$$u(x,y) = u(x,y)\Big|_{C_0} + \frac{\partial u}{\partial x}\Big|_{C_0}(x-x_0) + \frac{\partial u}{\partial y}\Big|_{C_0}(y-y_0)$$

$$+ \frac{1}{2}\frac{\partial^2 u}{\partial x^2}\Big|_{C_0}(x-x_0)^2 + \frac{\partial^2 u}{\partial x \partial y}\Big|_{C_0}(x-x_0)(y-y_0)$$

$$+ \frac{1}{2}\frac{\partial^2 u}{\partial y^2}\Big|_{C_0}(y-y_0)^2 + \cdots \quad (3.32)$$

The boundary conditions and the requirement that u be a solution of (3.4) will be used for the evaluation of the partial derivatives in (3.32). Once these quantities are evaluated along C_0, the solution $u(x,y)$ is uniquely determined within the radius of convergence of the series.

The first term on the right-hand side of (3.32) is simply the value of u taken for points lying along the base curve C_0, i.e. $u[x_0(s), y_0(s)]$. This forms one of the boundary conditions of (3.4), and so may be considered to be a known function.

To evaluate the two first-order terms in (3.32), we observe that the derivative of u along C_0 is, according to the chain rule, given by

$$\frac{du}{ds} = \frac{\partial u}{\partial x}\Big|_{C_0}\frac{dx_0}{ds} + \frac{\partial u}{\partial y}\Big|_{C_0}\frac{dy_0}{ds} \quad (3.33)$$

Since $x_0(s)$, $y_0(s)$ and $u[x_0(s), y_0(s)]$ are known functions that are assumed to have derivatives of all orders, du/ds, dx_0/ds and dy_0/ds are also known functions. Thus, Equation (3.33) provides a relation involving the two unknown functions u_x and u_y evaluated along C_0. Since we cannot determine both of these functions from one equation, an additional relation is needed that specifies u_x, u_y or a linear combination of u_x and u_y along C_0 that is independent of (3.33). This relation is a requirement for solving the differential equation and forms the second boundary condition. Thus, up to first-order terms in (3.32), the expansion of u about C_0 is determined completely by the boundary conditions. We have not yet made use of the fact that u is a solution of a particular differential equation. Had the quasi-linear equation being solved been of first order, then the equation itself would have provided the required second condition for evaluating u_x and u_y along C_0 (Problem 8).

Proceeding to the second-order terms, the determination of the three unknown quantities u_{xx}, u_{xy} and u_{yy} along C_0 requires three equations. Two equations are obtained by differentiating the *known* functions u_x and u_y with respect to s (along C_0):

$$\frac{du_x}{ds} = \left.\frac{\partial^2 u}{\partial x^2}\right|_{C_0}\frac{dx_0}{ds} + \left.\frac{\partial^2 u}{\partial x \partial y}\right|_{C_0}\frac{dy_0}{ds} \qquad (3.34a)$$

$$\frac{du_y}{ds} = \left.\frac{\partial^2 u}{\partial x \partial y}\right|_{C_0}\frac{dx_0}{ds} + \left.\frac{\partial^2 u}{\partial y^2}\right|_{C_0}\frac{dy_0}{ds} \qquad (3.34b)$$

The third equation is the differential equation (3.4) evaluated along C_0:

$$a(s)\left.\frac{\partial^2 u}{\partial x^2}\right|_{C_0} + b(s)\left.\frac{\partial^2 u}{\partial x \partial y}\right|_{C_0} + c(s)\left.\frac{\partial^2 u}{\partial y^2}\right|_{C_0} = F(s) \qquad (3.35)$$

where the notation $a(s)$, $b(s)$, $c(s)$ and $F(s)$ means that the indicated quantities are evaluated along C_0, e.g. $a(s) \equiv a[u(s), v(s)]$. Using a prime to indicate differentiation with respect to s, Equations (3.34) and (3.35) together may be written in matrix form as

$$\begin{pmatrix} x'_0 & y'_0 & 0 \\ 0 & x'_0 & y'_0 \\ a & b & c \end{pmatrix} \begin{pmatrix} u_{xx} \\ u_{xy} \\ u_{yy} \end{pmatrix} = \begin{pmatrix} u'_x \\ u'_y \\ F \end{pmatrix} \qquad (3.36)$$

Equations (3.36) have unique solutions for the second partial derivatives of u evaluated along C_0 provided that the determinant of the matrix of coefficients,

$$\Delta = \begin{vmatrix} x'_0 & y'_0 & 0 \\ 0 & x'_0 & y'_0 \\ a & b & c \end{vmatrix} = c(x'_0)^2 - bx'_0 y'_0 + a(y'_0)^2 \qquad (3.37)$$

is nonvanishing. If $\Delta \neq 0$, then the solutions of (3.36) for u_{xx}, u_{xy} and u_{yy} can be obtained using Cramer's rule:

$$u_{xx} = \frac{1}{\Delta} \begin{vmatrix} u'_x & y'_0 & 0 \\ u'_y & x'_0 & y'_0 \\ F & b & c \end{vmatrix}, \quad u_{xy} = \frac{1}{\Delta} \begin{vmatrix} x'_0 & u'_x & 0 \\ 0 & u'_y & y'_0 \\ a & F & c \end{vmatrix}, \quad u_{yy} = \frac{1}{\Delta} \begin{vmatrix} x'_0 & y'_0 & u'_x \\ 0 & x'_0 & u'_y \\ a & b & F \end{vmatrix}$$

(3.38)

Before studying the implications of the condition $\Delta = 0$, we examine the requirements for evaluating the higher-order derivatives in (3.32). Determining the four third-order derivatives of u requires four equations. Three equations are obtained by differentiating the three second-order partial derivatives with respect to s, and a fourth may be obtained by differentiating (3.4) with respect to either x or y and evaluating the resulting equation along C_0. These equations have unique solutions again provided that the matrix of coefficients is nonvanishing, which yields the condition $\Delta \neq 0$, with Δ given by (3.37). We will demonstrate below that the kth-order derivatives may be evaluated also subject to the condition that $\Delta \neq 0$. Thus, if $\Delta \neq 0$, the Taylor series (3.32) uniquely specifies the solution of (3.4) within the radius of convergence of the series.

The appearance of the condition $\Delta \neq 0$ in the evaluation of the higher-order terms in (3.32) can be demonstrated explicitly using *Mathematica*. The kth-order terms in (3.32) involve the $k+1$ distinct partial derivatives of u with respect to x and y, since the assumed analyticity of u implies that the order of differentiation is unimportant. The function u and the second-order terms in the differential equation are first defined:

```
x=x0[s];
y=y0[s];
f=u[x,y];

Eqn=a[s] D[f,{x,2}] + b[s] D[f,x,y] + c[s] D[f,{y,2}];
```

The $k+1$ equations for the kth-order derivatives of u are obtained from differentiating the derivatives of order $k-1$ of f with respect to s, which yields k equations, and differentiating (3.4) $k-2$ times. The resulting matrix of coefficients, M[k_] is constructed as a Table:

```
M[k_]:=Table[If[i<k+1,Coefficient[D[f,s,{x,k-i},{y,i-1}],
    Derivative[k+1-j,j-1][u][x,y]],
    Coefficient[D[Eqn,{x,k-2}],
    Derivative[k+1-j,j-1][u][x,y]]],
    {i,1,k+1},{j,1,k+1}];
```

88 Partial Differential Equations with Mathematica

The argument **k** has been defined to correspond to the order of the partial derivatives that are required. The coefficients contained in **M** can be displayed as a matrix using **MatrixForm** and the determinant of **M** can be evaluated using **Det**. Examples are shown for the second and fourth-order terms. In both cases, the determinant is proportional to (3.37). This can be seen from the structure of **M** as a function of k and the evaluation of the determinant using minors. In this way, the condition for obtaining solutions of kth-order derivatives in the expansion (3.32) can be extended to all orders.

MatrixForm[M[2]]

x0'[s]	y0'[s]	0
0	x0'[s]	y0'[s]
a[s]	b[s]	c[s]

Simplify[Det[M[2]]]

$$c[s]\, x0'[s]^2 - b[s]\, x0'[s]\, y0'[s] + a[s]\, y0'[s]^2$$

MatrixForm[M[4]]

x0'[s]	y0'[s]	0	0	0
0	x0'[s]	y0'[s]	0	0
0	0	x0'[s]	y0'[s]	0
0	0	0	x0'[s]	y0'[s]
a[s]	b[s]	c[s]	0	0

Simplify[Det[M[4]]]

$$y0'[s]\, (c[s]\, x0'[s]^2 - b[s]\, x0'[s]\, y0'[s] + a[s]\, y0'[s]^2)$$

We now consider the consequences of the vanishing of the determinant in (3.37). The condition $\Delta = 0$ means that Equations (3.34) and (3.35) for u_{xx}, u_{xy} and u_{yy} are not linearly independent and so do not have a unique solution because they form an underspecified set of equations. Since Δ is constructed from a matrix equation involving both the differential equation (3.4) and the base curve C_0, the vanishing of Δ can be used to find an equation for x_0 and y_0. From the definition (3.37), the equation $\Delta = 0$ may be interpreted as a quadratic equation for either y_0' or x_0'. Choosing the former, we have the solutions

$$y_0' = \frac{1}{2a} \left(b \pm \sqrt{b^2 - 4ac} \right) x_0' \qquad (3.39)$$

Regarding y_0 as a function of x_0, the chain rule yields $y_0' = x_0' dy_0/dx_0$, and the solution (3.39) reduces to

$$\frac{dy_0}{dx_0} = \frac{1}{2a(x_0, y_0)} \left[b(x_0, y_0) \pm \sqrt{b^2(x_0, y_0) - 4a(x_0, y_0)c(x_0, y_0)} \right] \qquad (3.40)$$

For each unique sign of the discriminant, (3.40) represents a first-order ordinary differential equation whose solution is a one-parameter family of lines in the x-y plane, called the **characteristic base curves** of the partial differential equation (3.4). The one parameter is the arbitrary constant obtained by integrating the differential equation (3.40). Since these curves were determined by solving the equation $\Delta = 0$, where the solutions for the partial derivatives of u are not unique, specifying C_0 to be a characteristic base curve does *not* lead to a unique solution of (3.4).

The characteristic base curves are seen to depend only on the quantities a, b and c and the number and type of solutions depend upon the quantity $\sqrt{b^2 - 4ac}$. Thus, the classification of partial differential equations as hyperbolic, parabolic and elliptic discussed in Section 3.1 enters into the determination of the characteristic base curves, and therefore into the specification of the appropriate boundary conditions for a given equation. Depending upon the sign of the discriminant, there are either two families of real characteristic base curves (hyperbolic), one family of real characteristic base curves (parabolic), or no real characteristic base curves (elliptic). We will see the implications of this for the solutions of different types of equation in Sections 3.4–3.6.

3.3 Geometry of Cauchy Boundary Conditions

To put the boundary conditions required for obtaining a solution to (3.32) into more geometric terms, we first rewrite the derivative of u along C_0 in Equation (3.33) as

$$\frac{du}{ds} = \left(\frac{\partial u}{\partial x} \mathbf{i} + \frac{\partial u}{\partial y} \mathbf{j} \right) \cdot \left(\frac{dx_0}{ds} \mathbf{i} + \frac{dy_0}{ds} \mathbf{j} \right) \qquad (3.41)$$

where **i** and **j** are unit vectors along the x and y directions, respectively. The first factor on the right-hand side of (3.41) is the two-dimensional gradient of u, $\nabla u = u_x \mathbf{i} + u_y \mathbf{j}$, while the second factor is the tangent, or 'velocity' vector along $\mathcal{C}_0$, $\mathbf{t} = x_0' \mathbf{i} + y_0' \mathbf{j}$. If the parameter s is chosen to be the arc length, this tangent vector is a unit vector, as shown in Problem 9. Thus, Equation (3.41) states that the derivative of u along $\mathcal{C}_0$ is the projection of the gradient of u on to $\mathbf{t}$:

$$\frac{du}{ds} = \mathbf{t} \cdot \nabla u \tag{3.42}$$

i.e. (3.41) is the **directional derivative** of u along $\mathbf{t}$. A quantity independent of (3.42) is therefore the directional derivative of u along any direction that has a component orthogonal to $\mathbf{t}$. A natural choice for an independent direction is the direction normal to $\mathcal{C}_0$. We signify the unit vector along this direction by $\mathbf{n}$. The requirements of orthogonality, $\mathbf{t} \cdot \mathbf{n} = 0$, and normalization, $\mathbf{n} \cdot \mathbf{n} = 1$, yield two possibilities (Problem 9):

$$\mathbf{n}^{\pm} = \pm \left[\frac{dv}{ds} \mathbf{i} - \frac{du}{ds} \mathbf{j} \right] \tag{3.43}$$

The two signs in this equation correspond to 'inward' and 'outward' normals to the base curve. In the following discussion, we will use $\mathbf{n}^+ \equiv \mathbf{n}$. The normal derivative of u along $\mathcal{C}_0$, denoted by $\partial u/\partial n$, is thus the directional derivative of u along $\mathbf{n}$, which is the gradient of u projected along $\mathbf{n}$:

$$\frac{\partial u}{\partial n} = \nabla u \cdot \mathbf{n} \tag{3.44}$$

In terms of the vector components defined in (3.41) and (3.44), this equation can be written as

$$\frac{\partial u}{\partial n} = \left(\frac{\partial u}{\partial x} \mathbf{i} + \frac{\partial u}{\partial y} \mathbf{j} \right) \cdot \left(\frac{dy_0}{ds} \mathbf{i} - \frac{dx_0}{ds} \mathbf{j} \right) = \frac{\partial u}{\partial x} \frac{dy_0}{ds} - \frac{\partial u}{\partial y} \frac{dx_0}{ds} \tag{3.45}$$

Using the fact that both $\mathbf{t}(s)$ and $\mathbf{n}(s)$ are unit vectors, (3.33) and (3.45) may be solved for u_x and u_y in terms of $\partial u/\partial n$ and du/ds to obtain the expressions:

$$\frac{\partial u}{\partial x} = \frac{du}{ds}\frac{dx_0}{ds} + \frac{\partial u}{\partial n}\frac{dy_0}{ds}, \quad \frac{\partial u}{\partial y} = \frac{du}{ds}\frac{dy_0}{ds} - \frac{\partial u}{\partial n}\frac{dx_0}{ds} \tag{3.46}$$

The problem of specifying both u along $\mathcal{C}_0$ and the derivative of u along a direction normal to $\mathcal{C}_0$ is thus seen to be equivalent to specifying the normal derivative $\partial u/\partial n$ and the tangential derivative du/ds.

To represent these boundary conditions geometrically, we write the solution surface of (3.4) with u regarded as a function of x and y:

$f[x,y,u(x,y)]=0$. The normal to the solution surface lies along the gradient of u:

$$\nabla f = \frac{\partial f}{\partial x}\mathbf{i} + \frac{\partial f}{\partial y}\mathbf{j} + \frac{\partial f}{\partial u}\mathbf{k} \qquad (3.47)$$

Expressions for the partial derivatives f_x and f_y can be obtained as in Equations (2.7) and (2.8) from the differential of f, with the result

$$\frac{\partial f}{\partial x} = -\frac{\partial f}{\partial u}\frac{\partial u}{\partial x}, \qquad \frac{\partial f}{\partial y} = -\frac{\partial f}{\partial u}\frac{\partial u}{\partial y} \qquad (3.48)$$

Substituting these expressions into (3.47), we obtain

$$\nabla f = -\frac{\partial f}{\partial u}\left(\frac{\partial u}{\partial x}\mathbf{i} + \frac{\partial u}{\partial y}\mathbf{j} - \mathbf{k}\right) \qquad (3.49)$$

which shows that the gradient of the solution surface lies along a direction determined by the partial derivatives of u with respect to x and y. Since the gradient is normal to the solution surface at each point, the gradient is also normal to planes that are tangent to the solution surface at each point. Thus, specifying u along C_0 and a partial derivative of u along a direction normal to C_0 is equivalent to specifying a *line* through which the solution surface must pass, C, as well as the orientations of the planes tangent to the solution surface along C. The geometrical relationships among these quantities is illustrated in Figure 3.1.

Boundary conditions where a function and the normal derivative of the function are specified along a curve C_0 are called **Cauchy boundary conditions**. Determining the solution of (3.4) with Cauchy boundary conditions is called the **Cauchy problem**. There are also other types of boundary conditions that we will encounter: **Dirichlet boundary conditions** occur when only u is specified along a boundary C_0, while **Neumann boundary conditions** occur when only $\partial u/\partial n$ is specified along C_0. We will examine the interrelationships between the classification of a second-order partial differential equation and the most appropriate boundary conditions in the next three sections.

3.4 Hyperbolic Equations

If the coefficients a, b and c of the derivatives in a partial differential equation of the general form (3.4) take on values such that $b^2 - 4ac > 0$ everywhere (recall that $a = a(x,y)$, $b = b(x,y)$ and $c = c(x,y)$) then the equation is called **hyperbolic**. As discussed in Section 3.1, the characteristics form a 'natural' coordinate system for hyperbolic equations in the sense that the left-hand side of (3.4) takes the simplest possible form. To show this for the

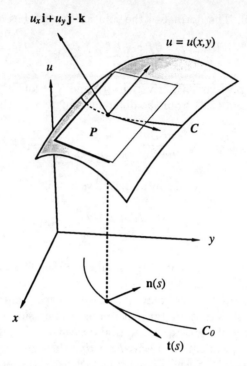

Figure 3.1 Geometric representation of the solution surface $f(x,y)$ together with the base curve C_0 and the curve C. The curve C_0 is shown with the unit normal vector **n** and the unit tangent vectors **t**. Specifying of f along a curve C_0 and the derivative of u along the normal **n** to C_0 is equivalent to specifying the curve C through which the solution surface must pass, as well as the orientation of the plane P intersecting C and tangent to the solution surface at each point along C.

more general equation (3.4), we represent the solutions to (3.40), which are the two families of characteristic base curves, as:

$$\lambda(x,y) = c_1, \qquad \mu(x,y) = c_2 \qquad (3.50)$$

where c_1 and c_2 are constants of integration. Each of these equations can be interpreted either as giving x as a function of y or y as a function of x along the curves. Regarding y as a function of x, the differential of each equation in (3.50) is

$$d\lambda = \left(\frac{\partial \lambda}{\partial x} + \frac{\partial \lambda}{\partial y}\frac{dy}{dx}\right) dx = 0, \qquad d\mu = \left(\frac{\partial \mu}{\partial x} + \frac{\partial \mu}{\partial y}\frac{dy}{dx}\right) dx = 0 \qquad (3.51)$$

Solving for dy/dx in each equation yields

$$\frac{dy}{dx} = -\frac{\lambda_x}{\lambda_y}, \qquad \frac{dy}{dx} = -\frac{\mu_x}{\mu_y} \qquad (3.52)$$

According to (3.40), the quantity dy/dx satisfies the quadratic equation

$$a\left(\frac{dy}{dx}\right)^2 - b\frac{dy}{dx} + c = 0 \qquad (3.53)$$

along a characteristic base curve. Substituting in turn the two expressions for dy/dx in Equation (3.52) produces the two equations

$$a\left(\frac{\partial\lambda}{\partial x}\right)^2 + b\frac{\partial\lambda}{\partial x}\frac{\partial\lambda}{\partial y} + c\left(\frac{\partial\lambda}{\partial y}\right)^2 = 0 \qquad (3.54a)$$

$$a\left(\frac{\partial\mu}{\partial x}\right)^2 + b\frac{\partial\mu}{\partial x}\frac{\partial\mu}{\partial y} + c\left(\frac{\partial\mu}{\partial y}\right)^2 = 0 \qquad (3.54b)$$

Using Equations (3.54), we may now make the change of variables from (x,y) to (λ,μ) to transform (3.4) into a simpler form. The required derivatives are obtained from (3.24) by identifying ξ with λ and η with μ. Substituting the second derivatives into (3.4), the transformed equation reads

$$au_{xx} + bu_{xy} + cu_{yy}$$
$$= (a\lambda_x^2 + b\lambda_x\lambda_y + c\lambda_y^2)u_{\lambda\lambda} + (2a\lambda_x\mu_x + b\lambda_x\mu_y$$
$$+ b\lambda_y\mu_x + 2c\lambda_y\mu_y)u_{\lambda\mu} + (a\mu_x^2 + b\mu_x\mu_y + c\mu_y^2)u_{\mu\mu} + \cdots \qquad (3.55)$$

where the ellipses again denote terms that contain the first-order derivatives u_λ and u_μ. The coefficients of $u_{\lambda\lambda}$ and $u_{\mu\mu}$ are immediately identified as the left-hand sides of (3.54a) and (3.54b), respectively, and therefore vanish. Referring now to Equation (3.28), the coefficient of $u_{\lambda\mu}$ must be positive, since $b^2 - 4ac > 0$, and the equation must remain hyperbolic under a change of coordinates. Thus, we may divide both sides by this quantity to obtain the normal form of a hyperbolic differential equation:

$$\frac{\partial^2 u}{\partial\lambda\partial\mu} = G(\lambda,\mu,u,u_\lambda,u_\mu) \qquad (3.56)$$

The function G depends on the particular equation both through the coefficients a, b and c and the function F.

We have seen in Section 3.2 that a unique solution to (3.4) could be obtained only if the initial conditions were not specified along one of the characteristic curves given in (3.50). If the initial curve does coincide with one of the characteristics, then a solution with two arbitrary functions cannot be constructed away from the initial curve and the Cauchy problem is *underspecified*. On the other hand, if any of the characteristics are intersected more than once, then the problem is *overspecified*, since the value along any characteristic is determined by the Cauchy initial conditions at any *one* point along that characteristic. Thus, a unique solution to the Cauchy problem for a hyperbolic equation can be obtained only if the initial curve intersects each characteristic once and only once.

Example 3.1. The one-dimensional wave equation

$$\frac{\partial^2 u}{\partial x^2} - \frac{1}{v^2}\frac{\partial^2 u}{\partial t^2} = 0 \tag{3.57}$$

was seen in Section 3.1 to be a hyperbolic equation ($b^2 - 4ac = 4/v^2 > 0$). The characteristics of this equation are determined by $dx/dt = \pm v$, which has the solutions

$$\lambda(x,t) = x - vt = c_1, \qquad \mu(x,t) = x + vt = c_2 \tag{3.58}$$

where c_1 and c_2 are constants. These are thus straight lines in the x-t plane with slopes $\pm v$ and intersections c_1 and c_2 with the x-axis. The general solution of (3.57) is given by (3.13):

$$u(x,t) = \mathcal{F}(x+vt) + \mathcal{G}(x-vt) \tag{3.59}$$

As an example of Cauchy boundary conditions for the wave equation, consider the initial-value problem where the initial displacement and the initial velocity are given by two functions f and g:

$$u(x,0) = f(x), \qquad u_t(x,0) = g(x) \tag{3.60}$$

Since $\mathcal{C}_0$ in this case corresponds to the x-axis, each characteristic is intersected once and only once, so the solution is determined uniquely everywhere in the x-t plane. The solution to (3.59) and (3.60) can be obtained in closed form by solving for the arbitrary functions $\mathcal{F}$ and $\mathcal{G}$. The boundary conditions (3.60) imply

$$u(x,0) = \mathcal{F}(x) + \mathcal{G}(x) = f(x) \tag{3.61a}$$

$$u_t(x,0) = v\mathcal{F}'(x) - v\mathcal{G}'(x) = g(x) \tag{3.61b}$$

where the prime indicates differentiation with respect to the arguments of $\mathcal{F}$ and $\mathcal{G}$. Equation (3.61b) can be integrated directly as follows:

$$v\int_{x_0}^{x}\mathcal{F}'(s)\,ds - v\int_{x_0}^{x}\mathcal{G}'(s)\,ds = \int_{x_0}^{x} g(s)\,ds \tag{3.62}$$

where x_0 here signifies any initial point of integration. Carrying out the integrations in (3.62), we obtain

$$\mathcal{F}(x) - \mathcal{G}(x) = \mathcal{F}(x_0) - \mathcal{G}(x_0) + \frac{1}{v}\int_{x_0}^{x} g(s)\,ds \tag{3.63}$$

Expressions for $\mathcal{F}(x)$ and $\mathcal{G}(x)$ are obtained by solving (3.61a) and (3.63):

$$\mathcal{F}(x) = \tfrac{1}{2}\mathcal{F}(x_0) - \tfrac{1}{2}\mathcal{G}(x_0) + \tfrac{1}{2}f(x) + \frac{1}{2v}\int_{x_0}^{x} g(s)\,ds$$

$$\mathcal{G}(x) = \tfrac{1}{2}\mathcal{G}(x_0) - \tfrac{1}{2}\mathcal{F}(x_0) + \tfrac{1}{2}f(x) - \frac{1}{2v}\int_{x_0}^{x} g(s)\,ds \quad (3.64)$$

Substituting the respective arguments into $\mathcal{F}$ and $\mathcal{G}$ as indicated in (3.59),

$$\mathcal{F}(x+vt) = \tfrac{1}{2}\mathcal{F}(x_0) - \tfrac{1}{2}\mathcal{G}(x_0) + \tfrac{1}{2}f(x+vt) + \frac{1}{2v}\int_{x_0}^{x+vt} g(s)\,ds$$

$$\mathcal{G}(x-vt) = \tfrac{1}{2}\mathcal{G}(x_0) - \tfrac{1}{2}\mathcal{F}(x_0) + \tfrac{1}{2}f(x-vt) - \frac{1}{2v}\int_{x_0}^{x-vt} g(s)\,ds \quad (3.65)$$

and combining the two integrals in (3.65) into a single term,

$$\int_{x_0}^{x+vt} g(s)\,ds - \int_{x_0}^{x-vt} g(s)\,ds = \int_{x-vt}^{x+vt} g(s)\,ds \quad (3.66)$$

produces **d'Alembert's solution** of the initial-value problem for the wave equation:

$$u(x,t) = \tfrac{1}{2}[f(x+vt) + f(x-vt)] + \frac{1}{2v}\int_{x-vt}^{x+vt} g(s)\,ds \quad (3.67)$$

The form of this solution reveals several important features of the wave equation. The terms involving f show that disturbances are propagated along the characteristics with velocity v. Both terms taken together show that the value of the solution at position x and at time t depends only upon the initial values of f at $x-vt$ and $x+vt$ and the values of g *between* these points. Finally, d'Alembert's solution is seen to be a continuous function of the initial conditions. In other words, a small change in either f of g results in a correspondingly small change in $u(x,t)$. The significance of this will become apparent when we investigate solution of Laplace's equation in Section 3.6.

The influence of the two terms on the form of the solution in (3.64) can be seen by considering two cases:

$$f(x) = \exp(-x^2), \quad g(x) = 0 \quad (3.68a)$$

$$f(x) = 0, \quad g(x) = \exp(-x^2) \quad (3.68b)$$

The profiles of these two solutions with $v=1$ that have been generated with **Plot** are shown in Figure 3.2.

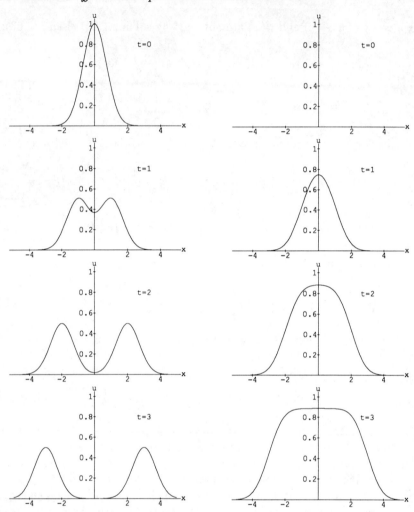

Figure 3.2 The time development of solutions of the initial-value problem for the wave equation with initial conditions given by $f(x) = \exp(-x^2)$ and $g(x) = 0$ (left) and with $f(x) = 0$ and $g(x) = \exp(-x^2)$ (right). The time sequence begins with $t=0$ and increases by one unit downward, as indicated.

For the initial conditions in (3.68a), the initial form is seen to break up into two travelling waves moving in opposite directions with unit speed, i.e. along the characteristics. The solution for the initial conditions (3.68b) evolves by the value at each point rising to the value

$$\frac{1}{2}\int_{-\infty}^{\infty} \exp(-x^2)\,dx = \tfrac{1}{2}\sqrt{\pi} \approx 0.886 \qquad (3.69)$$

with a 'wavefront' that also moves with unit speed.

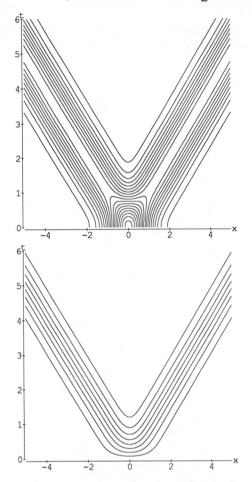

Figure 3.3 Contour plots of solutions of the initial-value problem (3.60) for the wave equation for $f(x) = \exp(-x^2)$ and $g(x) = 0$ (top) and for $f(x) = 0$ and $g(x) = \exp(-x^2)$ (bottom).

A direct correspondence between the time development of these solutions and the characteristics can be more easily seen by using ContourPlot, as shown in Figure 3.3. For both initial conditions in (3.68), the contours of the solutions eventually coincide with the characteristics.

∎

3.5 Parabolic Equations

If $b^2 - 4ac = 0$ everywhere in the x-y plane, then the partial differential equation in (3.4) is called **parabolic**. We will assume below that $a \neq 0$, $b \neq 0$ and $c \neq 0$. Analogous steps can be carried out for other combinations of values taken by a, b and c appropriate to this case as was done in Section 3.1 for the special case of Equation (3.8). The discriminant in Equation (3.40) vanishes, so there is only only one set of characteristics, which correspond to the solution of $dy/dx = b/2a$. We signify this family of characteristics by

$$\lambda(x, y) = c_1 \tag{3.70}$$

where c_1 is a constant. A 'natural' coordinate system for a parabolic equation analogous to that for hyperbolic equations is in terms of λ and x (or y). As in the hyperbolic case, the equation assumes the simplest form when expressed in terms of characteristic coordinates. Thus, making the change of variables from x and y to x and $\lambda(x, y)$ and using the chain rule to calculate the required second derivatives yields

$$\begin{aligned} au_{xx} &+ bu_{xy} + cu_{yy} \\ &= a(u_{xx} + 2u_{x\lambda}\lambda_x + u_{\lambda\lambda}\lambda_x^2 + u_\lambda \lambda_{xx}) \\ &\quad + b(u_{x\lambda}\lambda_y + u_{\lambda\lambda}\lambda_x\lambda_y + u_\lambda \lambda_{xy}) + c(u_{\lambda\lambda}\lambda_y^2 + u_\lambda \lambda_{yy}) \\ &= au_{xx} + u_{x\lambda}(2a\lambda_x + b\lambda_y) + u_{\lambda\lambda}(a\lambda_x^2 + b\lambda_x\lambda_y + c\lambda_y^2) \\ &\quad + u_\lambda(a\lambda_{xx} + b\lambda_{xy} + c\lambda_{yy}) \end{aligned} \tag{3.71}$$

We have from (3.51) and the characteristic equation for λ that the coefficient of $u_{x\lambda}$ vanishes. Furthermore, according to (3.54a) the coefficient of $u_{\lambda\lambda}$ vanishes as well. Thus, since we can express y in terms of x and λ and u_x and u_y in terms of x, λ, u_x and u_λ by using the chain rule, (3.71) implies the following normal form of (3.4) for parabolic equations:

$$\frac{\partial^2 u}{\partial x^2} = G(x, \lambda, u, u_x, u_\lambda) \tag{3.72}$$

As in Equation (3.56), the G is determined by the function F and coefficients a, b and c in (3.4). The diffusion equation,

$$D\frac{\partial^2 u}{\partial x^2} = \frac{\partial u}{\partial t} \tag{3.73}$$

is an example of a parabolic equation which is already in normal form. The quantity D is a constant which is the *thermal diffusivity* in the case of heat diffusion or simply the *diffusivity* in the case of particle diffusion.

Parabolic equations are encountered frequently in applications with the variable λ in (3.72) usually corresponding to the time. Solutions are sought to an initial-value problem where u is the specified function at $t=0$: $u(x, 0) = f(x)$, i.e. along the base curve $t = 0$, which corresponds to a characteristic of (3.73). Cauchy boundary conditions would overspecify the problem because if u is determined along a characteristic, then the derivatives u_x, u_{xx},... are also determined along this characteristic. According to (3.72), there is no freedom remaining to then specify u_λ. Thus, although the Cauchy boundary conditions do not yield a solution in this case, *Dirichlet* or *Neumann* boundary conditions for the initial-value problem do.

Example 3.2. The expansion use in Equation (3.32) to derive the characteristic equations can be used to illustrate some general features of solutions to the initial-value problem for the diffusion equation in (3.73). For notational convenience we set $D = 1$. The initial condition is taken as $u(x,0) = f(x)$ and, in accordance with the discussion in Section 3.2, we require that f have a convergent Taylor series in a neighborhood around the origin. The base curve $\mathcal{C}_0$ for the initial-value problem is the x axis, $y=0$, so the Taylor series of the solution of (3.71) is an expansion in powers of y:

$$u(x,y) = u(x,0) + \left.\frac{\partial u}{\partial y}\right|_{y=0} y + \frac{1}{2!}\left.\frac{\partial^2 u}{\partial y^2}\right|_{y=0} y^2 + \frac{1}{3!}\left.\frac{\partial^3 u}{\partial y^3}\right|_{y=0} y^3 + \cdots \quad (3.74)$$

The first term on the right-hand side of (3.74) is simply the initial condition, $u(x,0) = f(x)$. The first derivative evaluated along the x axis is determined directly from the diffusion equation to be $u_y(x,0) = u_{xx}(x,0) = f''(x)$. The second-order derivative is evaluated in two steps. The diffusion equation is first differentiated with respect to y: $u_{yy} = u_{xxy}$. An expression for the third-order derivative can be obtained by differentiating the diffusion equation twice with respect to x: $u_{xxy} = u_{xxxx}$. Combining these two expressions, we obtain $u_{yy} = u_{xxxx}$ which, when evaluated along the x axis, yields

$$\left.\frac{\partial^2 u}{\partial y^2}\right|_{y=0} = \left.\frac{\partial^4 u}{\partial x^4}\right|_{y=0} = \frac{d^4 f}{dx^4} \quad (3.75)$$

In fact, we deduce by induction that if

$$\left.\frac{\partial^n u}{\partial y^n}\right|_{y=0} = \left.\frac{\partial^{2n} u}{\partial x^{2n}}\right|_{y=0} = \frac{d^{2n} f(x)}{dx^{2n}} \quad (3.76)$$

then the diffusion equation implies

$$\frac{\partial^{n+1} u}{\partial y^{n+1}} = \frac{\partial^n}{\partial y^n}\frac{\partial u}{\partial y} = \frac{\partial^n}{\partial y^n}\frac{\partial^2 u}{\partial x^2} = \frac{\partial^2}{\partial x^2}\frac{\partial^n u}{\partial y^n} \quad (3.77)$$

Thus,

$$\left.\frac{\partial^{n+1} u}{\partial y^{n+1}}\right|_{y=0} = \frac{\partial^2}{\partial x^2}\frac{d^{2n} f}{dx^{2n}} = \frac{d^{2(n+1)} f}{dx^{2(n+1)}} \quad (3.78)$$

With a general expression for the nth-order derivative of u evaluated along the x axis, the expansion in Equation (3.74) takes the form

$$u(x,y) = \sum_{n=0}^{\infty} \frac{d^{2n} f(x)}{dx^{2n}} \frac{y^n}{n!} \tag{3.79}$$

The solution of the initial-value problem of the diffusion is therefore completely determined in terms of the initial function $f(x)$. As discussed above, specifying the initial function along the characteristic $y=0$ results in there being only one arbitrary function in the solution. Notice that this way of deriving solutions to the diffusion equation explicitly requires f to have derivatives of all orders. We will see in later chapters that this solution of the initial-value problem can be derived under much less restrictive conditions.

As a particular example of (3.79), consider the initial-value problem with $f(x) = \cos kx$. Then, since

$$\frac{d^{2n}}{dx^{2n}} \cos kx = (-1)^n k^{2n} \cos kx \tag{3.80}$$

we obtain the closed-form solution

$$u(x,y) = \left[\sum_{n=0}^{\infty} (-1)^n \frac{(k^2 y)^n}{n!}\right] \cos kx = e^{-k^2 y} \cos kx \tag{3.81}$$

The solution is seen to retain the form of a cosine function, but with an amplitude that decays with increasing y. In fact, the larger the value of k, and so the more rapidly varying (as a function of x) the cosine function, the quicker the decay. This is one of the key features of the behavior of solutions to the diffusion equation which will be explored in the problems at the end of the chapter and in later chapters.

Mathematica can be used to investigate solutions of the diffusion equation for a wide range of initial conditions by using the symbolic capabilities to carry out the differentiations and the summation in the expansion in (3.79). This is not the most efficient way of solving the diffusion equation, but it does provide an instance where the power series method provides an explicit exact solution of this equation by using only elementary operations.

Consider the initial-value problem with $f(x) = \exp(-x^2)$. The solution (3.79) is obtained by first defining the initial function f, performing the required derivatives of f and assembling the terms in (3.79) as a partial sum using Sum:

Second-Order Partial Differential Equations

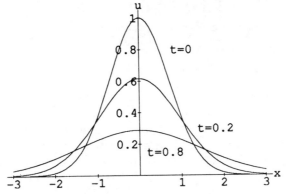

Figure 3.4 The time development of the solution to the initial-value problem of the diffusion equation (3.73) with the initial condition $u(x,0) = \exp(-x^2)$. With increasing time, the value of the solution at the origin decreases as the profile broadens.

```
f[x_]:=Exp[-x^2]

a[x_,k_]:=D[f[s],{s,2k}]/.s->x

u[x_,y_,n_]:=Sum[(1/k!)a[x,k]y^k,{k,0,n}]
```

The profiles of this solution for $y = 0$, 0.2 and 0.8 obtained by retaining the required number of terms to achieve an accuracy of 1% are shown in Figure 3.4. With increasing time, the solution is seen to decrease in value near the origin and the profile broadens. This is a characteristic feature of the diffusion equation and is related to the behavior of the solution in (3.81) as a function of the quantity k (Problem 19).

3.6 Elliptic Equations

If $b^2 - 4ac < 0$ everywhere, then the partial differential equation in Equation (3.4) is termed **elliptic**. The solution of Equation (3.40) yields characteristics that are complex conjugates: $\mu(x,y) = \lambda^*(x,y)$. As in Section 3.1, we can define real variables ξ and η that are the real and imaginary parts of λ^* and λ:

$$\lambda = \xi + i\eta, \qquad \lambda^* = \xi - i\eta \tag{3.82}$$

In terms of λ and λ^*, the normal form of elliptic equations can be transcribed from that shown in Equation (3.56) for hyperbolic equations:

$$\frac{\partial^2 u}{\partial \lambda \partial \lambda^*} = F(\lambda, \lambda^*, u, u_\lambda, u_{\lambda^*}) \tag{3.83}$$

102 Partial Differential Equations with Mathematica

Alternatively, with ξ and η as the independent variables, the normal form is

$$\frac{\partial^2 u}{\partial \xi^2} + \frac{\partial^2 u}{\partial \eta^2} = G(\xi, \eta, u, u_\xi, u_\eta) \tag{3.84}$$

Since the characteristics of elliptic equations are complex, specifying the boundary curves would appear not to present any difficulties, since *any* base curve could be chosen without the problem of intersecting a characteristic. However, the complex characteristics mean that careful attention must be paid to the analytic behavior of the solution in the complex plane if spurious singularities are to be avoided, as the following example illustrates.

Example 3.3. Consider the two-dimensional Laplace equation

$$\frac{\partial^2 u}{\partial x^2} + \frac{\partial^2 u}{\partial y^2} = 0 \tag{3.85}$$

with Cauchy boundary conditions:

$$u(x, 0) = f(x), \qquad u_y(x, 0) = g(x) \tag{3.86}$$

The general solution to (3.85) was found in Section 3.1 to be

$$u(x, y) = \mathcal{F}(x + iy) + \mathcal{G}(x - iy) \tag{3.87}$$

Combining (3.86) with (3.87), the solution can be written in the d'Alembertian form (3.64) by identifying the speed v with i:

$$u(x, y) = \tfrac{1}{2}[f(x + iy) + f(x - iy)] - \frac{i}{2}\int_{x-iy}^{x+iy} g(s)ds \tag{3.88}$$

This way of writing the solution to (3.86) and (3.87) is very revealing, since it shows that the values taken by u in the *real* x-y plane are determined by the functions f and g evaluated with *complex* arguments. To see the implications of this we consider two sets of Cauchy boundary conditions. Suppose first that

$$f(x) = \frac{1}{1 + x^2}, \qquad g(x) = 0 \tag{3.89}$$

Along the real line, f is a well-behaved function, but when viewed as functions of the complex variable $x+iy$, f has simple poles at $x=\pm i$, which are transformed to singularities in the solution u at the points $(0, \pm 1)$ (Figure 3.5). Since the derivatives of u do not exist at these singular points, the solution fails to satisfy Laplace's equation there. On the other hand, if the boundary conditions are given by

$$f(x) = \exp(-x^2), \qquad g(x) = 0 \tag{3.90}$$

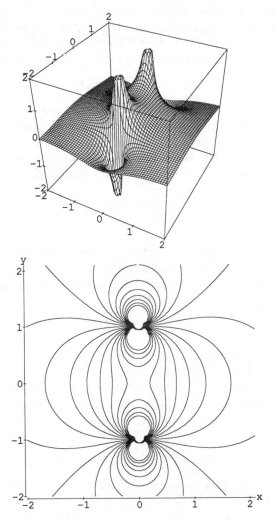

Figure 3.5 Surface plot of the solution of Laplace's equation in two dimensions with the boundary conditions $u(x,0) = 1/(1 + x^2)$ and $u_y(x,0) = 0$ in the region $-2 \leq x \leq 2$ and $-2 \leq y \leq 2$. The singularities of the solution at $(0, \pm 1)$ occur because the initial function in (3.90) has simple poles at $\pm i$ in the complex plane.

then the solution is regular in any finite region of the x-y plane, but increases without bound away from the origin along the y axis.

This example shows that although a solution to Laplace's equation with Cauchy boundary conditions can be constructed as in (3.88), singularities can appear that cannot be controlled by the boundary conditions. In fact, since the only function that is analytic *everywhere* is a constant, u will have a singularity somewhere in the x-y plane (though possibly not in the region of interest to the problem at hand) unless the initial conditions are chosen such that F is a constant and G vanishes. These considerations, and others (Problem 23), suggest that the analytic behavior of solutions to Laplace's equation is simply too sensitive to Cauchy boundary conditions. ∎

To identify more appropriate boundary conditions for Laplace's equation, we use some well-known vector relations. Suppose we have a vector function $\mathbf{F}$ within a region $\mathcal{V}$ bounded by a surface $\mathcal{S}$. The divergence theorem relates the integral of the divergence of $\mathbf{F}$ over $\mathcal{V}$ to the integral of the normal component of $\mathbf{F}$ over the bounding surface $\mathcal{S}$:

$$\int_\mathcal{V} \nabla \cdot \mathbf{F}\, d\mathcal{V} = \int_\mathcal{S} \mathbf{F} \cdot \mathbf{n}\, dA \tag{3.91}$$

where $\mathbf{n}$ is the unit normal of the surface, pointing outwards, and dA is the element of surface area of $\mathcal{S}$. By defining $\mathbf{F}$ in terms of two functions ϕ and ψ as $\mathbf{F} = \phi \nabla \psi$, and substituting into (3.91), we arrive at an identity due to Green:

$$\int_\mathcal{V} [\phi \nabla^2 \psi + \nabla \phi \nabla \psi]\, d\mathcal{V} = \int_\mathcal{S} \phi [\nabla \psi \cdot \mathbf{n}]\, dA \tag{3.92}$$

The quantity $\nabla \psi \cdot \mathbf{n}$ on the right-hand side of (3.92) is the normal derivative of ψ and, as in (3.44), will be written as $\partial \psi / \partial n$.

Suppose that ψ and ϕ are both solutions of Laplace's equation within $\mathcal{V}$ and either the values of ψ and ϕ, or the normal derivatives of these functions are specified on $\mathcal{S}$. Then, the function

$$\varphi = \phi - \psi \tag{3.93}$$

satisfies Laplace's equation within $\mathcal{V}$, and either φ or $\partial \varphi / \partial n$ vanishes on $\mathcal{S}$. Substituting φ for both ϕ and ψ in (3.92) yields

$$\int_\mathcal{V} [\varphi \nabla^2 \varphi + (\nabla \varphi)^2]\, d\mathcal{V} = \int_\mathcal{S} \varphi \frac{\partial \varphi}{\partial n}\, dA \tag{3.94}$$

Since $\nabla^2 \varphi = 0$ within $\mathcal{V}$ and either $\varphi = 0$ or $\partial \varphi / \partial n = 0$ on $\mathcal{S}$, (3.94) reduces to

$$\int_\mathcal{V} (\nabla \varphi)^2\, d\mathcal{V} = 0 \tag{3.95}$$

Since the integrand in (3.95) is a nonnegative quantity, we deduce that φ is a constant within $\mathcal{V}$, i.e. that $\phi - \psi =$ constant. Thus, to within an additive constant, specifying either the value of a function (Dirichlet boundary conditions) or its normal derivative (Neumann boundary conditions) over a bounding surface, uniquely determines the solution to Laplace's equation within $\mathcal{V}$. Notice that our argument is not restricted in the number of independent variables, so the conclusions are equally applicable in two and three spatial dimensions. The extension of this argument to unbounded regions is a slightly more delicate matter but can be accomplished provided the solution decays sufficiently rapidly at infinity (Sneddon, 1957).

Example 3.6. To examine the behavior of solutions of Laplace's equation with Dirichlet boundary conditions along a closed boundary, we construct solutions in polar coordinates using the general solution (3.79). Polar coordinates (r, ϕ) are related to rectangular coordinates (x, y) by

$$x = r \cos \phi, \qquad y = r \sin \phi \tag{3.96}$$

where the range of the variables is $0 \leq r$ and $0 \leq \phi < 2\pi$. Then, $x \pm iy = r \exp(\pm i\phi)$ and (3.79) becomes

$$u(r, \phi) = \mathcal{F}\left(re^{i\phi}\right) + \mathcal{G}\left(re^{-i\phi}\right) \tag{3.97}$$

Since the real and imaginary parts of (3.97) are separately solutions of Laplace's equation (Problem 2), we take $\mathcal{G} = \mathcal{F}$ and obtain as particular solutions the quantities $r^k \cos(k\phi)$ and $r^k \sin(k\phi)$, where k is a real number. The requirement that u be single-valued, i.e. $u(r, \phi+2\pi) = u(r, \phi)$, means that k must be an integer.

Suppose a solution of Laplace's equation is required within the circular region $r < 1$ that on the unit circle reduces to $\cos \phi$: $u(1, \phi) = \cos \phi$. From our discussion in the preceding paragraph, the solution that is regular over the interior of the unit circle and satisfies the boundary condition is the cosine solution with $k=1$:

$$u(r, \phi) = r \cos \phi \tag{3.98}$$

Alternatively, suppose the solution is required in the region $r > 1$, with the same boundary conditions. The corresponding regular solution is

$$u(r, \phi) = r^{-1} \cos \phi \tag{3.99}$$

The solution (3.98) is singular at $r = \infty$ and the solution (3.99) is singular at the origin. A solution that is regular over the entire plane and satisfies the boundary condition can be constructed by defining the quantities $r_>$ as the maximum of r and 1, and $r_<$ as the minimum of r and 1. Then (3.98) and (3.99) can be combined into a single expression by writing

$$u(r, \phi) = \frac{r_<}{r_>} \cos \phi \tag{3.100}$$

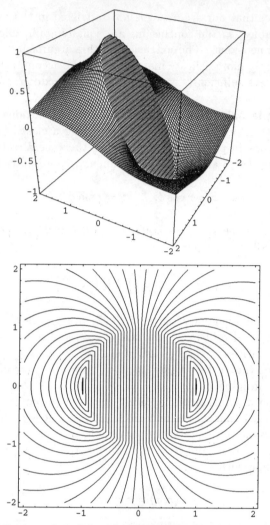

Figure 3.6 Surface plot of the solution (3.98) of Laplace's equation generated by the Plot3D command of *Mathematica* (top) and a contour plot of the same solution generated with the ContourPlot command (bottom). The matching of the two solutions across the unit circle generates a discontinuity in the first derivative of the solution across the boundary.

Surface and contour plots of this solution are shown in Figure 3.6. The matching of the solutions for the interior region of the circle in (3.98) and that for the exterior region in (3.99) produces a solution which is continuous but with a discontinuous first derivative across the boundary. The surface plot also shows that neither a maximum nor a minimum of the solution is attained within the region bounded by the circle. This is a general feature of solutions to Laplace's equation which is discussed in Problem 26. ∎

Further Reading

The books by Morse and Feshbach (1953) and Hildebrand (1962) provide additional discussions of second-order equations. Morse and Feshbach (1953) in particular provide a much more detailed treatment of the origin of the differential equations and the consequences of various types of boundary condition. A comprehensive mathematical treatment of second-order equations may be found in Garabedian (1964).

An excellent introductory discussion on the Laplace equation is given by Jackson (1975) and the use of complex analysis for solving problems in two-dimensional electrostatics is covered by Panofsky and Phillips (1962). Milne-Thomson (1973) provides an analogous treatment of problems in two-dimensional fluid flow. The Laplace equation is discussed from the standpoint of harmonic analysis in Ahlfors (1966), where other topics on complex analysis are also covered. Introductory treatments of conformal mapping may also be found in Morse and Feshbach (1953) and Hildebrand (1962).

References

Ahlfors L. V. (1966). *Complex Analysis* 2nd edn. New York: McGraw-Hill.
Garabedian P. R. (1964). *Partial Differential Equations.* New York: Wiley.
Hadamard J. (1952). *Lectures on Cauchy's Problem in Linear Partial Differential Equations.* New York: Dover.
Hildebrand F. B. (1962). *Advanced Calculus for Applications.* Englewood Cliffs NJ: Prentice-Hall.
Milne-Thomson L. M. (1973). *Theoretical Aerodynamics* 4th edn. New York: Dover.
Jackson J. D. (1975). *Classical Electrodynamics* 2nd edn. New York: Wiley.
Morse P. M. and Feshbach H. (1953). *Methods of Theoretical Physics* Vol. 1. New York: McGraw-Hill.
Panofsky W. K. H. and Phillips M. (1962). *Classical Electricity and Magnetism* 2nd edn. Reading MA: Addison-Wesley.
Sneddon I. N. (1957). *Elements of Partial Differential Equations.* New York: McGraw-Hill.

Problems

1. The technique used in Section 3.1 for solving second-order equations can be extended to the corresponding types of higher-order equations. By assuming a trial solution of the form $u(x,y) = f(x+my)$ obtain the general solutions to the biharmonic partial differential equation:

$$u_{xxxx} + 2u_{xxyy} + u_{yyyy} = 0$$

2. The general solution of Laplace's equation

$$u_{xx} + u_{yy} = 0$$

was derived in Section 3.1 as

$$u(x,y) = \mathcal{F}(x+iy) + \mathcal{G}(x-iy)$$

where $\mathcal{F}$ and $\mathcal{G}$ are any twice-differentiable functions. By appropriate choices for the functions $\mathcal{F}$ and $\mathcal{G}$ show that the real and imaginary parts of the general solution are solutions of Laplace's equation.

By making the choices $\mathcal{F}(s) = s^k$, where k is a positive integer, and taking real and imaginary parts, obtain particular polynomial solutions of Laplace's equation. For example, $k=1$ yields the solutions x and y, while $k=2$ yields $x^2 - y^2$ and xy as solutions.

3. Repeat the procedure in Problem 2 using polar coordinates, $x = r \cos \phi$ and $y = r \sin \phi$. Then, use the *Mathematica* command `ParametricPlot` to display polar plots of these solutions of Laplace's equation for a selection of values of k.

4. Classify each of the differential equations below as either hyperbolic, parabolic or elliptic, determine the characteristics, transform the equations to characteristic coordinates, and express the general solutions in the original variables.

(a) $u_{xx} + 3u_{xy} + 2u_{yy} - u_x - u_y = 0$

(b) $u_{xx} + 3u_{xy} + 2u_{yy} - 2u_x - 4u_y = 0$

(c) $u_{xx} + 2u_{xy} + 2u_{yy} = 0$

(d) $u_{xx} + 2u_{xy} + u_{yy} + u_x + u_y = 0$

(e) $u_{xx} + 4u_{xy} - 5u_{yy} + u_x - u_y = 0$

(f) $u_{xx} + 4u_{xy} - 5u_{yy} + u_x + 5u_y = 0$

(g) $u_{xx} + 4u_{xy} + 5u_{yy} = 0$

(h) $2u_{xx} + 4u_{xy} + 2u_{yy} + 2u_x + 2u_y = 0$

5. Eliminate the arbitrary functions in the following expressions to obtain the partial differential equations satisfied by $u(x,y)$. The quantities $\mathcal{F}_1, \mathcal{F}_2, \mathcal{F}_3$ and $\mathcal{F}_4$ are arbitrary functions that have derivatives up to the required order.

(a) $u(x,y) = \mathcal{F}_1(x+y) + \mathcal{F}_2(x-2y)$

(b) $u(x,y) = \mathcal{F}_1(x) + \mathcal{F}_2(y)$

(c) $u(x,y) = \mathcal{F}_1(x+2i) + \mathcal{F}_2(x-2i)$

(d) $u(x,y) = \mathcal{F}_1(x+y) + y\mathcal{F}_2(x+y)$

(e) $u(x,y) = \mathcal{F}_1(x+y) + \mathcal{F}_2(x+iy) + \mathcal{F}_3(x-iy)$

(f) $u(x,y) = \mathcal{F}_1(x+y) + \mathcal{F}_2(x-y) + \mathcal{F}_3(2x+y)$

(g) $u(x,y) = \mathcal{F}_1(x+y) + \mathcal{F}_2(x-y) + \mathcal{F}_3(x+iy) + \mathcal{F}_4(x-iy)$

(h) $u(x,y) = \mathcal{F}_1(x+y) + \mathcal{F}_2(x-y) + \mathcal{F}_3(2x+y) + \mathcal{F}_4(2x-y)$

6. Construct a sequence of commands in *Mathematica* that employs the command `Eliminate` to eliminate arbitrary functions from an expression of the type in Problem 5 to obtain the partial differential equation satisfied by an expression involving the arbitrary functions. For example, for expressions involving n arbitrary functions of arguments linear in the independent variables, only the nth-order derivatives need be considered. For $n = 2$, the following commands eliminate f_1 and f_2

```
F[x,y] := f1[g1[x,y]]+f2[g2[x,y]];

g1[x,y] = x + I y;
g2[x,y] = x - I y;

Eqn1 = uxx == D[F[x,y],{x,2}];
Eqn2 = uxy == D[F[x,y],x,y];
Eqn3 = uyy == D[F[x,y],{y,2}];

Eliminate[{Eqn1,Eqn2,Eqn3},{f1''[g1[x,y]],f2''[g2[x,y]]}]
```

to produce Laplace's equation:

```
uyy == -uxx
```

Modify these commands for expressions with more than two arbitrary functions and for more complex arguments.

7. Using a sequence of commands in *Mathematica*, or by other means, eliminate the arbitrary functions in the expressions below to obtain the partial differential equations satisfied by $u(x,y)$. The quantities $\mathcal{F}$ and $\mathcal{G}$ are arbitrary functions that are twice-differentiable.

(a) $u(x,y) = \mathcal{F}(xy) + \mathcal{G}(x/y)$

(b) $u(x,y) = \mathcal{F}(xy) + y\mathcal{G}(xy)$

(c) $u(x,y) = \mathcal{F}(ye^x) + \mathcal{G}(ye^{-x})$

(d) $u(x,y) = \mathcal{F}(y\cos x) + \mathcal{G}(y\sin x)$

8. Apply the power series method described in Section 3.2 to a first-order quasi-linear equation of the form

$$P(x,y)\frac{\partial u}{\partial x} + Q(x,y)\frac{\partial u}{\partial y} = R(x,y,u)$$

Show that a unique solution can be obtained provided that the base curve is not determined by the equations

$$\frac{dx_0}{P(x_0,y_0)} = \frac{dy_0}{Q(x_0,y_0)}$$

where x_0 and y_0 are the coordinate functions of the base curve. From this relation obtain the equation determining the characteristics.

9. Consider a curve $\mathcal{C}$ given in terms of the parameter s by the equations

$$\mathcal{C}(s) = [x_0(s), y_0(s)]$$

Using the fact that s is an arc length, i.e.

$$ds^2 = dx^2 + dy^2$$

show that the tangent vector $\mathbf{t}$ to the curve

$$\mathbf{t} = \frac{dx_0}{ds}\mathbf{i} + \frac{dy_0}{ds}\mathbf{j}$$

is a unit vector.

By writing $\mathbf{n} = n_x\mathbf{i} + n_y\mathbf{j}$ (where n_x and n_y are vector components, *not* partial derivatives), show that the requirement of normalization, $\mathbf{n}\cdot\mathbf{n}=1$, and orthogonality to the tangent vector $\mathbf{t}$, $\mathbf{t}\cdot\mathbf{n}=0$, yield the following expression for the unit normals to $\mathcal{C}$:

$$\mathbf{n}^{\pm} = \pm\left(\frac{dy_0}{ds}\mathbf{i} - \frac{dx_0}{ds}\mathbf{j}\right)$$

10. Consider the solution of the wave equation

$$\frac{\partial^2 u}{\partial x^2} - \frac{\partial^2 u}{\partial y^2} = 0$$

with Cauchy boundary conditions along an open curve C_0:

$$u(x,y)\Big|_{C_0} = f(x), \qquad \frac{\partial u}{\partial y}(x,y)\Big|_{C_0} = g(x)$$

Specialize the d'Alembertian general solution in Equation (3.67) to the case where C_0 is the straight line $y = mx$ and show that this solution can be written as

$$u(x,y) = \tfrac{1}{2}[(1+m)f(x+y) + (1-m)f(x-y)] + \tfrac{1}{2}(1-m^2)\int_{x-y}^{x+y} g(s)\,ds$$

Describe what happens to the solution and why if $m = \pm 1$. In particular, describe the uniqueness of the solution in relation to the characteristics.

11. Consider the Cauchy problem for the wave equation

$$\frac{\partial^2 u}{\partial x^2} - \frac{\partial^2 u}{\partial y^2} = 0$$

where the solution is required to satisfy the boundary conditions

$$u(x,0) = f(x), \qquad u_y(x,0) = g(x)$$

The functions $f(x)$ and $g(x)$ are assumed to have derivatives of all orders. By expanding the solution in Taylor series about the line $y=0$,

$$u(x,y) = \sum_{n=0}^{\infty} \frac{\partial^n u}{\partial y^n}\Big|_{y=0} \frac{y^n}{n!}$$

use the method of Example 3.2 to obtain

$$u(x,y) = \sum_{n=0}^{\infty} \frac{d^{2n} f}{dx^{2n}} \frac{y^{2n}}{(2n)!} + \sum_{n=0}^{\infty} \frac{d^{2n+1} h}{dy^{2n+1}} \frac{y^{2n+1}}{(2n+1)!}$$

where h and g are related by

$$\frac{dh(x)}{dx} = g(x)$$

By writing the first summation on the right-hand side of the solution as

$$\sum_{n=0}^{\infty} \frac{d^{2n} f}{dx^{2n}} \frac{y^{2n}}{(2n)!} = \frac{1}{2}\sum_{n=0}^{\infty} \frac{d^n f}{dx^n} \frac{y^n}{n!} + \frac{1}{2}\sum_{n=0}^{\infty} (-1)^n \frac{d^n f}{dx^n} \frac{y^n}{n!}$$

deduce that each of the two terms can be summed to obtain

$$\tfrac{1}{2}\left[f(x+y)+f(x-y)\right]$$

Perform an analogous sequence of steps for the second term in the solution to obtain d'Alembert's form of the general solution of the wave equation in (3.67).

12. The **telegraph equation** is given by:

$$\frac{\partial^2 u}{\partial x^2} = \frac{\partial^2 u}{\partial y^2} + \frac{\partial u}{\partial y}$$

The right-hand side of this equation is composed of both a propagating part, u_{yy}, and a diffusive part, u_y. The power series method of Section 3.2 can be applied to solve the initial value problem for this equation:

$$u(x,0) = f(x), \qquad u_y(x,0) = g(x)$$

Show that the coefficients in the power series expansion of $u(x,y)$,

$$u(x,y) = \sum_{k=0}^{\infty} \frac{1}{n!} u^{(k)}(x,0) y^k$$

where $u^{(k)}(x,0) \equiv \partial^k u/\partial x^k|_{y=0}$ are related by

$$u^{(k)} = u_{xx}^{(k-2)} - u^{(k-1)}$$

Use *Mathematica* to construct this series solution symbolically with the commands

```
a[0,x_]:=f[x];
a[1,x_]:=g[x];
a[n_,x_]:=If[n<0,0,D[a[n-2,s],{s,2}]-a[n-1,s]/.s->x

u[x_,y_,n_]:=Sum[(1/k!)a[k,x]y^k,{k,0,n}]
```

13. Specialize the procedure in Problem 12 to solving the initial-value problem of the telegraph equation with

$$u(x,0) = \cos kx, \qquad u_y(x,0) = 0$$

By observing that only even-order derivatives of f enter into the solution, solve this initial-value problem by substituting the trial solution

$$u(x,t) = A(t)\cos kx$$

14. Find the solution of the wave equation with the following Cauchy boundary conditions $u(x,0) = f(x)$ and $u_y(x,0) = g(x)$ where $f(x)$ and $g(x)$ are given by:

$$f(x) = \begin{cases} 1 & \text{if } x < 1 \\ 0 & \text{if } x \geq 1 \end{cases} \qquad g(x) = 0$$

Use *Mathematica* to plot this solution in two ways: with `Plot` to see the evolution of the initial wave form, and with `ContourPlot` to see the evolution of the solution according to the characteristics.

15. Repeat the procedure in Problem 14 for the Cauchy initial conditions of the wave equation given by:

$$f(x) = \begin{cases} 0 & \text{if } x < 0 \\ x & \text{if } 0 < x < \frac{1}{2} \\ 1-x & \text{if } \frac{1}{2} < x < 1 \\ 0 & \text{if } 1 < x \end{cases} \qquad G(x) = 0$$

16. Consider a function $u(x, y)$ that is a solution of Equation (3.4). We will examine the behavior of the derivatives of $u(x, y)$ across a curve $\mathcal{C}$ given by $\xi(x, y) = 0$ in the x-y plane. This function is first transformed into an orthogonal coordinate system (ξ, η) such that $u(\xi, \eta)$ has continuous first and second derivatives in a region containing $\mathcal{C}$, with the possible exception of $u_{\xi\xi}$, which may exhibit a discontinuity across $\mathcal{C}$.

By transforming Equation (3.4) into the (ξ, η) coordinate system, subtracting the two equations evaluated on either side of $\mathcal{C}$ and using the continuity of all quantities, except $u_{\xi\xi}$, along and across $\mathcal{C}$, obtain

$$(a\xi_x^2 + b\xi_x\xi_y + c\xi_y^2)\Delta u_{\xi\xi} = 0$$

where $\Delta u_{\xi\xi}$ is the 'jump' of $u_{\xi\xi}$ across $\mathcal{C}$. Thus, deduce that such a discontinuity can occur only across a characteristic.

By differentiating this expression the required number of times, show that a discontinuity in $\partial^n u/\partial \xi^n$, with all other required derivatives being continuous, can also only occur across a characteristic.

17. Use the method of Example 3.2 to obtain the solution of the initial-value problem of the diffusion equation with a source term $J(t)$:

$$\frac{\partial u}{\partial t} = \frac{\partial^2 u}{\partial x^2} + J(t)$$

In the particle interpretation of the diffusion equation, $J(t)$ represents a spatially uniform source of particles that depends only on the time. The initial value of u is given by
$$u(x,0) = f(x)$$
Show that expanding the solution of this equation as in Equation (3.74) yields the expression
$$u(x,t) = \sum_{n=0}^{\infty} \frac{d^{2n} f}{dx^{2n}} \frac{t^n}{n!} + \sum_{n=1}^{\infty} \frac{d^{n-1} J}{dx^{n-1}} \frac{t^n}{n!}$$
Show that the second summation in this expression can be written as
$$J(t) = \int_0^t J(s)\,ds$$

18. Use the method of Example 3.2 to obtain the solution of the initial-value problem
$$\frac{\partial u}{\partial t} = \frac{\partial^2 u}{\partial x^2} + J(x), \qquad u(x,0) = f(x)$$
in the form
$$u(x,t) = \sum_{n=0}^{\infty} \frac{d^{2n} f}{dx^{2n}} \frac{t^n}{n!} + \sum_{n=1}^{\infty} \frac{d^{2n-2} J}{dx^{2n-2}} \frac{t^n}{n!}$$
The term $J(x)$ in this equation represents a time-independent but spatially-varying source of particles (or heat).

19. Deduce from the result of Example 3.2 that the solution of the diffusion equation with the initial value
$$u(x,0) = n^{-1}[1 + \cos x + \cos 2x + \cdots + \cos(n-1)x]$$
is given by
$$u(x,t) = n^{-1} \sum_{k=0}^{n-1} \exp(-k^2 x) \cos kx$$
Use *Mathematica* to plot this solution for $n = 10$. In particular, observe that the initial function is sharply peaked near the origin and, as time increases, the distribution broadens, with the value at the origin decreasing continuously. Explain the origin of this effect in terms of the time development of each of the contributions to $u(x,0)$.

20. Consider two functions N and Q of the arguments x, t, u, u_x, u_t, ..., for some function $u(x,t)$, where $-\infty < x < \infty$. If N and Q are related by the equation
$$\frac{\partial N}{\partial t} + \frac{\partial Q}{\partial x} = 0$$

then show that
$$\frac{\partial}{\partial t}\int N\,dx = 0$$
if $\lim_{x\to\pm\infty} Q = 0$. In other words the quantity $\int N\,dx$ is a constant of the motion of the equation, and the local density N corresponds to the local current Q. Identify these quantities for the diffusion equation
$$\frac{\partial u}{\partial t} = \frac{\partial^2 u}{\partial x^2}$$
How is this conservation property reflected in plots of the solutions of these equations as a function of time?

21. As a particular application of the result of Problem 20, consider the solution of the diffusion equation with the initial condition $u(x,0) = f(x)$ shown in (3.79):
$$u(x,y) = \sum_{n=0}^{\infty} \frac{d^{2n} f(x)}{dx^{2n}} \frac{y^n}{n!}$$
Suppose that the initial function $f(x)$ is characterized by the value of the function and all of its derivatives vanishing as $x \to \pm\infty$. Show that
$$\int_{-\infty}^{\infty} u(x,t)\,dx = \int_{-\infty}^{\infty} f(x)\,dx$$

22. Consider the Cauchy problem for Laplace's equation, where the solution is required to satisfy the boundary conditions
$$u(x,0) = f(x), \qquad u_y(x,0) = g(x)$$
By expanding the solution in a Taylor series about the line $y=0$,
$$u(x,y) = \sum_{n=0}^{\infty} \left.\frac{\partial^n u}{\partial y^n}\right|_{y=0} \frac{y^n}{n!}$$
deduce from the result of Problem 11 that the application of the procedure in Example 3.2 yields
$$u(x,y) = \sum_{n=0}^{\infty} (-1)^n \frac{d^{2n} f}{dx^{2n}} \frac{y^{2n}}{(2n)!} + \sum_{n=0}^{\infty} (-1)^n \frac{d^{2n+1} h}{dy^{2n+1}} \frac{y^{2n+1}}{(2n+1)!}$$
The function $h(x)$ is given in terms of $g(x)$ as in Problem 11. By following the steps in Problem 11, show that these series can be summed to obtain the expression in Equation (3.88).

23. Another way to understand why Cauchy boundary conditions are inappropriate for Laplace's equation is provided by an example due to Hadamard

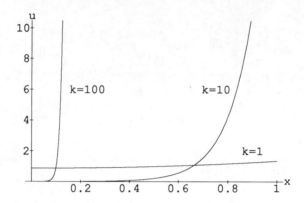

Figure 3.7 An illustration of Hadamard's example discussed in Problem 23 to show why Cauchy boundary conditions are unsuitable for Laplace's equation. Shown are slices of the solution obtained in Problem 23 taken at the point $x = \frac{1}{2}$ for $k = 1$, 10 and 100 generated with the Plot command. As k increases the solution attains arbitrarily large values arbitrarily close to the line $y=0$, while approaching zero along that line.

(1952). Consider the d'Alembert form of the solution of Laplace's equation in Equation (3.88). If the functions determining the initial conditions are given by $f(x) = k^{-2} \cos kx$ and $g(x) = 0$, show that the solution is

$$u(x,y) = \frac{1}{k^2} \cos kx \cosh ky$$

Along the initial line $y=0$, this solution is bounded by the value k^{-2}. Suppose now that $y > 0$. By examining the behavior of the initial conditions and the solution as $k \to \infty$, deduce that the initial conditions can be made arbitrarily close to zero, while the solution can be made arbitrarily large near the initial line $y = 0$. This example shows that an arbitrarily small change in the boundary conditions can produce an arbitrarily large change in the solution as close as desired near the boundary. In this sense, the solution is not a continuous function of the Cauchy boundary conditions.

Use Plot to graph this solution near $y=0$ for a fixed value of x with increasing values of k. An example is shown in Figure 3.7.

24. Using the arguments leading to (3.95), show that given a region $\mathcal{R}$ in which the function $u(x,y)$ satisfies Laplace's equation, the integral of the normal derivative of u around the boundary, $\mathcal{C}$, vanishes:

$$\int_\mathcal{C} \frac{\partial u}{\partial n} ds$$

where ds is the element of arc length along $\mathcal{C}$.

Consider a point (x_0, y_0) within $\mathcal{R}$. A circular contour of radius r centered at this point can be written as

$$(x_0 + r\cos\phi, y_0 + r\sin\phi)$$

where $0 \leq \phi < 2\pi$. Thus, construct the average value of $u(x, y)$ evaluated along such a contour contained entirely within $\mathcal{R}$:

$$\bar{u}(x_0, y_0; r) = \frac{1}{2\pi} \int_0^{2\pi} u(x_0 + r\cos\phi, y_0 + r\sin\phi) \, d\phi$$

By differentiating this expression with respect to r show that $\bar{u}(x_0, y_0; r)$ is, in fact, independent of r. Thus, deduce that the value of $u(x_0, y_0)$ is equal to the average of its values taken on the circumference of a circle centered at (x_0, y_0).

25. Using solutions of Laplace's equation obtained in Problem 2, verify the mean-value theorem proven in Problem 24. For example, taking the solution $u(x, y) = x^2 - y^2$, the following commands calculate the appropriate integral:

```
u[x_,y_]:=x^2-y^2

U[x_,y_,r_]:=N[(1/(2 Pi))NIntegrate[u[x+r Cos[phi],y+r Sin[phi]],
   {phi,0,2Pi}]]
```

26. Use the mean-value theorem proven in Problem 24 to show that a non-constant function satisfying Laplace's equation in a region $\mathcal{R}$ can attain neither a maximum nor a minimum within $\mathcal{R}$. Thus, within a closed and bounded region $\mathcal{R}$, the extrema of $u(x, y)$ are attained on the boundary of $\mathcal{R}$.

Construct the argument as follows. Suppose that a point (x_0, y_0) is the maximum of $u(x, y)$ within $\mathcal{R}$. By evaluating the average value of $u(x, y)$ using the mean-value theorem, show that a contradiction is reached. Use the same line of reasoning applied to $-u(x, y)$ to show that a minimum cannot be attained within $\mathcal{R}$.

27. Identify the regions in the x-y plane where the differential equations are either hyperbolic, parabolic or elliptic. Determine the characteristics of each equation and, in regions where the equations are hyperbolic, find the general solutions in the original variables.

(a) $x^2 u_{xx} - y^2 u_{yy} + x u_x - y u_y = 0$

(b) $y^3(x u_{xx} - u_x) - x^3(y u_{yy} + u_y) = 0$

(c) $u_{xx} - y^2 u_{yy} - y u_y = 0$

(d) $x^2 u_{xx} - u_{yy} + x u_x = 0$

(e) $x u_{xx} - u_{yy} + \frac{1}{2} u_x = 0$

(f) $u_{xx} - y u_{yy} - \frac{1}{2} u_y = 0$

28. For each of the equations in Problem 27, determine the Jacobian for the transformation between the original coordinates (x, y), and the characteristic

coordinates (λ, μ) in regions where the equations are hyperbolic. Show that the transformation between these two coordinate systems becomes singular at the boundaries where the classification of the equation changes. Use ContourPlot to examine the characteristics and to rationalize the behavior of the Jacobian.

29. For each of the equations in Problem 27, determine the general solution to an appropriate Cauchy problem in the region where the equation is hyperbolic. For example, consider the equation in 27(a). By applying Cauchy boundary conditions

$$u(x,y)\Big|_{C_0} = f(x), \qquad u_x(x,y)\Big|_{C_0} = g(x)$$

along the line $y=x^3$, i.e.

$$C_0(s) = s\mathbf{i} + s^3\mathbf{j}$$

show that the solution of this equation can be written as

$$u(x,y) = \tfrac{1}{3}f[\xi(x,y)] + \tfrac{2}{3}f[\eta(x,y)] - \frac{4}{3}\int_{x_0}^{\xi(x,y)} g(t)\,dt + \frac{4}{3}\int_{x_0}^{\eta(x,y)} g(t)\,dt$$

where $\xi(x,y) = (y/x)^{1/2}$ and $\mu(x,y) = (xy)^{1/4}$. Identify any restrictions on the quantity x_0.

Chapter 4

Separation of Variables and the Sturm–Liouville Problem

In the preceding chapter we derived some general properties of quasi-linear second-order partial differential equations. We were able to obtain general solutions in certain special cases and to make specific statements concerning the appropriate boundary conditions for each type of equation. However, the determination of general solutions in a simple closed form is not always straightforward, and is not necessarily the best way to solve problems where boundaries are involved. The next several chapters are devoted to developing techniques for finding solutions to initial-value and boundary-value problems from linear combinations of elementary particular solutions. We will find that these methods have natural extensions in higher dimensions, so their application is not limited to the case of two independent variables, which has been the basis of most of our discussion up to now.

We begin this chapter with the technique of separation of variables. The idea behind this method is to transform a partial differential equation in n independent variables into n ordinary differential equations by separating the functional dependence of the solution into a product of functions of each of the independent variables. Solving these ordinary differential equations leads to eigenvalue problems and the general solution is then constructed as a superposition of the eigenfunctions, with the coefficients determined by the initial conditions and the boundary conditions. The method of Fourier series, where the solution is written as an infinite sum of trigonometric functions, is one of the best-known methods of solution with this approach. We will use Fourier series in conjunction with the method of separation of variables to solve several different types of boundary-value problem.

The representation of solutions of partial differential equations as an infinite series of functions is central to the method of separation of variables. Many of the properties of these functions are determined by the general form of the ordinary differential equations after a separation of variables has been performed. The ordinary differential equations are known as **Liouville** equations and determining the solutions and the properties of solutions is known as the **Sturm–Liouville** problem. Many of the familiar features of special functions and orthogonal polynomials are based upon the properties of solutions of Liouville's equation. Some of the important general properties of the solutions will be discussed in this chapter, while properties of particular functions will be examined in Chapter 6.

4.1 The Method of Separation of Variables

The method of separation of variables is one of the oldest techniques for solving partial differential equations and provides an important illustration of the power of the superposition principle in constructing general solutions from linear combinations of elementary particular solutions. The basic idea behind this method is very simple. Given a partial differential equation for a function F of n independent variables, $F(x_1, x_2, \ldots, x_n)$, the functional dependence of F is factored into separate functions of each of the independent variables:

$$F(x_1, x_2, \ldots, x_n) = f_1(x_1) f_2(x_2) \cdots f_n(x_n) \tag{4.1}$$

This product is then substituted into the partial differential equation and, if the equation is separable in the $(x_1, x_2, \ldots, x_n)$ coordinate system, the equation can be reduced to a sum of n terms each of which is composed only of the jth variable, the corresponding function f_j and its derivatives. Since the sum of these terms must vanish and since each of the variables x_j can be varied independently, then each of the n terms in the sum must be equal to a constant. This leads to n equations each involving only x_j, $f_j(x_j)$ and the (ordinary) derivatives of f_j, i.e. n ordinary differential equations for the f_j. The solution of the original *partial* differential equation has thereby been reduced to finding the solution of n *ordinary* differential equations.

To illustrate the method of separation of variables for the construction of solutions to initial and boundary-value problems, we will solve the wave equation,

$$\frac{\partial^2 u}{\partial x^2} - \frac{\partial^2 u}{\partial t^2} = 0 \tag{4.2}$$

The solution u is first written as the product of a function X of the variable x and a function T of t:

$$u(x,t) = X(x)T(t) \tag{4.3}$$

This expression is then substituted into (4.2) and the required derivatives are carried out. Dividing the resulting equation by the product XT yields

$$\frac{1}{X}\frac{d^2 X}{dx^2} - \frac{1}{T}\frac{d^2 T}{dt^2} = 0 \tag{4.4}$$

The first term on the left-hand side of this equation is a function of x only and the second term is a function of t only, with their difference vanishing for *all* values of x and t. Since x and t may be varied independently, Equation (4.4) can be satisfied only if each of the two terms is equal to the *same* constant:

$$\frac{1}{X}\frac{d^2 X}{dx^2} = \frac{1}{T}\frac{d^2 T}{dt^2} = -k^2 \tag{4.5}$$

The quantity k is called a **separation constant**; the negative square of k is introduced only for convenience in writing the solutions of (4.5). Equation (4.5) can be written as ordinary differential equations for X and T:

$$\frac{d^2 X}{dx^2} + k^2 X = 0, \qquad \frac{d^2 T}{dt^2} + k^2 T = 0 \tag{4.6}$$

The solutions to these equations that satisfy (4.5) can be obtained easily and written either as complex exponentials

$$X(x) = e^{\pm ikx}, \qquad T(t) = e^{\pm ikt} \tag{4.7}$$

or, in terms of real functions, as

$$X(x) = \begin{cases} \cos kx \\ \sin kx \end{cases}, \qquad T(t) = \begin{cases} \cos kt \\ \sin kt \end{cases} \tag{4.8}$$

These solutions in Equations (4.7) and (4.8) each yield four linearly independent products XT that solve (4.2). Thus, for a given separation constant, the superposition principle can be used to construct a solution to (4.2) that is a linear combination of products of the solutions in (4.7),

$$u(x,t) = a e^{+ik(x+t)} + b e^{+ik(x-t)}$$
$$+ c e^{-ik(x-t)} + d e^{-ik(x+t)} \tag{4.9}$$

or a linear combination of products of the solutions in (4.8),

$$u(x,t) = A \cos kx \cos kt + B \cos kx \sin kt$$
$$+ C \sin kx \cos kt + D \sin kx \sin kt \tag{4.10}$$

The coefficients a,b,c,d and A,B,C,D are constants that will be determined below. The arguments of the exponentials in (4.9) are readily identified as the characteristics of (4.2), $x+t$ and $x-t$. The solutions in (4.9) and (4.10) are thus seen to be sinusoidal waves propagating in the positive and negative x directions with unit speed.

Although the expressions given in (4.9) and (4.10) are not the most general solutions of the wave equation, they do provide some guidance about the way to proceed to construct more flexible solutions. Suppose that a solution of (4.2) is sought on the closed boundary $[0, L]$, which at the endpoints is required to vanish at all times,

$$u(0,t) = u(L,t) = 0 \qquad (4.11)$$

An example of such a problem is a vibrating string of length L whose ends are fixed. Applying the boundary conditions in (4.11) to the solution in (4.10) yields

$$u(0,t) = A \cos kt + B \sin kt = 0 \qquad (4.12a)$$

$$u(L,t) = A \cos kL \cos kt + B \cos kL \sin kt$$
$$+ C \sin kL \cos kt + D \sin kL \sin kt = 0 \qquad (4.12b)$$

At $t=0$, Equation (4.12a) yields $A=0$, while for $kt=\frac{1}{2}\pi$, we obtain $B=0$. Thus, Equation (4.12a) can be satisfied for all times only if $A=B=0$, in which case (4.12b) reduces to

$$u(L,t) = \sin kL (C \cos kt + D \sin kt) = 0 \qquad (4.13)$$

There are two ways to satisfy this equation. We can set $C = D = 0$, but this produces the trivial solution, $u(x,t)=0$, which satisfies only the trivial boundary conditions $u(x,0) = 0$ and $u_t(x,0) = 0$. Alternatively, we can require that $\sin kL = 0$. This necessitates choosing kL to be an integral multiple of π: $k_n L = n\pi$, where n is an integer. For any value of k satisfying this condition, the solution in (4.10) becomes

$$u(x,t) = \sin k_n x \, (C \cos k_n t + D \sin k_n t) \qquad (4.14)$$

This is a solution of the wave equation (4.2) with the boundary conditions (4.11) and with the initial conditions given by

$$u(x,0) = C \sin k_n x, \qquad u_t(x,0) = k_n D \sin k_n x \qquad (4.15)$$

In other words, (4.14) describes the time development of the solution to the wave equation (4.2) from an initial sinusoidal displacement with an initial sinusoidal velocity.

Although the solution in Equation (4.14) solves a very restricted initial-value problem, it can be generalized to any initial functions that can be expressed as sums of terms proportional to $\sin k_n x$. Thus, suppose that $u(x,0)$ and $u_t(x,0)$ are given by

$$u(x,0) = f_1 \sin k_1 x + f_2 \sin k_2 x + \cdots + f_N \sin k_N x$$
$$u_t(x,0) = g_1 \sin k_1 x + g_2 \sin k_2 x + \cdots + g_N \sin k_N x$$
(4.16)

where the coefficients f_n and g_n are known quantities. The solution of the wave equation with these initial conditions and with the boundary conditions in Equation (4.11) is then constructed as a superposition of solutions of the form shown in (4.14), with the result

$$u(x,t) = \sum_{n=1}^{N} \sin k_n x \left(f_n \cos k_n t + \frac{g_n}{k_n} \sin k_n t \right) \quad (4.17)$$

Thus, the time development from any initial conditions of the form in (4.16) with the boundary conditions (4.11) has been decomposed into the superposition of elementary solutions given in (4.14). These solutions evolve independently of one another from initial conditions determined by the coefficients f_n and g_n in (4.16). The solution to the wave equation at a later time is determined by summing the amplitudes of the individual components.

The solution in Equation (4.17) shows that if the functions characterizing the initial conditions can be represented as a trigonometric sum, then the solution of the initial-value problem for the wave equation can also be represented as a trigonometric sum. In fact, since any pair of functions $\sin k_n x$ with different values of n are linearly independent, as a calculation of the Wronskian shows, there is no reason to expect the sums in (4.16) to contain only a finite number of terms for any given functions $u(x,0)$ and $u_t(x,0)$. However, a question then naturally arises: which functions can be represented by an infinite or finite series of trigonometric functions? Specifically, for particular functions that determine the initial conditions $u(x,0)$ and $u_t(x,0)$, a procedure is needed both for calculating the coefficients f_n and g_n in (4.16) and for determining whether the sums on the right-hand sides of the equations in (4.16) are equal to the functions on the left-hand sides.

The series obtained by letting $N \to \infty$ in (4.16), where the separation constant takes *discrete* values are examples of **Fourier series**, and the conditions under which functions have Fourier series representations are examined in the next two sections. As we have seen in applying the boundary conditions in Equation (4.11) to the wave equation, the discreteness of the separation constant arises from the solution being confined to a bounded region, such as a closed interval. The case where k takes continuous values and the sums in Equations (4.16) and (4.17) are replaced by integrals are examples of **Fourier transforms**. Continuous values of the separation

4.2 Orthogonality of Functions

We have already seen in Chapter 1 that the trigonometric functions corresponding to different separation constants are linearly independent. In this section, we will carry this property one step further and show that in representations such as those in Equations (4.16) and (4.17), we can regard the trigonometric functions are 'orthogonal vectors' in a space composed of functions. This has many important practical and conceptual consequences for calculating solutions of differential equations from expansions in terms of trigonometric functions and other functions. First, however, the notion of orthogonality of functions must be made more precise.

We begin by considering two functions $\phi_1(x)$ and $\phi_2(x)$ defined over an open interval (a, b). To test for the linear independence of these functions, we construct the equation

$$A\phi_1(x) + B\phi_2(x) = 0 \tag{4.18}$$

where A and B are constants. If ϕ_1 and ϕ_2 are linearly independent, then the only choice of A and B that insures (4.18) is satisfied for *every* value of x in (a, b) is $A = B = 0$. If, on the other hand, ϕ_1 and ϕ_2 are linearly dependent, then there are nonzero values of A and B that satisfy (4.18) at all points x within (a, b). In this case ϕ_1 and ϕ_2 are seen to be related by a proportionality constant, which enables (4.18) to be solved for A in terms of B or vice versa.

In Section 1.4, the constants A and B were determined using the Wronskian. While this approach provides an adequate way of assessing the linear independence of two or more functions, it is not especially well suited to providing a 'measure' of the extent to which two functions are linearly independent. Another way to test for linear independence in Equation (4.18), and one that leads naturally to the notion of the orthogonality of functions, is to multiply (4.18) in turn by ϕ_1 and ϕ_2 and integrate over (a, b) to obtain a matrix equation involving the quantities A and B and the integrals

$$S_{ij} = \int_a^b \phi_i(x)\phi_j(x)\,dx \tag{4.19}$$

For $i \neq j$, $S_{12} = S_{21}$ represents the overlap between ϕ_1 and ϕ_2 integrated over (a, b). Similarly, for $i = j$, S_{11} and S_{22} are the integrals of ϕ_1^2 and ϕ_2^2 over (a, b). The resulting matrix equation is

$$\begin{pmatrix} S_{11} & S_{12} \\ S_{21} & S_{22} \end{pmatrix} \begin{pmatrix} A \\ B \end{pmatrix} = \begin{pmatrix} 0 \\ 0 \end{pmatrix} \tag{4.20}$$

This equation can be solved for A and B only if the determinant of the matrix of coefficients vanishes, i.e. only if

$$\Delta \equiv S_{11}S_{22} - S_{12}S_{21} = 0 \tag{4.21}$$

Thus, if ϕ_1 and ϕ_2 are linearly *dependent* functions, then $\Delta = 0$ and (4.20) can be solved for A in terms of B. If, on the other hand, ϕ_1 and ϕ_2 are linearly *independent*, then $\Delta \neq 0$ and the only solution to (4.20) is $A = B = 0$.

If ϕ_1 and ϕ_2 are linearly dependent over (a, b), then solving, say, the first of the equations in (4.20) and substituting into (4.18) produces

$$\phi_2 = \frac{S_{12}}{S_{11}} \phi_1 \tag{4.22}$$

Even if ϕ_1 and ϕ_2 are linearly *independent*, the quantities S_{12} and S_{11} can still be calculated, though the equality in Equation (4.22) no longer holds. However, we can introduce the quantity ψ, which is the difference between the two sides of (4.22):

$$\psi = \phi_2 - \frac{S_{12}}{S_{11}} \phi_1 \tag{4.23}$$

Multiplying both sides of this equation by ϕ_1 and integrating over (a, b) shows that the integrated overlap between ψ and ϕ_1 vanishes:

$$\int_a^b \phi_1(x)\psi(x)\,dx = 0 \tag{4.24}$$

Similarly, multiplying both sides of (4.23) by ϕ_2 and integrating yields, after some rearrangement, the expression

$$\int_a^b \phi_2(x)\psi(x)\,dx = S_{22}\left(1 - \frac{S_{12}S_{21}}{S_{11}S_{22}}\right) \tag{4.25}$$

Two functions ϕ_1 and ψ, neither of which vanishes over an entire interval (a, b), that satisfy (4.24) are said to be **orthogonal** over that interval. The decomposition of ϕ_2 in Equation (4.23) can now be interpreted as an elimination of the contribution to ϕ_2 that is proportional to ϕ_1. The remaining contribution, represented by ψ, is orthogonal to ϕ_1, as Equation (4.24) shows. A measure of the contribution of ϕ_2 to ψ is shown in Equation (4.25). In particular, if ϕ_1 and ϕ_2 are linearly dependent, then the right-hand side of (4.25) vanishes, while if the two functions are already orthogonal, then S_{12} and S_{21} both vanish, and the right-hand side is equal to S_{22}. In effect, the decomposition (4.23) is equivalent to the **Gram–Schmidt orthogonalization** for two vectors **u** and **v**, whereby **v** is written as a linear combination of a vector 'colinear' with **u** and a vector 'orthogonal' to **u**. This analogy is carried out in Problem 1.

> **Example 4.1.** Integrate can be used to evaluate symbolically the overlap integrals of the functions $\sin k_n x$, which were determined in the preceding section to be solutions of the wave equation with the boundary conditions (4.11). With $k_n = n\pi/L$, where n is an integer, we obtain
>
> ```
> Simplify[Integrate[Sin[(n Pi x)/L]Sin[(m Pi x)/L],{x,0,L}]]
> ```
>
> $$\frac{L \sin[\mathrm{Pi}\,(m-n)]}{2\,(\mathrm{Pi}\,m - \mathrm{Pi}\,n)} - \frac{L \sin[\mathrm{Pi}\,(m+n)]}{2\,\mathrm{Pi}\,(m+n)}$$
>
> If $m \neq n$, both terms in this expression vanish, which shows that $\sin k_n x$ and $\sin k_m x$ are orthogonal. If $m = n$, the corresponding commands yield
>
> ```
> Integrate[(Sin[(n Pi x)/L])^2,{x,0,L}]
> ```
>
> $$\frac{L}{2} - \frac{L \cos[\mathrm{Pi}\,n] \sin[\mathrm{Pi}\,n]}{2\,\mathrm{Pi}\,n}$$
>
> The second term vanishes, so the integral takes the value $\frac{1}{2}L$, independent of n. Thus, the orthogonality relation of the sine functions over the interval $(0, L)$ can be summarized as
>
> $$\int_0^L \sin k_n x \sin k_m x \, dx = \begin{cases} \frac{1}{2}L & \text{if } m = n \\ 0 & \text{if } m \neq n \end{cases} \quad (4.26)$$

■

The orthogonality in (4.26) suggests a representation of solutions of the initial-value problem of the wave equation with the boundary conditions in (4.11) as a series of sine functions. In order to establish the conditions under which this representation and those based on other trigonometric functions are meaningful, we need to study some general properties of trigonometric series. This will be done in the next section.

4.3 Fourier Series

The representation of solutions of the initial-value problem of the wave equation as a series of sine functions is a direct result of the boundary conditions in Equation (4.11). Other types of boundary condition lead to solutions in the form of cosine series or series that involve both sine and cosine functions, as discussed in the problems at the end of the chapter. Therefore, to make our discussion of trigonometric series as widely applicable as possible, we will consider a function $f(x)$ with a period of 2π and examine the properties of the following series representation of $f(x)$:

$$f(x) = \tfrac{1}{2}a_0 + \sum_{n=0}^{\infty}(a_n \cos nx + b_n \sin nx)$$

$$= \tfrac{1}{2}a_0 + a_1 \cos x + a_2 \cos 2x + \cdots + b_1 \sin x + b_2 \sin 2x + \cdots \quad (4.27)$$

This expansion is called a **Fourier series**. The idea of such expansions first arose in the study of physical problems involving oscillations of vibrating strings, but it was Fourier, in a comprehensive study of heat conduction, who first showed that a function could be expanded in this way. A brief historical sketch of Fourier series may be found in the book by Carslaw (1950).

There are several observations to make concerning the expansion in (4.27). First, the functions $\sin nx$ and $\cos nx$ have a *common* period of 2π, as does $f(x)$. Second, it is easy to verify that the functions $\sin nx$ and $\cos nx$ are mutually orthogonal over any interval of length 2π. We will use this property shortly to obtain expressions for the expansion coefficients a_n and b_n. Third, even though the left-hand and right-hand sides of this equation have been set equal to one another, it must be stressed that we as yet have only written a *formal* expansion of $f(x)$; we will identify the conditions under which equality is obtained later in this section. Finally, the reason for writing the first term on the right-hand side of this equation as $\tfrac{1}{2}a_0$ is simply one of convenience, as will become clear below.

The expansion coefficients a_n and b_n, which are called **Fourier coefficients**, are calculated by multiplying both sides of (4.29) by $\sin nx$ or $\cos nx$, integrating over $(-\pi, \pi)$ (the choice of this interval is largely a matter of convention) and using the orthogonality properties of the trigonometric functions:

$$a_n = \frac{1}{\pi}\int_{-\pi}^{\pi} f(x)\cos nx \, dx$$

$$b_n = \frac{1}{\pi}\int_{-\pi}^{\pi} f(x)\sin nx \, dx$$

(4.28)

for $n \geq 1$. By performing the corresponding steps to determine a_0, we obtain

$$a_0 = \frac{1}{2\pi}\int_{-\pi}^{\pi} f(x)\, dx \quad (4.29)$$

The reason for the factor $\tfrac{1}{2}$ in front of this term in (4.27) is now seen to be merely a consolidation of the expressions for a_0 and for a_n, with $n \neq 0$.

The theorems that determine the conditions under which the representation (4.27) exists are stated in many forms. One set of conditions requires f to be finite over $(-\pi, \pi)$ with the possible exception of a finite number of points, and the integral

$$\int_{-\pi}^{\pi} |f(x)|\, dx \tag{4.30}$$

to exist. If these conditions are satisfied, then at points $-\pi < x < \pi$ where the limits

$$f(x^+) = \lim_{\varepsilon \to 0} f(x+\varepsilon), \qquad f(x^-) = \lim_{\varepsilon \to 0} f(x-\varepsilon) \tag{4.31}$$

exist, the series in (4.27) converges to

$$\tfrac{1}{2}[f(x^+) + f(x^-)] \tag{4.32}$$

Thus, at points x where f is continuous, the value of $f(x)$ is the same whether x is approached from the left or the right, and the series converges to $f(x)$. On the other hand, if f is discontinuous at x, the series converges to the mean of the two limits obtained when the point is approached from the left and the right. A clear and concise discussion of the proofs of these statements may be found in Rudin (1964).

For the class of functions fulfilling the criteria for a convergent Fourier series, the trigonometric functions are said to form a **complete** basis set. For functions that do not satisfy these criteria, the convergence of the Fourier series cannot be guaranteed. Completeness must be established both for a given set of functions and for the conditions that must be satisfied by the functions to be expanded. Thus, theorems that prove the existence of the representation (4.27) are essentially completeness theorems for the trigonometric functions. A discussion of the criteria for completeness of the trigonometric functions and other functions has been given by Birkhoff and Rota (1989).

Example 4.2. To illustrate some of the points just discussed, we consider the Fourier series representation of two different types of function. Consider first the function

$$f(x) = \begin{cases} \pi^{-1}(\pi + x) & \text{if } -\pi < x \leq 0 \\ \pi^{-1}(\pi - x) & \text{if } 0 < x \leq \pi \end{cases} \tag{4.33}$$

This function is continuous everywhere, but has discontinuities in its first derivative at $x = \pm\pi$ and $x = 0$. To generate the Fourier series of this function within *Mathematica*, we first construct the function in (4.33),

```
f[x_]:=If[x<=0,(x+Pi)/Pi,(Pi-x)/Pi]
```

The Fourier series is now constructed by first defining the expansion coefficients as in (4.28) and (4.29). When we do so, we find that the b_n all vanish for reasons that will be explained below. Thus, we need only consider the a_n, which are calculated using Integrate:

```
a[n_]:=(1/Pi)Integrate[((Pi+t)/Pi) Cos[n t],{t,-Pi,0}]+
       (1/Pi)Integrate[((Pi-t)/Pi) Cos[n t],{t,0,Pi}]
```

The partial sums of the Fourier series are now constructed using Sum. The distinction between a_0 and the a_n for $n \neq 0$ as shown in (4.28) and (4.29) can be taken into account by defining new coefficients A[k] using If:

```
A[k_]:=If[k>0,a[k],(1/2)a[0]]

F[x_,n_]:=Sum[A[k]Cos[k x],{k,0,n}]
```

The partial sums of the Fourier series can now be generated up to any desired order. For example, the first five terms are obtained as

```
F[x,5]
```

$$\frac{1}{2} + \frac{4 \cos[x]}{2 \, Pi} + \frac{4 \cos[3\,x]}{9 \, Pi} + \frac{4 \cos[5\,x]}{25 \, Pi}$$

The Fourier representations obtained from F[x,1] and F[x,5] are shown in Figure 4.1. Even the first term reproduces the profile of the function quite closely. The only marked differences between the partial sum and the function are at the points of discontinuity of the first derivative. Such features cannot be reproduced with a finite series of smooth functions, since any finite sum of functions with continuous first derivatives will also have a continuous first derivative.

Consider now the function

$$f(x) = \begin{cases} 0 & \text{if } -\pi < x \leq \tfrac{1}{8}\pi \\ 1 & \text{if } \tfrac{1}{8}\pi < x \leq \tfrac{7}{8}\pi \\ 0 & \text{if } \tfrac{7}{8}\pi < x \leq \pi \end{cases} \qquad (4.34)$$

which has discontinuities at $x = \tfrac{1}{8}\pi$ and $x = \tfrac{7}{8}\pi$. The Fourier series for (4.34) is generated by following similar steps to those used for obtaining the series in (4.33) and the partial sums including the first 5, the first 10 and the first 20 terms are shown in Figure 4.2.

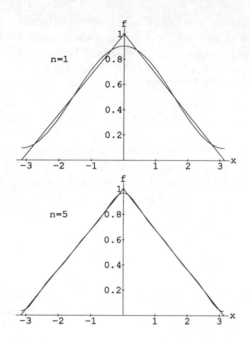

Figure 4.1 The Fourier representation of the function in (4.33), including the first term (top) and the first five terms in the partial sum (bottom). The function in (4.33) is also shown with each partial sum. The only points where the convergence is slow is near the discontinuities in the first derivative at the points $x = \pm \pi$ and $x = 0$.

Although the general profile of the function is reproduced even by including only first five terms, appreciable differences are evident between the Fourier representation and the function even after the first twenty terms have been included in the partial sum. The slow convergence is due to the rapid variation of the function near the discontinuities, which requires the higher-order trigonometric functions to be included in the partial sum because of their short periods.

However, including more terms in the Fourier expansion of this function does not lead to an improved representation in all respects. While the value of the function is better reproduced at points *near* the discontinuities, the series representation *at* the discontinuities shows marked deviations from the profile of the function regardless of how many terms are included in the Fourier series. This effect, which is known as the **Gibbs phenomenon** and is a general feature of Fourier series of discontinuous functions, is discussed in Problem 8.

In constructing the Fourier series for the function in (4.33) we found that only the cosine terms constributed. Alternatively, we have seen in the

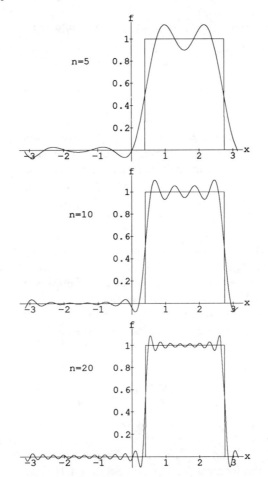

Figure 4.2 The partial sums of the Fourier series of the function in (4.34) including, from the top, the first five terms, the first ten terms, and the first twenty terms. The function in (4.34) is shown with each of the partial sums. The gross shape of the function is reproduced with relatively few terms, though the behavior near the discontinuities converges slowly. This behavior should be contrasted with that shown in Figure 4.1 for the function in (4.33).

preceding section that the sine functions form a natural basis for representing solutions to the wave equation that vanish at the end-points of a finite interval. Other types of boundary conditions will be seen to lead to Fourier cosine series or trigonometric series with both sine and cosine contributions. To relate the type of Fourier series to the different types of function and boundary condition, we first observe that any function $f(x)$ can be written as:

$$f(x) = \tfrac{1}{2}[f(x) + f(-x)] + \tfrac{1}{2}[f(x) - f(-x)]$$
$$\equiv f_+(x) + f_-(x)$$

(4.35)

By their construction $f_+(x)$ and $f_-(x)$ have the properties

$$f_+(-x) = \tfrac{1}{2}[f(-x) + f(x)] = f_+(x)$$
$$f_-(-x) = \tfrac{1}{2}[f(-x) - f(x)] = -f_-(x)$$
(4.36)

i.e. $f_+(x)$ has even parity about the origin and is thus called an *even* function, while $f_-(x)$ has odd parity about the origin and is called an *odd* function. Substitution of the decomposition in Equation (4.35) into the integrals in (4.28) shows that only the *odd* component contributes to a_n and only the *even* component contributes to b_n. In other words, if $f(x)$ is an even function, then the trigonometric series in (4.27) reduces to a Fourier *cosine* series, while if $f(x)$ is an odd function, as in (4.33), then (4.27) reduces to a Fourier *sine* series.

Suppose now that we are given a function $f(x)$ on the interval $(0, \pi)$. Then, using the constructions in (4.35) we can extend this function to the interval $(-\pi, \pi)$ to form either an even function or an odd function. In the former case, the Fourier series of the extended function contains only terms proportional to $\cos nx$ and in the latter case only terms proportional to $\sin nx$. In fact, we need not confine ourselves to the interval $(0, \pi)$, since a simple change of variables,

$$x = \frac{\xi - a}{b - a}\pi$$
(4.37)

changes the range of the independent variable from $-\pi < x \leq \pi$ to $a \leq \xi \leq b$, with the corresponding changes in functions forming the trigonometric series being $\cos[k_n(\xi-a)]$ and $\sin[k_n(\xi-a)]$, where $k_n = n\pi/(b-a)$. In particular, if we consider the interval $(0, L)$, as in Section 4.1, then the trigonometric basis for odd functions is given by the functions $\sin(k_n\xi)$, with $k_n = n\pi/L$, which are precisely the functions obtained in (4.14) and (4.15) by separating variables. In other words, the boundary conditions in (4.11) dictate that the solution $u(x,t)$ and the initial functions $u(x,0)$ and $u_t(x,0)$ must be *odd* functions of x. Similarly, if the boundary conditions in (4.11) are replaced by $u_x(0,t) = u_x(L,t)$, then these functions would be required to be *even* functions of x (Problem 16). If the boundary conditions require a linear combination of the function and its first derivative to take a particular value at the end-points, then the series solution has both sine and cosine terms.

These statements have consequences that go far beyond simply determining whether the solution to a particular boundary-value problem can be expressed as a Fourier sine series, a cosine series or a more general trigonometric series. We will see in Chapter 7 that more general methods for obtaining solutions based on the method of images can be traced to the results obtained in this section. For the moment, however, we will use the results of this section to obtain solutions to the wave equation, the diffusion equation and Laplace's equation.

4.4 Fourier Series Solutions of Differential Equations

In Section 4.1, we found that by requiring the solutions of the wave equation to satisfy the fixed end-point boundary conditions at $x=0$ and $x=L$

$$u(0,t) = 0, \qquad u(L,t) = 0 \qquad (4.38)$$

the method of separation of variables provided solutions of the form

$$u(x,t) = \sin k_n x (A \cos k_n t + B \sin k_n t) \qquad (4.39)$$

where A and B are constants and $k_n = n\pi/L$, with n being an integer. This solution satisfies the initial conditions

$$u(x,0) = A \sin k_n x, \qquad u_t(x,0) = k_n B \sin k_n x \qquad (4.40)$$

From the discussion of the preceding section, we now know how to proceed to solve the more general initial-value problem where the initial displacement and initial velocity are given by

$$u(x,0) = f(x), \qquad u_t(x,0) = g(x) \qquad (4.41)$$

We use the particular solutions in (4.39) as a basis and represent the solution as the infinite series

$$u(x,t) = \sum_{n=1}^{\infty} \sin k_n x (A_n \cos k_n t + B_n \sin k_n t) \qquad (4.42)$$

Since each term in the summation is a solution of the wave equation that satisfies the boundary conditions in (4.38), so is the left-hand side of the equation, due to the wave equation being linear. To satisfy the initial values in (4.41), we calculate $u(x,0)$ and $u_t(x,0)$ from Equation (4.42) and equate the resulting series to f and g:

$$u(x,0) = \sum_{n=0}^{\infty} A_n \sin k_n x = f(x) \qquad (4.43a)$$

$$u_t(x,0) = \sum_{n=0}^{\infty} k_n B_n \sin k_n x = g(x) \qquad (4.43b)$$

We can now use the orthogonality of the functions $\sin k_n x$ over the interval $(0, L)$ to solve for the coefficients A_n and B_n, as was done in obtaining the expressions in (4.28). Carrying out the analogous steps in (4.43), we obtain

$$A_n = \frac{2}{L} \int_0^L f(x) \sin k_n x \, dx \qquad (4.44a)$$

$$k_n B_n = \frac{2}{L} \int_0^L g(x) \sin k_n x \, dx \qquad (4.44b)$$

In fact, recognizing the the right-hand sides of Equations (4.44) as the Fourier coefficients of the sine series of $f(x)$ and $g(x)$,

$$f(x) = \sum_{n=1}^\infty f_n \sin k_n x, \qquad g(x) = \sum_{n=1}^\infty g_n \sin k_n x \qquad (4.45)$$

we can make the identifications $A_n = f_n$ and $k_n B_n = g_n$ and write the solution to the wave equation with the initial values in (4.41) and the boundary conditions in (4.38) as

$$u(x,t) = \sum_{n=1}^\infty \sin k_n x \left(f_n \cos k_n t + \frac{g_n}{k_n} \sin k_n t \right) \qquad (4.46)$$

The Fourier representations of the functions $f(x)$ and $g(x)$ in (4.45) are valid not only in the interval $(0, L)$, but over the entire real line. In fact, it is often useful to make a distinction between the original functions f and g, which are defined only for $0 \leq x \leq L$, and the periodic extensions of these functions, which have a periodicity of $2L$ and are defined for $-\infty < x < \infty$. We will signify these periodic extensions by $\bar{f}$ and $\bar{g}$. The Fourier sine series for f and g in (4.45) show that the $\bar{f}$ and $\bar{g}$ are odd functions of x.

We will use the periodic extensions to show how the Fourier series representation of the solution to the wave equation in (4.46) can be transformed into a closed-form general solution analogous to d'Alembert's solution in Equation (3.64). Standard trigonometric identities for the sines and cosines of the sum and difference of two angles are first employed to rewrite the solution in (4.46) in a more suggestive form:

$$u(x,t) = \sum_{n=1}^\infty f_n \sin k_n x \cos k_n t + \sum_{n=1}^\infty \frac{g_n}{k_n} \sin k_n x \sin k_n t$$

$$= \tfrac{1}{2} \sum_{n=1}^\infty f_n \sin[k_n(x+t)] + \tfrac{1}{2} \sum_{n=1}^\infty f_n \sin[k_n(x-t)]$$

$$+ \tfrac{1}{2} \sum_{n=1}^\infty \frac{g_n}{k_n} \cos[k_n(x-t)] - \tfrac{1}{2} \sum_{n=1}^\infty \frac{g_n}{k_n} \cos[k_n(x+t)] \qquad (4.47)$$

It is now evident that this solution is in fact a function of the characteristics of the wave equation, $x+t$ and $x-t$, which was not readily apparent from the original form in (4.46). Referring now to (4.45), the first two terms on the right-hand side of (4.47) are seen to be proportional to the function $\bar{f}$ evaluated with the arguments $x+t$ and $x-t$:

$$\bar{f}(x \pm t) = \sum_{n=1}^{\infty} f_n \sin[k_n(x \pm t)] \tag{4.48}$$

The last two terms on the right-hand side of (4.47) can be combined by first writing:

$$\sum_{n=1}^{\infty} \frac{f_n}{k_n}\{\cos[k_n(x-t)] - \cos[k_n(x+t)]\} = \sum_{n=1}^{\infty} f_n \left(\int_{x-t}^{x+t} \sin k_n s \, ds \right) \tag{4.49}$$

Then, interchanging the order of summation and integration on the right-hand side of this equation, we obtain

$$\sum_{n=1}^{\infty} f_n \left(\int_{x-t}^{x+t} \sin k_n s \, ds \right) = \int_{x-t}^{x+t} \left(\sum_{n=1}^{\infty} f_n \sin k_n s \right) ds$$

$$= \int_{x-t}^{x+t} \bar{g}(s) \, ds \tag{4.50}$$

Thus, combining (4.48) and (4.50), the solution in (4.46) can be written in the form of d'Alembert's solution in (3.64):

$$u(x,t) = \tfrac{1}{2}[\bar{f}(x+t) + \bar{f}(x-t)] + \tfrac{1}{2} \int_{x-t}^{x+t} \bar{g}(s) \, ds \tag{4.51}$$

Although this expression was derived from the Fourier series representation in Equation (4.46), it is a much less restrictive form of solution since functions with slowly-convergent Fourier series can be handled within the same framework as those with rapidly-convergent series. It should also be apparent that the solution in Equation (4.51) could have been obtained directly once we realized that the boundary conditions *required* the periodic extensions of f and g to be odd functions. Examples of other types of boundary conditions are given in the problems at the end of the chapter.

Example 4.3. To see how boundary conditions of the type in Equation (4.38) affect the behavior of solutions of the wave equation, we will obtain the solution with the initial conditions

$$f(x) = \begin{cases} 0 & \text{if } 0 \leq x < \tfrac{3}{8} \\ 1 & \text{if } \tfrac{3}{8} < x < \tfrac{5}{8}, \\ 0 & \text{if } \tfrac{5}{8} \leq x \leq 1 \end{cases} \qquad g(x) = 0 \tag{4.52}$$

and with the requirement that the solution vanish at $x=0$ and $x=1$:

$$u(0,t) = 0, \qquad u(1,t) = 0 \tag{4.53}$$

The time development of this solution is shown in Figure 4.3 both in the region between the boundaries, $0 \leq x \leq 1$, and in the extended region outside of the boundaries. The profile of $\bar{f}$ clearly shows it to be an odd function of x. The initial form of the solution is seen to split into two pulses that propagate toward the boundaries. The same behavior is seen in neighboring pulses in the extended region, though with opposite sign due to $\bar{f}$ being an odd function. At $t = \frac{1}{2}$ pulses of opposite sign annihilate one another at the boundary, which causes the solution to vanish at the boundaries, in keeping with (4.53). For $t > \frac{1}{2}$, the pulses on either side of the boundary have passed through one another and the pulses originally in the physical region are now in neighboring extended regions and vice versa.

The behavior of the characteristics in the region of interest show the trajectories of the pulses moving toward the boundaries, then reflected toward the center, meeting the boundaries again and so on. This periodic process corresponds to successively more distant pulses in the extended region passing through the region between the boundaries. In this way, the presence of the boundaries is contained *entirely* in the construction of the periodic extension $\bar{f}$ from the original function f. ∎

We now turn our attention to finding solutions for the diffusion equation

$$\frac{\partial u}{\partial t} = \frac{\partial^2 u}{\partial x^2} \tag{4.54}$$

in a region between two boundaries. The boundaries are placed at $x = 0$ and $x = L$ where the solution is required to vanish at all times:

$$u(0,t) = 0, \qquad u(L,t) = 0 \tag{4.55}$$

Viewed from the standpoint of heat conduction, these boundary conditions correspond to heat being totally absorbed at the boundaries. For this reason, the boundary conditions in (4.55) are sometimes referred to as 'absorbing boundary conditions.' The initial value of u is given by

$$u(x,0) = f(x) \tag{4.56}$$

We apply the method of separation of variables as in Section 4.1 by writing u as the product $u(x,t) = X(x)T(t)$, substituting into the diffusion equation, and dividing the resulting expression by XT:

$$\frac{1}{T}\frac{dT}{dt} = \frac{1}{X}\frac{d^2 X}{dx^2} \tag{4.57}$$

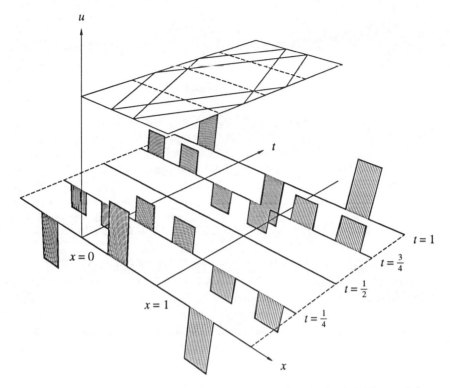

Figure 4.3 The solution of the wave equation with the initial conditions in Equation (4.52) and with fixed boundary conditions at $x=0$ and $x=1$: $u(0,t)=u(1,t)=0$. The time development of the solution is shown both in the region between the boundaries, $0 \le x \le 1$, and in the neighboring region outside of the boundaries. The characteristics of this solution are shown on the plane above the solution.

The left-hand side is a function of t only, the right-hand side is a function of x only, with the two expressions being equal for all independent variations of x and t. Thus, equality can be maintained for all such variations only if the two terms are separately equal to the same constant, which we again write as $-k^2$:

$$\frac{1}{T}\frac{dT}{dt} = \frac{1}{X}\frac{d^2 X}{dx^2} = -k^2 \tag{4.58}$$

These equations imply the following differential equations for X and T:

$$\frac{dT}{dt} + k^2 T = 0, \qquad \frac{d^2 X}{dx^2} + k^2 X = 0 \tag{4.59}$$

The solutions to (4.56) that satisfy (4.55) are

$$T(t) = e^{-k^2 t}, \qquad X(x) = \begin{cases} \cos kx \\ \sin kx \end{cases} \tag{4.60}$$

if $k \neq 0$, and
$$T(t) = a_0, \qquad X(x) = b_0 + c_0 x \tag{4.61}$$

if $k = 0$, where a_0, b_0 and c_0 are constants. The solution for X in (4.60) can also be written in complex form as in (4.7). The solution of the initial-value problem of the diffusion equation is now written as products XT of the expressions in (4.60) and (4.61). By forming these products, we notice that the contribution from (4.61) is the only part of the solution that is nonvanishing as $t \to \infty$. Thus, we can consider fitting this solution to the boundary conditions separately from the contributions in (4.60). The solution formed from (4.61) is of the form $A_0 + B_0 x$, where A_0 and B_0 are constants. From the boundary conditions in Equation (4.55), we can see immediately that $A_0 = 0$ and $B_0 = 0$, so there is no contribution from the solution corresponding to $k = 0$ (see, however, the discussion in Problem 21).

We now turn our attention to the contribution from (4.60). By analogy with the discussion in Section 4.1, the boundary conditions (4.55) mean that only the sine functions in (4.60) are allowed solutions and that the separation constant k is restricted to the values $k_n = n\pi/L$, where n is an integer. Thus, the solution of the boundary-value problem of the diffusion equation can be written as the Fourier series

$$u(x,t) = \sum_{n=1}^{\infty} A_n e^{-k_n^2 t} \sin k_n x \tag{4.62}$$

By setting $t=0$ in this equation and comparing with (4.56), we obtain

$$u(x,0) = \sum_{n=1}^{\infty} A_n \sin k_n x = f(x) \tag{4.63}$$

This again shows that the boundary conditions require the periodic extension of $f(x)$ to be an odd function, and that the coefficients A_n in (4.58) are in fact the Fourier coefficients of the sine series of $f(x)$. Thus, we can write the solution in (4.62) as

$$u(x,t) = \sum_{n=1}^{\infty} f_n e^{-k_n^2 t} \sin k_n x \tag{4.64}$$

where the expansion coefficients f_n are given by

$$f_n = \frac{2}{L} \int_0^L f(x) \sin k_n x \, dx \tag{4.65}$$

Example 4.4. The form of the solution in (4.64) provides some interesting insights into the way solutions of the diffusion equation develop in time. The important point is that the rate of decay of the nth term in the sum in (4.64) is proportional to n^2. As we have seen in Example 4.2, the higher-order terms in a Fourier series represent the sharp features of a function. Thus, this solution shows that the sharpest features of the initial function have the fastest rate of decay. In particular, any sharp features such as discontinuities in the initial function are quickly smoothed out.

To see this behavior in a specific example consider the solution (4.64) with the initial function $f(x)$ given by a form similar to that in (4.33):

$$f(x) = \begin{cases} x & \text{if } 0 \leq x < \frac{1}{2} \\ 1-x & \text{if } \frac{1}{2} \leq x < 1 \end{cases} \tag{4.66}$$

This function is continuous but has a discontinuous first derivative at $x = \frac{1}{2}$. The Fourier series in Equation (4.65) is constructed by following steps analogous to those in Example 4.3. The time development of the solution from the initial profile in (4.66) is shown in Figure 4.4 for the times $t = 0$, $t = 0.025$ and $t = 0.1$. The first twenty-five nonvanishing terms were included in the Fourier sine series. As expected, the sharply-peaked structure in $f(x)$ near $x = \frac{1}{2}$ is lost quickly, since the higher-order Fourier components that are required to reproduce this behavior decay most rapidly. As $t \to \infty$, the solution is seen to approach $u = 0$, as all of the heat contained in the interior region has diffused to the boundaries, where it has been absorbed.

∎

Our final example of the use of the method of separation of variables and Fourier series will be to solve Laplace's equation,

$$\frac{\partial^2 u}{\partial x^2} + \frac{\partial^2 u}{\partial y^2} = 0 \tag{4.67}$$

in polar coordinates, $x = r\cos\phi$, $y = r\sin\phi$, where the ranges of r and θ are $0 \leq r < \infty$ and $0 \leq \phi < 2\pi$. The transformation of Equation (4.67) into polar coordinates can be accomplished with the techniques described in Section 1.4, with the result

$$r\frac{\partial}{\partial r}\left(r\frac{\partial u}{\partial r}\right) + \frac{\partial^2 u}{\partial \phi^2} = 0 \tag{4.68}$$

To apply the method of separation of variables, we first write the solution to Laplace's equation as $u(r, \phi) = R(r)\Phi(\phi)$. We then substitute this form into (4.68), and divide by $R\Phi$ to obtain

$$\frac{r}{R}\frac{d}{dr}\left(r\frac{dR}{dr}\right) = -\frac{1}{\Phi}\frac{d^2\Phi}{d\phi^2} \tag{4.69}$$

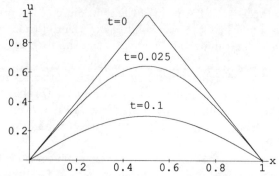

Figure 4.4 The time development of the solution of the diffusion equation with the initial conditions in Equation (4.66) and with the boundary conditions $u(0,t) = u(1,t) = 0$. The solution is shown at the times $t = 0$, $t = 0.025$ and $t = 0.1$. The discontinuity in the first derivative at $t = \frac{1}{2}$ decays rapidly and the solution is seen to approach zero everywhere within the interval $[0,1]$.

The two sides of this equation are functions of different independent variables, so they must be equal to the same constant, which we again denote by k^2. This implies the following differential equations for R and Φ:

$$r^2 \frac{d^2 R}{dr^2} + r \frac{dR}{dr} - k^2 R = 0, \qquad \frac{d^2 \Phi}{d\phi^2} + k^2 \Phi = 0 \qquad (4.70)$$

The solutions of these equations that satisfy (4.69) are

$$R(r) = \begin{cases} r^k \\ r^{-k} \end{cases}, \qquad \Phi(\phi) = \begin{cases} e^{ik\phi} \\ e^{-ik\phi} \end{cases} \qquad (4.71)$$

for $k \neq 0$, and

$$R(r) = \begin{cases} 1 \\ \ln r \end{cases}, \qquad \Phi(\phi) = \begin{cases} 1 \\ \phi \end{cases} \qquad (4.72)$$

for $k = 0$. The solution of Laplace's equation is then written as a superposition of solutions in (4.71) and (4.72).

We will apply this procedure to obtain the solution to Laplace's equation within a circle of radius ϱ where on the boundary the solution is specified in terms of a function f:

$$u(\varrho, \phi) = f(\phi) \qquad (4.73)$$

Since a solution of Laplace's equation is sought in the region $r < \varrho$, some of the terms in (4.71) and (4.72) must be excluded if unphysical singularities in the solution are to be avoided. For this reason, the terms $\ln r$ and r^{-k} for $k > 0$ must be omitted. Furthermore, since the range of ϕ is unrestricted, the coordinates (r, ϕ) and $(r, \phi + 2\pi)$ correspond to the same point, so u

Separation of Variables and the Sturm–Liouville Problem

must take on the same value at the point regardless of how that point is represented, i.e. $u(r,\phi) = u(r, \phi + 2\pi)$. In other words, u must be single-valued. This requirement means that terms proportional to ϕ in (4.72) must be omitted and the separation constant k must be set equal to an integer: $k = n$. Thus, the appropriate form for the solution of Laplace's equation within a circular region that includes the origin is

$$u(r,\phi) = \sum_{n=-\infty}^{\infty} A_n r^n e^{in\phi} \qquad (4.74)$$

The expansion coefficients A_n are determined by the boundary condition (4.73), as we will see below. The type of expansion in Equation (4.74) is the complex form of a Fourier series. Any Fourier series can be written in this form with an appropriate choice of the expansion coefficients (Problem 9).

By evaluating the solution (4.74) at $r = \varrho$ and comparing the resulting expression with the boundary condition in Equation (4.73), we obtain

$$u(\varrho,\phi) = \sum_{n=-\infty}^{\infty} A_n \varrho^n e^{in\phi} = f(\phi) \qquad (4.75)$$

This equation represents a complex Fourier series representation of the function $f(\phi)$. The requirement that f be a real function means that the coefficients A_n and A_{-n} are required to be complex conjugates: $A_{-n} = A_n^*$. Then, the orthogonality of the functions $e^{in\phi}$ over the interval $(0, 2\pi)$ (Problem 2) enables the expansion coefficients in (4.74) to be determined in terms of f:

$$A_n = \frac{1}{2\pi \varrho^n} \int_0^{2\pi} f(\phi) e^{-in\phi} d\phi \equiv \frac{1}{\varrho^n} f_n \qquad (4.76)$$

where the f_n are the coefficients of the complex Fourier series of $f(x)$. Thus, the solution of Laplace's equation within a circle of radius ϱ is given by

$$u(r,\phi) = \sum_{n=-\infty}^{\infty} f_n \left(\frac{r}{\varrho}\right)^n e^{in\phi} \qquad (4.77)$$

The solution is seen to be determined entirely by the values prescribed on the circumference of the circular region by the function $f(x)$.

Example 4.5. The Fourier series solution in Equation (4.77) can be used to verify a particular case of the result obtain in Problem 3.24, i.e. that the value taken by a solution to Laplace's equation is equal to an average taken by that function on the circumference of a circle centered at that point. We first separate the terms corresponding to $n=0$ from the other terms in (4.77):

$$u(r,\phi) = f_0 + \sum_{n\neq 0} f_n \left(\frac{r}{\varrho}\right)^n e^{in\phi} \tag{4.78}$$

Then, taking the limit of this expression as $r \to 0$, we obtain

$$\lim_{r\to 0} u(r,\phi) = f_0 = \frac{1}{2\pi} \int_0^{2\pi} f(\phi) d\phi \tag{4.79}$$

as required. In fact, we can carry this result one step further by taking the average of (4.77) over *any* circle of radius $0 < r \leq 1$. Signifying this average by $\bar{u}(r)$, we find

$$\bar{u}(r) \equiv \frac{1}{2\pi} \int_0^{2\pi} u(r,\phi) \, d\phi$$

$$= f_0 + \sum_{n\neq 0} f_n \left(\frac{r}{\varrho}\right)^n \int_0^{2\pi} e^{in\phi} d\phi$$

$$= f_0 \tag{4.80}$$

again as expected from Problem 3.24. ∎

Just as we were able to manipulate the Fourier series solution of the wave equation into the form of d'Alembert's solution, we can perform analogous operations to obtain a more flexible expression for the solution of (4.68) with the boundary condition (4.73). By substituting (4.76) into (4.74), a few simple steps produce

$$u(r,\phi) = \frac{1}{2\pi} \sum_{n=-\infty}^{\infty} \left(\frac{r}{\varrho}\right)^n e^{in\phi} \int_0^{2\pi} f(\phi') e^{-in\phi'} d\phi'$$

$$= \frac{1}{2\pi} \int_0^{2\pi} f(\phi') \left\{ \sum_{n=-\infty}^{\infty} \left[\frac{r}{\varrho} e^{i(\phi-\phi')}\right]^n \right\} d\phi' \tag{4.81}$$

The summation in (4.81) can be carried out by observing that in the interior of the circular region, $r < \varrho$, so

$$\left| \frac{r}{\varrho} e^{i(\phi-\phi')} \right| = \frac{r}{\varrho} < 1 \tag{4.82}$$

Thus, we can use the result

$$\sum_{n=1}^{\infty} x^n = \frac{x}{1-x} \tag{4.83}$$

which is valid for $x<1$, to evaluate the summation in (4.81):

$$\sum_{n=-\infty}^{\infty}\left[\frac{r}{\varrho}e^{i(\phi-\phi')}\right]^n = 1 + 2\operatorname{Re}\left\{\sum_{n=1}^{\infty}\left[\frac{r}{\varrho}e^{i(\phi-\phi')}\right]^n\right\}$$

$$= 1 + 2\frac{\varrho r\cos(\phi-\phi') - r^2}{\varrho^2 + r^2 - 2\varrho r\cos(\phi-\phi')}$$

$$= \frac{\varrho^2 - r^2}{\varrho^2 + r^2 - 2\varrho r\cos(\phi-\phi')} \tag{4.84}$$

where the notation $\operatorname{Re}\{f\}$ is used to indicate the real part of the complex function f. Substituting (4.84) into (4.81), we obtain

$$u(r,\phi) = \frac{1}{2\pi}\int_0^{2\pi}\frac{\varrho^2 - r^2}{\varrho^2 + r^2 - 2\varrho r\cos(\phi-\phi')}f(\phi')\,d\phi' \tag{4.85}$$

This expression, which is known as **Poisson's integral formula**, gives the value of the solution of Laplace's equation within the circle of radius ϱ in terms of values specified on the circle. In regions not too close to the boundary (because of the restriction in (4.82)) this formula provides a practical alternative to the series representation in (4.77) in the same way that the d'Alembert form of the solution in (4.51) provides a conceptually and computationally simpler alternative to the Fourier series solution of the wave equation in (4.46).

4.5 The Sturm–Liouville Problem

Expansions in terms of orthogonal functions, of which the Fourier series is the simplest and best-known example, are a common way of representing solutions of many types of differential equations. In fact, many of the properties that we used in the last section can be shown to be valid for a general class of differential equations and boundary conditions. The equation we will consider in this section encompasses many ordinary differential equations that arise from the separation of variables of a higher-dimensional partial differential equation. This equation, which is called the **Liouville equation**, is

$$\frac{d}{dx}\left[p(x)\frac{dy}{dx}\right] + [q(x) + \lambda r(x)]y = 0 \tag{4.86}$$

The quantity λ is a separation constant, and p, q and r are real-valued functions of x with p and r taking only positive values. The function y is assumed to be defined over an interval $[a, b]$ (often the open interval (a, b)), and we take the boundary conditions to be

$$y(a) = y(b) = 0 \tag{4.87}$$

though the results derived below are valid for more general boundary conditions (Problem 33). The solutions y to (4.86) and (4.87) are called **eigenfunctions** and the associated λ are **eigenvalues**. The determination of the general properties of (4.86) and (4.87) and, in particular, the dependence of y upon λ is called the **Sturm–Liouville problem**. We will now derive several of the more familiar and useful properties of the eigenfunctions and eigenvalues of (4.86) and (4.87).

The eigenvalues are discrete. Let $u(x; \lambda)$ and $v(x; \lambda)$ be the fundamental solutions of (4.86) and (4.87), i.e. u and v are two linearly-independent solutions in terms of which all other solutions may be expressed (for fixed λ). Then there are constants A and B that allow any solution y to be expressed as a linear combination of u and v:

$$y(x; \lambda) = Au(x; \lambda) + Bv(x; \lambda) \tag{4.88}$$

These constants are determined by requiring $y(x; \lambda)$ to satisfy the boundary conditions in (4.87). Applying these boundary conditions to the expression in (4.88) leads to the following equations:

$$\begin{aligned} y(a; \lambda) = Au(a; \lambda) + Bv(a; \lambda) = 0 \\ y(b; \lambda) = Au(b; \lambda) + Bv(b; \lambda) = 0 \end{aligned} \tag{4.89}$$

Equations (4.89) can be regarded as two simultaneous equations for the (unknown) quantities A and B. To determine the conditions under which this system has a nontrivial solution, it is most convenient to write this system of two equations in matrix form as

$$\begin{pmatrix} u(a; \lambda) & v(a; \lambda) \\ u(b; \lambda) & v(b; \lambda) \end{pmatrix} \begin{pmatrix} A \\ B \end{pmatrix} = \begin{pmatrix} 0 \\ 0 \end{pmatrix} \tag{4.90}$$

Thus, we see that Equations (4.89) can be solved for A and B only if the determinant of the matrix of coefficients in (4.90) vanishes. Otherwise the only solution is $A = B = 0$, which yields the trivial solution $y = 0$. The condition for a nontrivial solution of (4.89) is therefore given by

$$\begin{vmatrix} u(a; \lambda) & v(a; \lambda) \\ u(b; \lambda) & v(b; \lambda) \end{vmatrix} = u(a; \lambda)v(b; \lambda) - u(b; \lambda)v(a; \lambda) = 0 \tag{4.91}$$

If we consider $u(x; \lambda)$ and $v(x; \lambda)$ to be analytic functions of λ, then the determinant itself is an analytic function of λ. Therefore the zeros of the determinant must be *isolated* (Ahlfors, 1966). Since the zeros of the determinant correspond to allowed solutions of the Sturm–Liouville problem, we conclude that the eigenvalues of (4.86) and (4.87) are discrete.

The eigenfunctions are orthogonal. We begin with the equations satisfied by eigenfunctions y_i and y_j with eigenvalues λ_i and λ_j:

$$\frac{\mathrm{d}}{\mathrm{d}x}\left[p(x)\frac{\mathrm{d}y_i}{\mathrm{d}x}\right] + [q(x) + \lambda_i r(x)]y_i = 0 \qquad (4.92a)$$

$$\frac{\mathrm{d}}{\mathrm{d}x}\left[p(x)\frac{\mathrm{d}y_j}{\mathrm{d}x}\right] + [q(x) + \lambda_j r(x)]y_j = 0 \qquad (4.92b)$$

We now multiply (4.92a) by $y_j(x)$ and (4.92b) by $y_i(x)$. Taking the difference between the resulting equations we obtain

$$y_j\frac{\mathrm{d}}{\mathrm{d}x}\left[p(x)\frac{\mathrm{d}y_i}{\mathrm{d}x}\right] - y_i\frac{\mathrm{d}}{\mathrm{d}x}\left[p(x)\frac{\mathrm{d}y_j}{\mathrm{d}x}\right] + (\lambda_i - \lambda_j)r(x)y_iy_j = 0 \qquad (4.93)$$

Integrating this equation over the interval $[a, b]$, performing an integration by parts and solving for the last term on the left-hand side yields

$$(\lambda_j - \lambda_i)\int_a^b ry_iy_j\,\mathrm{d}x = \int_a^b\left\{y_j\frac{\mathrm{d}}{\mathrm{d}x}\left[p\frac{\mathrm{d}y_i}{\mathrm{d}x}\right] - y_i\frac{\mathrm{d}}{\mathrm{d}x}\left[p\frac{\mathrm{d}y_j}{\mathrm{d}x}\right]\right\}\mathrm{d}x$$

$$= \left[y_jp\frac{\mathrm{d}y_i}{\mathrm{d}x} - y_ip\frac{\mathrm{d}y_j}{\mathrm{d}x}\right]\Big|_a^b - \int_a^b\left[p\frac{\mathrm{d}y_j}{\mathrm{d}x}\frac{\mathrm{d}y_i}{\mathrm{d}x} - p\frac{\mathrm{d}y_i}{\mathrm{d}x}\frac{\mathrm{d}y_j}{\mathrm{d}x}\right]\mathrm{d}x$$

$$= \left[y_jp\frac{\mathrm{d}y_i}{\mathrm{d}x} - y_ip\frac{\mathrm{d}y_j}{\mathrm{d}x}\right]\Big|_a^b \qquad (4.94)$$

The right-hand side of this equation vanishes because the boundary conditions insure that the quantity enclosed in brackets vanishes at $x = a$ and $x = b$. Thus, if $\lambda_i \neq \lambda_j$, the integral on the left-hand side of (4.94) must vanish:

$$\int_a^b r(x)y_i(x)y_j(x)\,\mathrm{d}x = 0 \qquad (4.95)$$

This expression is similar to the orthogonality relation in Equation (4.24) except for the presence of the function $r(x)$. This function is called the **weight function**. Two functions that satisfy Equation (4.95) are said to be orthogonal with respect to that weight function. In particular, we see that eigenfunctions of the Liouville equation corresponding to distinct eigenvalues are orthogonal with respect to the weight function of the governing equation.

The eigenvalues are real. We begin with the equation satisfied by the eigenfunction y_i with eigenvalue λ_i,

$$\frac{\mathrm{d}}{\mathrm{d}x}\left[p(x)\frac{\mathrm{d}y_i}{\mathrm{d}x}\right] + [q(x) + \lambda_i r(x)]y_i = 0 \qquad (4.96)$$

Taking the complex conjugate of this equation and recalling that p, q and r are real-valued functions of x produces the equation

$$\frac{d}{dx}\left[p(x)\frac{dy_i^*}{dx}\right] + [q(x) + \lambda_i^* r(x)]y_i^* = 0 \qquad (4.97)$$

We now multiply (4.96) by y_i^* and (4.97) by y_i, subtract one equation from the other and integrate over $[a, b]$ and perform an integration by parts to obtain

$$\begin{aligned}(\lambda_i^* - \lambda_i) \int_a^b r y_i y_i^* \, dx &= \int_a^b \left\{ y_i^* \frac{d}{dx}\left[p\frac{dy_i}{dx}\right] - y_i \frac{d}{dx}\left[p\frac{dy_i^*}{dx}\right]\right\} dx \\ &= \left[y_i^* p \frac{dy_i}{dx} - y_i p \frac{dy_i^*}{dx}\right]\Big|_a^b - \int_a^b \left[p\frac{dy_i^*}{dx}\frac{dy_i}{dx} - p\frac{dy_i}{dx}\frac{dy_i^*}{dx}\right] dx \\ &= \left[y_i^* p \frac{dy_i}{dx} - y_i p \frac{dy_i^*}{dx}\right]\Big|_a^b \qquad (4.98)\end{aligned}$$

The boundary conditions in (4.87) and their complex conjugates again insure that the right-hand side of this equation vanishes. Thus, equation (4.98) yields

$$(\lambda_i^* - \lambda_i) \int_a^b r(x) y_i^*(x) y_i(x) \, dx = \int_a^b r(x) |y_i(x)|^2 \, dx \qquad (4.99)$$

If $\lambda_i^* \neq \lambda_i$, then since $r > 0$, we must have that $y_i(x)$ reduces to the trivial solution $y_i(x) = 0$. Thus, for nontrivial eigenfunctions, the λ_i are real:

$$\lambda_i^* = \lambda_i \qquad (4.100)$$

The three properties just derived for the Liouville equation (4.86) with boundary conditions (4.87) can be shown to be valid under the more general boundary conditions

$$\left[\alpha y(x) + \beta \frac{dy(x)}{dx}\right]\Big|_{x=a} = 0, \qquad \left[\alpha y(x) + \beta \frac{dy(x)}{dx}\right]\Big|_{x=b} = 0 \qquad (4.101)$$

for constants α and β. This is taken up in the problems at the end of the chapter.

Further Reading

Morse and Feshbach (1953), Courant and Hilbert (1953), Hildebrand (1962) and Birkhoff and Rota (1989) provide general discussion on the material covered in this chapter. A comprehensive treatment of Fourier series may be found in the book by Carslaw (1950). Birkhoff and Rota (1989) provide a very readable discussion of the question of completeness of eigenfunctions in general and of the trigonometric functions in particular. A detailed discussion of the Sturm–Liouville theory and its extensions may be found in Ince (1956).

References

Ahlfors L. V. (1966). *Complex Analysis* 2nd edn. New York: McGraw-Hill.

Andrews L. C. (1985). *Special Functions for Engineers and Applied Mathematicians.* New York: Macmillan.

Birkhoff G. and Rota G.-C. (1989). *Ordinary Differential Equations* 4th edn. New York: Wiley.

Carslaw H. S. (1950). *An Introduction to the Theory of Fourier's Series and Integrals.* New York: Dover.

Courant R. and Hilbert D. (1953). *Methods of Mathematical Physics* Vol. 1. New York: Wiley-Interscience.

Hildebrand F. B. (1962). *Advanced Calculus for Applications.* Englewood Cliffs NJ: Prentice-Hall.

Ince E. L. (1956). *Ordinary Differential Equations.* New York: Dover.

Lin C. C. and Segel L. A. (1974). *Mathematics Applied to Deterministic Problems in the Natural Sciences.* New York: Macmillan.

Morse P. M. and Feshbach H. (1953). *Methods of Theoretical Physics* Vol. 1. New York: McGraw-Hill.

Rudin W. (1964). *Principles of Mathematical Analysis.* New York: McGraw-Hill.

Problems

1. The angle θ between two n-component vectors $\mathbf{u} = (u_1, u_2, \ldots, u_n)$ and $\mathbf{v} = (v_1, v_2, \ldots, v_n)$ is given in terms of the scalar product as

$$\cos\theta = \frac{\mathbf{u}\cdot\mathbf{v}}{\sqrt{\mathbf{u}\cdot\mathbf{u}}\sqrt{\mathbf{v}\cdot\mathbf{v}}}$$

or in terms of the vector components as

$$\cos\theta = \frac{\sum u_i v_i}{\sqrt{\sum u_i u_i}\sqrt{\sum v_i v_i}}$$

An analogous construction can be made for two functions $\phi_1(x)$ and $\phi_2(x)$, using the quantities $\mathcal{S}_{ij}$ defined in Equation (4.19):

$$\cos\theta = \frac{\mathcal{S}_{12}}{\sqrt{\mathcal{S}_{11}}\sqrt{\mathcal{S}_{22}}}$$

This equation is meaningful provided that the modulus of the right-hand side can be shown to be less than unity. To show this, consider the inequality

$$\int_a^b [\lambda\phi_1(x) - \mu\phi_2(x)]^2 \geq 0$$

where λ and μ are any two real constants. Since the integrand is nonnegative, the integral is also nonnegative. By making appropriate choices for λ and μ and observing that $\mathcal{S}_{12} = \mathcal{S}_{21}$, show that this inequality implies

$$\frac{\mathcal{S}_{12}}{\sqrt{\mathcal{S}_{11}}\sqrt{\mathcal{S}_{22}}} \leq 1$$

as required. Use this result to rewrite Equation (4.25) as

$$\int_a^b \phi_2(x)\psi(x)\,dx = \mathcal{S}_{22}[1 - \cos^2\theta]$$

This provides a geometrical interpretation of the orthogonalization procedure carried out in Section 4.2.

2. The orthogonality of the sine and cosine functions obtained in Section 4.2 can be used to deduce the orthogonality relation of the functions

$$\exp(ik_n x) = \cos k_n x + i\sin k_n x$$

over the interval $(0, L)$. The quantity k_n has its usual meaning: $k_n = n\pi/L$, where n is an integer. These functions differ from the trigonometric functions

Separation of Variables and the Sturm–Liouville Problem   149

in being complex quantities. If we view the real and imaginary parts of this expression as the components of a vector in the two-dimensional space of complex variables, then the magnitude is given by

$$\left|\exp(ik_n x)\right| = \exp(ik_n x)\exp(-ik_n x) = 1$$

Hence obtain the orthogonality relation

$$\int_0^L \exp(ik_n x)\exp(-ik_m x)\,dx = \begin{cases} L & \text{if } m = n \\ 0 & \text{if } m \neq n \end{cases}$$

3. Determine the Fourier sine series for the following functions defined on the interval $(0, \pi)$:

(a) $f(x) = x(\pi - x)$

(b) $f(x) = x$

(c) $f(x) = 1$

Use *Mathematica* to construct the partial sums of these series and to display the results using the command Plot. Comment on the relative rates of convergence of the series and indicate any regions where the convergence appears to be conspicuously slow.

4. Infinite product representations of functions provide a complementary, if not as widely used, representation to infinite series. In this problem, the infinite product representation of the sine function will be obtained (Andrews, 1985) and some of the properties will be studied in the following problem with *Mathematica*.

Show that the Fourier series representation of the function $f(x) = \cos kx$, where k is not an integer, is

$$\cos kx = \frac{\sin k\pi}{k\pi} + \frac{2k}{\pi}\sum_{n=1}^{\infty}\frac{(-1)^n}{k^2 - n^2}\sin k\pi \cos nx$$

By choosing $x = \pi$, show that the Fourier series can be written as

$$\frac{d}{dk}\left[\ln\left(\frac{\sin k\pi}{k\pi}\right)\right] = \sum_{n=1}^{\infty}\frac{d}{dk}\left[\ln(n^2 - k^2)\right]$$

Then by integrating this expression with respect to k from $k = 0$ to $k = x$, where $x < 1$, deduce the product representation

$$\sin x = x\prod_{n=1}^{\infty}\left(1 - \frac{x^2}{n^2 \pi^2}\right)$$

5. Construct the Fourier series for $\cos kx$ for noninteger k and use `Plot` to display the solution for several partial sums. Comment on the convergence of this series.

Construct the product representation using `Product` to expand the first several partial products. Compare this series with that obtained from the Taylor series expansions taken about $x=0$, which can be obtained with `Series`.

Plot the first several partial products together with the sine function using `Plot` over the domain $0 \le x \le \pi$. Identify the regions where the convergence is rapid and where the convergence is slow. How many terms are required to converge to within 1% at $x = \frac{1}{2}\pi$?

6. Consider the Fourier series representation of a function $f(x)$ defined on the interval $(-\pi, \pi)$ given in Equation (4.27):

$$f(x) = \tfrac{1}{2}a_0 + \sum_{n=0}^{\infty}(a_n \cos nx + b_n \sin nx)$$

Square both sides of this equation, integrate over $(-\pi, \pi)$ and use the orthogonality of the trigonometric functions to obtain

$$\frac{1}{\pi}\int_{-\pi}^{\pi}[f(x)]^2\,dx = \tfrac{1}{2}a_0^2 + \sum_{n=1}^{\infty}(a_n^2 + b_n^2)$$

This relation, which is a 'sum rule' for the Fourier coefficients, is known as **Parseval's theorem**. The rate of convergence of the sums on the right-hand side of this equation to the value on the left-hand side provides a measure of convergence of the Fourier series of $f(x)$.

7. Using Parseval's theorem, obtain the following relations for the functions in Problem 3:

(a) $\dfrac{\pi^6}{960} = \sum_{k=0}^{\infty}\dfrac{1}{(2k+1)^6}$

(b) $\dfrac{\pi^2}{6} = \sum_{k=1}^{\infty}\dfrac{1}{k^2}$

(c) $\dfrac{\pi^2}{8} = \sum_{k=0}^{\infty}\dfrac{1}{(2k+1)^2}$

Use *Mathematica* to investigate the convergence of these series. One way to see the relative rates of convergence of these series is to first normalize each series to unity by dividing both sides of the expressions given above by the left-hand sides. Then, form the partial sums into a table using the commands `Sum` and `Table` and use `ListPlot` to display the convergence of each series.

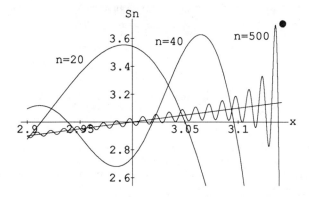

Figure 4.5 Illustration of the Gibbs phenomenon for the partial sums $S_n(x)$ of the Fourier sine series of the function $f(x)=x$ for $0 \le x \le \pi$. Shown are the partial sums for $n=20$, $n=40$ and $n=500$ in a neighborhood of the point $x=\pi$, where the function is discontinuous. The straight line represents $f(x)$ and the dot indicates the point to which the partial sums converge at $x=\pi$.

8. The **Gibbs phenomenon** is associated with the behavior of the Fourier series representation of a function $f(x)$ near points of discontinuity of that function. Gibbs observed that if the function $f(x)$ is continuous everywhere except for a finite number of points, then the sequence of partial sums, $S_n(x)$, of the Fourier representation approaches the curve consisting of $f(x)$ and a series of lines parallel to the y axis and passing through the points of discontinuity and *beyond* by a finite amount.

For the function $f(x)=x$ defined over the interval $(0, \pi)$, the points of discontinuity are the points $n\pi$ for every integer n, and the magnitude of the 'jump' is, according to Gibbs, given by

$$\lim_{n \to \infty} S_n(\pi) - \pi = 2 \int_0^\pi \frac{\sin x}{x}\, dx - \pi$$

Verify this by using *Mathematica* to evaluate the partial sums of the Fourier sine series of $f(x)=x$, and to evaluate numerically the jump at the discontinuity at $x=\pi$. A plot of the partial sums near $x=\pi$ is shown in Figure 4.5 for $n=20$, $n=40$ and $n=500$. Notice how the partial sums converge to the function except in a small neighborhood about $x=\pi$ and that with increasing terms in the partial sums, the local maximum of the partial sums moves toward the point $x=\pi$ from the left. A comprehensive discussion of this phenomenon has been given by Carslaw (1950).

9. Show that the representation of a function $f(x)$ over the interval $(0, L)$ in (4.27) can be written as a complex Fourier series,

$$f(x) = \sum_{n=-\infty}^{\infty} c_n \exp(ik_n x)$$

where the expansion coefficients c_n are given in terms of a_n and b_n by

$$c_n = \tfrac{1}{2}(a_n - ib_n) = \frac{1}{L}\int_0^L f(x)\exp(-ik_n x)\,dx$$

with

$$c_{-n} = c_n^* = \tfrac{1}{2}(a_n + ib_n)$$

10. Beginning with the complex Fourier series representation of a function $f(x)$ defined over the interval $(-\pi, \pi)$, we define $f_N(x)$ as the partial sum obtained by summing the complex Fourier series of $f(x)$ from $n = -N$ to $n = N$:

$$f_N(x) = \sum_{n=-N}^{N} c_n \exp(inx)$$

By substituting the integral expression for c_n into the summation show that this series can be written as

$$f_N(x) = \frac{1}{2\pi}\int_{-\pi}^{\pi} f(s) \sum_{n=-N}^{N} \exp[in(x-s)]\,ds$$

Notice that the order of summation and integration have been interchanged. By multiplying the summation in this expression by e^{ix} and performing a few simple manipulations, obtain the following closed-form expression:

$$\sum_{n=-N}^{N} \exp[in(x-s)] = \frac{\sin[(N+\tfrac{1}{2})(x-s)]}{\sin[\tfrac{1}{2}(x-s)]}$$

This summation, which is signified by $D_N(x-s)$, is known as **Dirichlet's kernel**. Thus, $f_N(x)$ can be written in terms of this kernel as

$$f_N(x) = \frac{1}{2\pi}\int_{-\pi}^{\pi} f(s) D_N(x-s)\,ds$$

11. Use the definition of $D_N(x)$ in the form of a summation to calculate the value of the integral

$$\frac{1}{2\pi}\int_{-\pi}^{\pi} D_N(x)\,dx$$

Then, use *Mathematica* to construct and plot the quantity

$$\frac{\sin[(N+\tfrac{1}{2})(x-s)]}{\sin[\tfrac{1}{2}(x-s)]}$$

for increasing values of N. How does the function behave near $x = s$?

Separation of Variables and the Sturm–Liouville Problem 153

12. Consider the Fourier sine series representations of the functions in Problem 3. Construct the integral representation in Problem 10 for these functions and plot the result for increasing values of N. Then construct the representation in (4.27) for the same values of N to verify explicitly that the same results are obtained as from the integral representation.

13. A quantity closely related to Dirichlet's kernel is **Fejér's kernel** (Rudin, 1965), which is defined by

$$K_N(x) = \frac{1}{n+1} \sum_{n=0}^{N} D_n(x)$$

Fejér's kernel is seen to be the average of the first N Dirichlet kernels. Using the same approach as was used in Problem 10 to obtain a closed-form expression for Dirichlet's kernel, show that Fejér's kernel is given by

$$K_N(x) = \frac{1}{n+1} \frac{1 - \cos[(N+1)x]}{1 - \cos x}$$

Verify that $K_N(x) \geq 0$ and that

$$\frac{1}{2\pi} \int_{-\pi}^{\pi} K_N(x)\,dx = 1$$

Use *Mathematica* to plot Fejér's kernel for increasing values of N. What are the primary differences between $K_N(x)$ and $D_N(x)$?

14. Fejér's kernel can be used to prove the following theorem due to Fejér (Rudin, 1965). We first introduce the quantity $\Sigma_N(x)$ as the arithmetic mean of the partial sums of the Fourier series of a function $f(x)$ defined on the interval $(-\pi, \pi)$:

$$\Sigma_N(x) = \frac{1}{N+1} [f_0(x) + f_1(x) + \cdots + f_N(x)]$$

Fejér's theorem states that if $f(x)$ is a continuous function, then $\Sigma_N(x)$ converges uniformly to $f(x)$ for all values of x:

$$\lim_{N \to \infty} \Sigma_N(x) = f(x)$$

Show that $\Sigma_N(x)$ can be written in terms of Fejér's kernel as

$$\Sigma_N(x) = \frac{1}{2\pi} \int_{-\pi}^{\pi} f(s) K_N(x - s)\,ds$$

Then use *Mathematica* to construct the quantity $|f(x) - \Sigma_N(x)|$ for the following functions:

(a) $f(x) = \pi^{-2}(\pi - x)(\pi + x)$

(b) $f(x) = \begin{cases} \pi - x & \text{if } x \geq 0 \\ \pi + x & \text{if } x < 0 \end{cases}$

Verify that $|f(x) - \Sigma_N(x)|$ decreases uniformly with increasing N for all values of x.

15. As an example for which Fejér's theorem does not apply, perform the analogous construction for the following function:

$$f(x) = \begin{cases} 1 & \text{if } |x| \leq \tfrac{1}{2}\pi \\ 0 & \text{if } |x| > \tfrac{1}{2}\pi \end{cases}$$

Calculate $|f(x) - \Sigma_N(x)|$ at $x = \tfrac{1}{2}\pi$ for increasing values of N and verify that this quantity does *not* approach zero as $N \to \infty$.

16. Apply the method of separation of variables to find the solution of the wave equation

$$\frac{\partial^2 u}{\partial x^2} - \frac{\partial^2 u}{\partial t^2} = 0$$

with the usual Cauchy initial conditions,

$$u(x, 0) = f(x), \qquad u_t(x, 0) = g(x)$$

within the interval $(0, L)$. At the end-points of the interval apply the boundary conditions

$$u_x(x, 0) = 0, \qquad u_x(x, L) = 0$$

By constructing the appropriate periodic extensions of $f(x)$ and $g(x)$ as implied by the Fourier series, write the solution of this boundary value problem in the form of d'Alembert's solution, as in Equation (4.51).

17. Repeat Problem 16, except with the boundary conditions

$$u(x, 0) = 0, \qquad u_x(x, L) = 0$$

18. Use *Mathematica* to construct the appropriate d'Alembert solutions of the wave equation on the interval $(0, 1)$ with the boundary conditions

(a) $u(0, t) = 0, \qquad u(1, t) = 0$

(b) $u_x(0, t) = 0, \qquad u_x(1, t) = 0$

(c) $u(0, t) = 0, \qquad u_x(1, t) = 0$

for initial conditions of the form in Equation (4.52):

$$f(x) = \begin{cases} 0 & \text{if } 0 \leq x < \frac{3}{8} \\ 1 & \text{if } \frac{3}{8} < x < \frac{5}{8} , \\ 0 & \text{if } \frac{5}{8} \leq x \leq 1 \end{cases} \qquad g(x) = 0$$

In assembling and plotting the solutions, you may find the following constructions useful for the appropriate periodic extensions for a function $f(x)$ defined on the interval $(0,1)$:

(a) `f[x_]:=(-1)^(x-Mod[x,1])F[Mod[x,1]]`

(b) `f[x_]:=F[Mod[x,1]]`

(c) `g[x_]:=If[x>=0,F[x],-F[Abs[-x]]]`
`f[x_]:=(-1)^((x-Mod[x+1,2]+1)/2)g[Mod[x+1,2]-1]`

Display the solutions using `Plot` to visualize the time development of the solution and `ContourPlot` for the characteristics.

19. The procedure used to obtain the d'Alembert solution of the wave equation in Equation (4.51) and in Problem 16, 17 and 18 can be applied to other types of boundary conditions without using the Fourier series representation. The idea is to identify whether the boundary conditions require the functions $f(x)$ and $g(x)$ to be odd or even with respect to reflections about the boundary. Thus, find the solution of the initial value problem of the wave equation in the region $x > 0$ in the presence of a *single* boundary at the origin in the cases where

(a) $u(0,t) = 0$ (fixed end-point)

(b) $u_t(0,t) = 0$ (free end-point)

20. Show that the solution of the diffusion equation

$$\frac{\partial u}{\partial t} = \frac{\partial^2 u}{\partial x^2}$$

obtained in the steady-state by setting $u_t = 0$ is given by

$$u_\infty(x) = A + Bx$$

where A and B are constants. By comparing this solution with those obtained in Equations (4.60) and (4.61), deduce that the separation constant in Equation (4.58) must be *negative* in order for the solution of the diffusion equation to approach $u_\infty(x)$ as $t \to \infty$.

21. Consider the initial value problem for the diffusion equation

$$\frac{\partial u}{\partial t} = \frac{\partial^2 u}{\partial x^2}, \qquad u(0,t) = u_0, \qquad u(L,t) = u_L$$

where u_0 and u_L are constants and with the initial condition

$$u(x,0) = f(x)$$

Show that fitting the boundary conditions requires the $k = 0$ solution of (4.61) to be chosen as

$$u_0 + (u_L - u_0)\frac{x}{L}$$

together with $k_n = n\pi/L$, where n is an integer. Thus, by writing the solution as

$$u(x,t) = u_0 + (u_L - u_0)\frac{x}{L} + \sum_{n=1}^{\infty} A_n \exp(-k_n^2 t) \sin k_n x$$

calculate the A_n and show that the solution to this initial-value problem is

$$u(x,t) = u_0 + (u_L - u_0)\frac{x}{L}$$
$$+ \frac{2}{\pi}\sum_{n=1}^{\infty}\frac{1}{n}\left[(-1)^n u_L - u_0\right]\exp(-k_n^2 t)\sin k_n x + \sum_{n=1}^{\infty} f_n \exp(-k_n^2 t)\sin k_n x$$

where f_n is given by (4.65). Determine the distribution attained in this region as $t \to \infty$. Verify your answer by setting $u_t = 0$ in the heat equation and solving the resulting *ordinary* differential equation with the boundary conditions at $x = 0$ and $x = L$.

22. Show that the solution in Problem 21 could have been obtained by writing the solution as

$$u(x,t) = u_0 + (u_L - u_0)\frac{x}{L} + \phi(x,t)$$

where $\phi(x,t)$ is a solution of

$$\frac{\partial \phi}{\partial t} = \frac{\partial^2 \phi}{\partial x^2}, \qquad \phi(0,t) = 0, \qquad \phi(L,t) = 0$$

and with a suitably chosen initial function

$$\phi(x,0) = g(x)$$

23. The boundary conditions (4.55) for the diffusion equation are sometimes called 'absorbing' boundary conditions because the initial distribution f is dissipated through the end-points and the system approaches a uniform zero distribution in the limit $t \to \infty$, as the solution in (4.64) shows.

Another type of boundary condition, called 'reflecting' boundary conditions, involves stipulating

$$u_x(0,t) = 0, \qquad u_x(L,t) = 0$$

at the end-points. In this case, the *net* flux emanating from the end-points is zero or, in terms of diffusion, any particle incident on either of the boundaries is reflected back into the region $(0, L)$. Show that solution of this boundary-value problem for the diffusion equation with $u(x,0) = f(x)$ is

$$u(x,t) = \frac{f_0}{2} + \sum_{n=1}^{\infty} f_n \exp(-k_n^2 t) \cos(k_n x)$$

where $k_n = n\pi/L$ and

$$f_n = \frac{2}{L} \int_0^L f(x) \cos(k_n x)\, dx$$

Show that

$$\int_0^L u(x,t)\, dx = \int_0^L f(x)\, dx$$

which demonstrates explicitly that no heat escapes through the end-points. Furthermore, show that the asymptotic behavior of $u(x,t)$ as $t \to \infty$ is given by

$$\lim_{t \to \infty} u(x,t) = \frac{1}{L} \int_0^L f(x)\, dx$$

which is a constant function whose value is equal to the average of the initial distribution f.

24. Show that the solution of the boundary-value problem of the diffusion equation

$$\frac{\partial u}{\partial t} = \frac{\partial^2 u}{\partial x^2}, \qquad u(0,t) = 0, \qquad u_x(L,t) = 0$$

with the initial condition

$$u(x,0) = f(x)$$

is given by

$$u(x,t) = \sum_{n=0}^{\infty} f_n \exp(-k_n^2 t) \sin(k_n x)$$

where

$$k_n = \frac{(2n+1)\pi}{2L}, \qquad f_n = \frac{2}{L} \int_0^L f(x) \sin k_n x\, dx$$

where n is a nonnegative integer.

25. Either solve directly or use the method described in Problem 22 to show that the solution of the diffusion equation in Problem 24 with the boundary conditions modified to

$$u(0,t) = u_0, \qquad u_x(L,t) = u_L'$$

is

$$u(x,t) = u_0 + xu'_L + \sum_{n=0}^{\infty}\left[f_n - \frac{4}{\pi^2}\frac{(2n+1)\pi u_0 + Lu'_L}{(2n+1)^2}\right]\exp(-k_n^2 t)\cos(k_n x)$$

26. A third type of boundary condition for the heat equation states that the flux across the bounding surface is proportional to the difference between the temperature at the surface and that of the surrounding medium:

$$K\frac{\partial u}{\partial n} = H(u_0 - u)$$

where K and H are positive constants, n is the *outward* normal of the boundary, and u_0 is the temperature of the surrounding medium.

Obtain the solution of the diffusion equation $u_t = u_{xx}$ with the initial condition

$$u(x,0) = f(x)$$

and with the boundary conditions

$$-u_x(0,t) = u_0 - u(0,t), \qquad u_x(L,t) = u_L - u(L,t)$$

using the method of Problem 22. In particular, show that the solution can be written as the sum of two contributions, $u(x,t) = v(x) + w(x,t)$, where

$$v(x) = \frac{1}{L+2}[(1+x)u_L + (L+1-x)u_0]$$

and

$$w(x,t) = \sum_{n=1}^{\infty} a_n \exp(-k_n^2 t)\left(\cos k_n x + \frac{1}{k_n}\sin k_n x\right)$$

Show that $k_n = \kappa_n/L$, where κ_n is the nth positive root of

$$\cot x = \frac{1}{2}\left(\frac{x}{L} - \frac{L}{x}\right)$$

and that the expansion coefficients a_n are given by

$$a_n = \frac{2k_n^2}{(k_n^2+1)^2 L + 2}\int_0^L [f(x) - v(x)]\left(\cos k_n x + \frac{1}{k_n}\sin k_n x\right)dx$$

27. Consider the boundary-value problem of Problem 26 for the case where $L=1$ and where $f(x)$ is given by

$$f(x) = x(1-x)$$

Use FindRoot to show that the first ten roots of

$$\cot x = \tfrac{1}{2}(x - x^{-1})$$

are given by

1.30654
3.67319
6.58462
9.63168
12.72324
15.834105
18.95497
22.08166
25.21203
28.34486

Then display the evolution of the solution using Plot and animate a time sequence to see the effect of these boundary conditions when compared with the other types of boundary conditions for the diffusion equation.

28. Given a Fourier series representation of a function $f(x)$ defined over the interval $(0, L)$,

$$f(x) = \sum_{n=1}^{\infty} f_n \sin k_n x$$

where $k_n = n\pi/L$ and the expansion coefficients are given by

$$f_n = \frac{2}{L} \int_0^L f(x) \sin k_n x$$

obtain a series representation for the derivatives of $f(x)$ with respect to x.

By expanding the exponential in the solution (4.64) of the diffusion equation,

$$u(x,t) = \sum_{n=1}^{\infty} f_n \exp(-k_n^2 t) \sin k_n x$$

show that this solution is equivalent to the solution obtained in Equation (3.79):

$$u(x,y) = \sum_{n=0}^{\infty} \frac{d^{2n} f(x)}{dx^{2n}} \frac{y^n}{n!}$$

In obtaining this result, you must assume that the initial function $f(x)$ is infinitely differentiable.

29. Consider the wave equation in three spatial dimensions:

$$\frac{\partial^2 u}{\partial x^2} + \frac{\partial^2 u}{\partial y^2} + \frac{\partial^2 u}{\partial z^2} - \frac{\partial^2 u}{\partial t^2} = 0$$

Transform this equation into spherical polar coordinates and show that the equation satisfied by the radial function for a spherically symmetric solution (no dependence on the polar or azimuthal angles) is

$$\frac{\partial^2 R}{\partial r^2} + \frac{2}{r}\frac{\partial R}{\partial r} - \frac{\partial^2 R}{\partial t^2} = 0$$

By defining $R(r,t) = r^{-1}\varrho(r,t)$, obtain the equation for $\varrho(r)$:

$$\frac{\partial^2 \varrho}{\partial r^2} - \frac{\partial^2 \varrho}{\partial t^2} = 0$$

Thus, if the initial conditions are given by

$$R(r,0) = f(r), \qquad R_r(r,0) = g(r)$$

show that the spherically symmetric solution to the three-dimensional wave equation is given by

$$R(r,t) = \frac{1}{2r}\Big[(r+t)f(r+t) + (r-t)f(r-t)\Big] + \frac{1}{2r}\int_{r-t}^{r+t} sg(s)\,ds$$

Notice that since $r \geq 0$, neither f nor g are defined for negative values of their arguments, so this solution is valid provided $r \geq t$.

30. The method of separation of variables can be used to obtain solutions of the two-dimensional wave equation in rectangular coordinates,

$$\frac{\partial^2 u}{\partial x^2} + \frac{\partial^2 u}{\partial y^2} - \frac{\partial^2 u}{\partial t^2} = 0$$

within the square region given by $0 \leq x \leq L$ and $0 \leq y \leq L$. Show that the Fourier series solution of this equation with the initial conditions

$$u(x,y,0) = f(x,y), \qquad u_t(x,y,0) = g(x,y)$$

and with the boundary conditions

$$u(x,0) = u(x,L) = 0, \qquad u(0,y) = u(L,y) = 0$$

is given by

$$u(x,y,t) = \sum_{m=1}^{\infty}\sum_{n=1}^{\infty} \sin k_m x \, \sin k_m x \left(f_{mn} \cos \kappa_{mn} t + \frac{g_{mn}}{\kappa_{mn}} \sin \kappa_{mn} t \right)$$

where $k_n = n\pi/L$, $k_m = m\pi/L$, $\kappa_{mn} = (k_n^2 + k_m^2)^{1/2}$ and

$$f_{mn} = \left(\frac{2}{L}\right)^2 \int_0^L \int_0^L f(x,y) \sin k_m x \sin k_n y \, dx \, dy$$

with an analogous expression for the g_{mn}.

Obtain the corresponding Fourier series solution for the boundary conditions

$$u_y(x,0) = u_y(x,L) = 0, \qquad u_x(0,y) = u_x(L,y) = 0$$

31. Use *Mathematica* to construct the solution in Problem 30 for the square region $0 \le x \le 1$ and $0 \le y \le 1$ with the initial conditions

$$f(x,y) = 16xy(1-x)(1-y), \qquad g(x,y) = 0$$

and with the boundary conditions

$$u(x,0) = u(x,1) = 0, \qquad u(0,y) = u(1,y) = 0$$

Use the command Plot3D to display this solution at various times during the oscillation period and then to animate the solution. Repeat this procedure for the boundary conditions

$$u_y(x,0) = u_y(x,1) = 0, \qquad u_x(0,y) = u_x(1,y) = 0$$

32. The methodology in Problems 30 and 31 can be applied to the two-dimensional diffusion equation:

$$\frac{\partial u}{\partial t} = \frac{\partial^2 u}{\partial x^2} + \frac{\partial^2 u}{\partial y^2}$$

within the square region given by $0 \le x \le L$ and $0 \le y \le L$. Show that the Fourier series solution of this equation with the initial conditions

$$u(x,y,0) = f(x,y)$$

and with the 'absorbing' boundary conditions

$$u(x,0) = u(x,L) = 0, \qquad u(0,y) = u(L,y) = 0$$

is given by

$$u(x,y,t) = \sum_{m=1}^{\infty} \sum_{n=1}^{\infty} f_{mn} \exp(-\kappa_{mn}^2 t) \sin k_m x \sin k_m x$$

where the quantities have the same meanings as in Problem 30. Obtain the corresponding solution for the 'reflecting' boundary conditions,

$$u_y(x,0) = u_y(x,L) = 0, \qquad u_x(0,y) = u_x(L,y) = 0$$

Use *Mathematica* to construct, display and animate these solutions for the square region $0 \le x \le 1$ and $0 \le y \le 1$ with the initial condition

$$f(x,y) = 16xy(1-x)(1-y)$$

33. Show that properties of eigenfunctions and eigenvalues of the Liouville equation defined on an interval (a,b),

$$\frac{d}{dx}\left[p(x)\frac{dy}{dx}\right] + [q(x) + \lambda r(x)]y = 0$$

that were obtained in Section 4.5, can also be obtained with the more general boundary conditions

$$\alpha y(a) + \beta y'(a) = 0, \qquad \gamma y(b) + \delta y'(b) = 0$$

where α, β, γ and δ are constants.

34. Determine the factors that are required to multiply the following equations in order to bring them into the form of a Liouville equation:

(Bessel) $\quad \dfrac{d^2y}{dx^2} + \dfrac{1}{x}\dfrac{dy}{dx} + \left(1 - \dfrac{n^2}{x^2}\right)y = 0$

(Chebyshev) $\quad x\dfrac{d^2y}{dx^2} + (1-x)\dfrac{dy}{dx} + ny = 0$

(Hermite) $\quad \dfrac{d^2y}{dx^2} - 2x\dfrac{dy}{dx} + 2ny = 0$

(Laguerre) $\quad x\dfrac{d^2y}{dx^2} + (1-x)\dfrac{dy}{dx} + ny = 0$

(ordinary Legendre) $\quad (1-x^2)\dfrac{d^2y}{dx^2} - 2x\dfrac{dy}{dx} + n(n+1)y = 0$

How are these multiplying factors related to the weight functions in the orthogonality integral of the eigenfunctions of each equation?

35. Show that the second-order ordinary differential equation

$$a(x)\frac{d^2y}{dx^2} + b(x)\frac{dy}{dx} + [c(x) + \lambda d(x)]y = 0$$

can be converted to Sturm–Liouville form by an appropriate change of variable $x \to u(x)$. Derive the differential equation satisfied by u and then obtain the solution in the form

$$u(x) = \int^x \frac{1}{a(s)} \exp\left[\int^s \frac{b(t)}{a(t)} dt\right] ds$$

What is the weight function as a function of u? By writing the orthogonality relation for the eigenfunctions of the Sturm–Liouville problem as a function of u, and then transforming to x, determine the weight function *as a function of x* for each of the equations in Problem 34. How are these multiplying factors related to the factors $u(x)$ obtained for the corresponding equations using the procedure of Problem 34?

36. Consider the Sturm–Liouville problem in Equations (4.86) and (4.87) with $p(x) = 1$, $q(x) = 0$, $r(x) = 1$ and $\lambda = k^2$, i.e.,

$$\frac{d^2 y}{dx^2} + k^2 y = 0$$

with $y(a) = 0$ and $y(b) = 0$. Using the notation of Equation (4.88), show that the fundamental solutions of this equation can be chosen to be

$$u(x; k) = \cos kx, \qquad v(x; k) = \sin kx$$

Use these solutions to impose the boundary conditions and to construct the determinant in Equation (4.91). Show that Equation (4.91) produces the following condition on the allowed values of k:

$$\sin[k(b-a)] = 0$$

With this criterion and the equations determining A and B, obtain (to within a constant of proportionality) the eigenfunctions of this Sturm–Liouville problem:

$$y_n(x) \propto \sin\left(n\pi \frac{x-a}{b-a}\right)$$

where n is an integer.

37. Repeat the calculation in Problem 36, but with the boundary conditions $y'(a) = 0$ and $y'(b) = 0$. Show that in this case, the eigenfunctions are given by

$$y_n(x) = \cos\left(n\pi \frac{x-a}{b-a}\right)$$

where n is again an integer.

38. Suppose that there are two solutions y_1 and y_2 that satisfy the Liouville equation with the general boundary conditions in Problem 33. Then, at $x = a$, both solutions must satisfy the same boundary condition:

$$\alpha y_1(a) + \beta y_1'(a) = 0, \qquad \alpha y_2(a) + \beta y_2'(a) = 0$$

164 *Partial Differential Equations with* Mathematica

Since α and β cannot *both* vanish, show that the determinant of the matrix of coefficients of this system must vanish. By observing that this determinant is the Wronskian of y_1 and y_2, deduce that these two solutions are proportional, so that every eigenvalue of the Sturm–Liouville problem is associated with only a single eigenfunction.

39. The restriction of only a single eigenfunction being associated with a given eigenvalue in the Sturm–Liouville problem can be relieved by considering boundary conditions called **periodic boundary conditions**. As an example, consider the equation in Problem 36 over the interval $(-\pi, \pi)$:

$$\frac{d^2 y}{dx^2} + k^2 y = 0$$

with the boundary conditions

$$y(-\pi) = y(\pi), \qquad y'(-\pi) = y'(\pi)$$

Apply these boundary conditions to the fundamental solutions in Problem 36 to show that the functions

$$u_n(x) = \cos nx, \qquad v_n(x) = \sin nx$$

both satisfy the Liouville equation with eigenvalue n. Notice that these functions are orthogonal.

40. In many applications, the behavior of eigenfunctions for large values of the eigenvalues provide considerable insight into the behavior of physical systems. For this purpose, it is useful to transform Equation (4.86) into a form that is more amenable to an asymptotic analysis. Thus, consider the transformation

$$y(x) = u(x)\phi[v(x)]$$

where the functions $u(x)$ and $v(x)$ are to be determined. By substituting this form into Equation (4.86), show that by requiring the coefficient of λ to be unity means that $v(x)$ must be chosen as

$$v(x) = \int_a^x \left[\frac{r(t)}{p(t)}\right]^{1/2} dt$$

Then combine this result with the requirement that the coefficient of $d\phi/dv$ vanish to determine $u(x)$ as

$$u(x) = [r(x)p(x)]^{-1/4}$$

Thus, obtain the reduced equation for $\phi(v)$:

$$\frac{d^2\phi}{dv^2} + (\lambda + \tilde{q})\phi = 0$$

where $\tilde{q}$ as a function of x is given by

$$\tilde{q}(x) = \frac{q}{r} + \frac{1}{ru}\frac{d}{dx}\left(p\frac{du}{dx}\right).$$

By regarding p, q, r and u as functions of v, show that this expression for $\tilde{q}$ can be written as

$$\tilde{q}(v) = \frac{q(v)}{r(v)} - u(v)\frac{d^2}{dv^2}\left[\frac{1}{u(v)}\right].$$

The utility of the equation for ϕ, which is known as **Liouville's normal form** (Lin and Segel, 1974), will be examined in Problem 41.

41. The results of Problem 40 can be used to obtain the asymptotic behavior of eigenfunctions and eigenvalues of Equation (4.86) with the boundary condition in (4.87). In the limit of large eigenvalues, Liouville's normal form becomes

$$\frac{d^2\phi}{dv^2} + k^2\phi = 0$$

where we have set $\lambda = k^2$. Show that the boundary conditions for $\phi(v)$ corresponding to (4.66) are

$$\phi[v(a)] = 0, \qquad \phi[v(b)] = 0.$$

By solving these equations, obtain the following asymptotic forms for the eigenvalues k_n and eigenfunctions y_n (Lin and Segel, 1974):

$$y_n(x) = [r(x)p(x)]^{-1/4}\sin[k_n v(x)], \qquad k_n = \frac{n\pi}{v(b)}.$$

Compare this result with that obtained for the Sturm–Liouville problem in Problem 36.

Chapter 5

Series Solutions of Ordinary Differential Equations

Most differential equations encountered in physics, engineering and applied mathematics do not have solutions that are given in terms of elementary functions, i.e. expressions that can be constructed from simple algebraic operations involving exponential functions, trigonometric functions and their inverses. Instead, the solutions to such equations are expressed as an infinite series of terms or a finite series with a large number of terms. Expressing solutions of differential equations as power series has already been used in Chapter 3 to establish some general properties of solutions of the different types of second-order partial differential equation for specific boundary conditions. Constructing the series solution was seen to put some very severe restrictions on the problem being solved, namely, that both the differential equation and the boundary conditions were required to be analytic in a region near the boundary curve.

In this chapter a similar procedure will be applied to finding series solutions to ordinary differential equations. The types of series solution that can be obtained will be seen to depend strongly on the analytical behavior of the differential equation at the point about which the expansion is made. However, the series method for ordinary differential equations will be found to be a much more versatile technique than that for partial differential equations. This is because the existence of a series solution near a particular point does not require the differential equation to be analytic there. Instead, the equation can have what is called a 'regular' singular point and still admit a series representation of at least one solution, though this solution need not be analytic at the point where the expansion is made. If the differential equation is analytic about the expansion point, then the method described in this chapter yields a series solution involving two arbitrary constants. This is the general solution of the equation within the radius of convergence of the series. Alternatively, if the differential equation has a regular singularity at the expansion point, then the Frobenius method can guarantee only one solution in the form of a power series. Auxiliary methods then must be applied to obtain an independent complementary solution. Since second-order equations turn out to exhibit all of the essential features of the Frobenius method, our discussion will focus on this class of equations.

In the last section of this chapter, we adopt a somewhat different point of view by considering the consequences that having a given number and type of singular points has on the form of the differential equation and on the solutions themselves. This leads to a classification scheme for differential equations that can be used to establish interrelationships among solutions to different equations that might not otherwise be obvious. Furthermore, on a more contemplative level, this approach highlights the role that singular points of all types have in providing the variety of special functions that will be studied in the next chapter.

5.1 Singular Points of Differential Equations

Consider a second-order differential equation of the form

$$\frac{d^2y}{dx^2} + p(x)\frac{dy}{dx} + q(x)y = 0 \qquad (5.1)$$

Points where $p(x)$ and $q(x)$ are analytic, i.e. where these functions can be expanded in power series that have nonzero radii of convergence, are said to be **regular points** of the differential equation. Points where either $p(x)$ is singular or $q(x)$ is singular are said to be **singular points** of the differential equation. Another way to characterize the analytic character of this differential equation is to multiply the equation by the appropriate factors to render the coefficients of $y'(x)$ and $y(x)$ regular at the point in question. A singular point is then a point where the coefficient of the second derivative vanishes.

The idea behind solving differential equations with the power series method is to first expand $p(x)$, $q(x)$ and the solution $y(x)$ about a particular point and to substitute these expansions into the differential equation in (5.1). This will be carried out in detail in the next section. In this section, we will focus on the consequences of the behavior of $p(x)$ and $q(x)$ near a given point for the analytic properties of the solutions of (5.1). We first expand $p(x)$, $q(x)$ and $y(x)$ as Taylor series about the desired point which, for simplicity and without loss of generality, we take to be the origin. A simple translation of coordinates can always be performed to transform any point into the origin. To encompass the possibilities that the origin may be a regular point or a singular point, we represent these functions as series of the form

$$p(x) = x^r \sum_{n=0}^{\infty} p_n x^n, \quad q(x) = x^t \sum_{n=0}^{\infty} q_n x^n, \quad y(x) = x^s \sum_{n=0}^{\infty} a_n x^n \quad (5.2)$$

No restriction is yet placed on the exponents r, t and s. If $r \geq 0$ and $t \geq 0$ then the origin is a regular point of (5.1). Alternatively, if $r < 0$ and/or $t < 0$, then the origin is a singular point of (5.1), as can be seen, fror example, from the series for $p(x)$:

$$p(x) = \frac{p_0}{x^r} + \frac{p_1}{x^{r-1}} + \cdots + p_r + p_{r+1} x + \cdots \quad (5.3)$$

It must be stressed that r, t, p_n and q_n are given as part of the differential equation, while the exponent s and the coefficients a_n are to be determined by solving the equation. We now take the required derivatives of $y(x)$,

$$\frac{dy}{dx} = a_0 s x^{s-1} + a_1(s+1) x^s + \cdots$$

$$\frac{d^2 y}{dx^2} = a_0 s(s-1) x^{s-2} + a_1 s(s+1) x^{s-1} + \cdots \quad (5.4)$$

and substitute these expansions together with those for $p(x)$ and $q(x)$ into Equation (5.1). The leading-order contributions to each of the three terms in (5.1) are

$$\left[a_0 s(s-1) x^{s-2} + \cdots \right] + \left(a_0 p_0 s x^{r+s-1} + \cdots \right) + \left(q_0 a_0 x^{s+t} + \cdots \right) = 0 \quad (5.5)$$

In order for the left-hand side of (5.5) to vanish for a range of values of x in an open interval, the function itself must reduce to zero identically, so each term in the series must vanish separately. This requirement determines the values that can be taken by the exponent s and the expansion coefficients a_n in the series for the solution $y(x)$ in (5.2). The term (or terms) in (5.5) with the smallest exponent determines the number and type of solutions of the differential equation. In particular, only if $r \geq -1$ and $t \geq -2$ is the leading-order term of (5.5) quadratic in s; otherwise, the leading contribution is at most linear in s. Furthermore, depending on the values of r and t, the roots of these equations for s can be either positive, negative or zero. Thus, the behavior of solutions to (5.1) in the neighborhood of the origin is seen to depend sensitively on the behavior of p and q there. There are three cases to consider.

Case I. If p and q are analytic in a neighborhood of $x=0$, i.e. if p and q have power series expansions as in (5.2) with $r=0$ and $t=0$, then the origin is a **regular point** of the differential equation. In this case, the leading-order contribution to (5.5) comes from the first term on the left-hand side, $a_0 s(s-1)x^{s-2}=0$, which yields the two solutions $s=0$ and $s=1$. Thus, by comparison with the expansion for $y(x)$ in (5.2), we conclude that at a regular point, the differential equation has two linearly independent solutions, each of which is regular at that point.

Case II. If the origin is not a regular point of the differential equation, then it is a singular point. If the quantities $xp(x)$ and $x^2 q(x)$ are both regular at $x=0$, which implies that $r=-1$ and that $t=-2$ or $t=-1$, then the origin is called a **regular singular point** of the differential equation. The leading-order terms in (5.5) are still quadratic in s, though the roots of the quadratic equation depend upon p_0 and possibly q_0, and so need neither be nonnegative nor even distinct. Thus, the differential equation does not necessarily have any (nontrivial) solutions that are regular at a regular singular point, though there is at least one solution of the form

$$y(x) = x^s \sum_{n=0}^{\infty} a_n x^n \qquad (5.6)$$

Such a solution is regular at the origin only if s is a nonnegative integer.

Case III. If the origin is neither a regular point nor a regular singular point, then it is an **irregular singular point** of the differential equation. In this case, $r < -1$ and/or $t < -2$, and the leading terms in (5.5) come from either or both of the second and third terms on the left-hand side, which yield at most a single solution for s. Thus, a (nontrivial) solution of the form (5.6) may or may not exist.

Example 5.1. Consider the problem of finding power series solutions near the origin of the following differential equation:

$$\frac{d^2 y}{dx^2} - \frac{2}{x}\frac{dy}{dx} - \frac{y}{x^4} = 0 \qquad (5.7)$$

The origin is seen to be an irregular singular point of this equation since, in the notation of Equation (5.2), $t=-4$. All other finite points are readily identified to be regular points. To check the behavior of the point at infinity, we make the transformation $\xi = 1/x$, which transforms the origin to infinity and vice versa. In terms of the new variable ξ, Equation (5.7) becomes

$$\frac{d^2 y}{d\xi^2} - y = 0 \qquad (5.8)$$

The general solution to this equation is easily obtained as

$$y(\xi) = Ae^{\xi} + Be^{-\xi} \tag{5.9}$$

where A and B are constants to be determined by the boundary conditions. In the original variable, the general solution to (5.7) is therefore given by

$$y(x) = Ae^{1/x} + Be^{-1/x} \tag{5.10}$$

The solution is seen to have an essential singularity at $x=0$ and so has no Taylor series expansion about the origin.

This example illustrates two important points. First, the transformation of the independent variable from x to $1/x$ allows the behavior of a differential equation at infinity to be determined using the methods developed in this section. This transformation will be used for this purpose several times in this chapter and in the next chapter. Second, the essential singularity of (5.10) at infinity is a general feature of the general solution at irregular singular points. One and possibly both of the solutions of a differential equation will have essential singularities at an irregular singular point (Morse and Feshbach, 1953). ∎

5.2 The Method of Frobenius

Although the form of Equation (5.1) is useful for identifying the analytic properties of differential equations, a more convenient form for obtaining power series solutions is

$$R(x)\frac{d^2y}{dx^2} + \frac{1}{x}P(x)\frac{dy}{dx} + \frac{1}{x^2}Q(x)y = 0 \tag{5.11}$$

The functions P, Q and R are assumed to be regular about the expansion point x_0, with $R(x_0) \neq 0$. In the following discussion, x_0 is again taken to be the origin without loss of generality. Thus, the expansions of R, P and Q can be written as

$$P(x) = \sum_{n=0}^{\infty} P_n x^n, \quad Q(x) = \sum_{n=0}^{\infty} Q_n x^n, \quad R(x) = \sum_{n=0}^{\infty} R_n x^n \tag{5.12}$$

where the requirement that $R(0) \neq 0$ means that $R_0 \neq 0$. Since P, Q and R are presumed to be regular at the origin, with R being nonzero as $x \to 0$, (5.11) can be written in the form of Equation (5.1) by dividing the equation by R. This shows that we can make the identifications

$$xp(x) = \frac{P(x)}{R(x)}, \quad x^2 q(x) = \frac{Q(x)}{R(x)} \tag{5.13}$$

172 Partial Differential Equations with Mathematica

Since P, Q and R are each analytic at the origin, with $R_0 \neq 0$, the origin is seen to be either a *regular point* of the equation or a *regular singular point* (Problem 2). Therefore, according to the discussion in the preceding section, we can expect at least one solution of the form (5.6).

The application of the method of Frobenius to obtain solutions of (5.11) proceeds by taking a trial solution y of the form given in (5.2):

$$y(x) = x^s \sum_{n=0}^{\infty} a_n x^n \qquad (5.14)$$

The exponent s and the coefficients a_n are to be determined by requiring that (5.14) is a solution of (5.11). To obtain the equations that determine s and a_n the expansions in (5.12) and (5.14) are first substituted into (5.11):

$$x^{s-2} \left[\sum_{n=0}^{\infty} R_n x^n \cdot \sum_{n=0}^{\infty} a_n (n+s)(n+s-1) x^n \right.$$
$$\left. + \sum_{n=0}^{\infty} P_n x^n \cdot \sum_{n=0}^{\infty} a_n (n+s) x^n + \sum_{n=0}^{\infty} Q_n x^n \cdot \sum_{n=0}^{\infty} a_n x^n \right] = 0 \qquad (5.15)$$

Each of the three terms in (5.15) is composed of a product of two power series. To combine these three products of series into an expression involving a single power series, we use the expression in Equation (1.37) for the product of two series. After some rearrangement to collect terms with common powers of x, and cancelling the inessential factor x^{s-2}, we obtain

$$\sum_{n=0}^{\infty} \left\{ \sum_{k=0}^{n} a_k \left[(k+s)(k+s-1) R_{n-k} + (k+s) P_{n-k} + Q_{n-k} \right] \right\} x^n = 0 \quad (5.16)$$

Equation (5.16) is the series representation of (5.11) and, where the series converges, is equivalent to the original differential equation.

The left-hand side of (5.16) is the Taylor series of a function whose expansion coefficients are contained in the braces, i.e. (5.16) has the general form $\sum_n c_n x^n = 0$, where $c_n = \{\cdots\}$. In order for this equation to be satisfied over an interval about the origin, this function must be identically zero. Thus, $c_n = 0$ for *every* value of n:

$$\sum_{k=0}^{n} a_k \left[(k+s)(k+s-1) R_{n-k} + (k+s) P_{n-k} + Q_{n-k} \right] = 0 \qquad (5.17)$$

Since this requirement cannot be met for every choice of s and a_k, the set of Equations (5.17) for $n = 0, 1, 2, \ldots$ determines the values of s and a_k that yield a *formal* solution of (5.11) that is of the form in (5.14). We say 'formal' because even though the expression (5.14) so obtained satisfies the differential equation, we must verify that this series does indeed have a nonzero radius of convergence in order to be a meaningful solution. Although there are theorems to deal with the various situations that arise (Birkhoff and Rota, 1989), we will content ourselves with examining the convergence of each series on a case-by-case basis.

The final step in the Frobenius method is to solve the equations in (5.17) for the unknown quantities s and a_n. We first consider the term in this equation for $n = 0$. The only term that contributes is the one obtained by setting $n = k = 0$:

$$a_0[s(s-1)R_0 + sP_0 + Q_0] = 0 \tag{5.18}$$

This equation implies that either $a_0 = 0$ or the quantity enclosed in brackets vanishes (or both). Referring to (5.14) we see that choosing $a_0 = 0$ simply involves changing the definition of s, but not the form of the series. Therefore, by requiring the coefficient of a_0 in (5.18) to vanish, we can satisfy (5.18) for *any* value of a_0. In other words, a_0 becomes an arbitrary constant in the solution of the differential equation. With this choice, (5.18) requires that

$$R_0 s^2 + (P_0 - R_0)s + Q_0 = 0 \tag{5.19}$$

This is a quadratic equation for the exponent s in (5.10) and is termed the **indicial equation**. The roots of the indicial equation, s_1 and s_2, are variously called the **exponents**, the **roots**, or the **indices** of the point about which the expansion is made. Depending upon the values of R_0, P_0 and Q_0 which, as we will see shortly, determine whether the origin is a regular point or a regular singular point, the roots s_1 and s_2 may not be distinct, nor even integers. The values of these roots will been seen to be the primary indicators of the behavior of the solutions near the origin. We will occasionally find it convenient to write the indicial equation in the following factored form that explicitly involves the roots:

$$R_0 s^2 + (P_0 - R_0)s + Q_0 = (s - s_1)(s - s_2) = 0 \tag{5.20}$$

For the contributions in (5.17) corresponding to $n > 0$, we separate the term for $k = n$ from the other terms to obtain

$$a_n[(n+s)(n+s-1)R_0 + (n+s)P_0 + Q_0]$$
$$= -\sum_{k=0}^{n-1} a_k \left[(k+s)(k+s-1)R_{n-k} + (k+s)P_{n-k} + Q_{n-k}\right] \tag{5.21}$$

The left-hand side of this equation can be simplified by observing that the coefficient of a_n reduces to the indicial equation (5.18) if we make the replacement $n+s \to s$. Thus, we can use (5.20) to write this expression as

$$a_n(n+s-s_1)(n+s-s_2)$$
$$= -\sum_{k=0}^{n-1} a_k \left[(k+s)(k+s-1)R_{n-k} + (k+s)P_{n-k} + Q_{n-k} \right] \quad (5.22)$$

This equation is seen to provide an expression for the coefficient a_n in terms of the quantities $a_0, a_1, \ldots, a_{n-1}$, i.e. the a_n are determined *recursively*: a_1 is given in terms of a_0, a_2 in terms of a_0 and a_1, and so on. For this reason, the hierarchy of equations in (5.22) is referred to as the **recurrence formulae** or the **recursion relations** for the differential equation. Expressions for $n=1$ and $n=2$ are shown below:

$$a_1(s+1-s_1)(s+1-s_2) = -a_0[s(s-1)R_1 + sP_1 + Q_1]$$
$$a_2(s+2-s_1)(s+2-s_2) = -a_0[s(s-1)R_2 + sP_2 + Q_2] \quad (5.23)$$
$$\qquad - a_1[s(s+1)R_1 + (s+1)P_1 + Q_1]$$

Equations (5.22) and (5.23) show that the recursion relations have a 'telescopic' structure, in the sense that if the functions R, P and Q have expansion coefficients of all orders, then the recursion relations will contain successively more terms with increasing n. In practice, however, the application of this method usually involves only one or two terms on the right-hand side of (5.22), enabling simple expressions to be obtained for the coefficient a_n as a function of n. In favorable circumstances, the series in (5.14) can be summed to obtain a closed-form expression for the solution.

Since Equation (5.18) is used to obtain the indicial equation (5.19), a_0 is arbitrary, so the solutions of the recursion relations are proportional to this constant, as can be seen from (5.22) and (5.23). Thus, if the two solutions obtained by solving these recursion relations with each of the two roots of the indicial equation are independent, then we have obtained a solution with two arbitrary constants, which is the general solution of the differential equation in (5.11). The conditions under which the Frobenius method yields two independent solutions will be discussed shortly.

The steps leading to the indicial equation in (5.19) and the recursion relations in (5.21) can be carried out symbolically in *Mathematica* and expressions for the expansion coefficients a_n can be obtained, however complex the recursion relations. This may prove useful if any of the functions P, Q or R are not simple algebraic expressions, resulting in complicated recursion relations that are difficult to solve in closed form.

The first step is to define the power series expansions for the quantities R, P, Q and y, as in (5.12) and (5.14):

```
R[x_,n_]:=Sum[r[k]x^k,{k,0,n}]

P[x_,n_]:=Sum[p[k]x^k,{k,0,n}]

Q[x_,n_]:=Sum[q[k]x^k,{k,0,n}]

y[x_,s_,n_]:=x^s Sum[a[k]x^k,{k,0,n}]
```

Equation (5.16) is then obtained by using `Collect` to group the terms according to powers of x. The resulting expression (the left-hand side of (5.16)) is called `Frobenius`:

```
Frobenius[n_]:=Collect[R[x,n]D[y[x,s,n],{x,2}] +
    x^(-1)P[x,n]D[y[x,s,n],x] + x^(-2)Q[x,n]y[x,s,n],x]
```

The indicial equation and the recursion relations can now be extracted from `Frobenius` by collecting the coefficient of the appropriate power of x. For example, the indicial equation is obtained from the coefficients of x^{s-2}, as in (5.16):

```
Simplify[Coefficient[Frobenius[0],x^(s-2)]]
```

$$a[0]\ (s\ p[0]\ +\ q[0]\ -\ s\ r[0]\ +\ s^2\ r[0])$$

Similarly, the recursion relation for a_n in (5.22) can be determined from the coefficient of x^{s-2+n} in `Frobenius` by using `Solve` to obtain an expression for a_n in terms of $a_{n-1}, \ldots, a_0$. For example, solving for a_1 yields an expression equivalent to that obtained by solving the first of Equations (5.23):

```
Simplify[Solve[Coefficient[Frobenius[1],x^(s-1)]==0,a[1]]]
```

$$a[1] \to -\left(\frac{a[0]\ (s\ p[1]\ +\ q[1]\ -\ s\ r[1]\ +\ s^2\ r[1])}{p[0]\ +\ s\ p[0]\ +\ q[0]\ +\ s\ r[0]\ +\ s^2\ r[0]}\right)$$

Our approach to finding series solutions to differential equations has so far been quite general. We have not yet made any arguments that distinguish the application of the method at a regular point of a differential equation from that at a regular singular point. We can examine the relationship between the solutions obtained from the Frobenius method and the analytic behavior of the differential equation near the origin by using (5.1), (5.12) and (5.13) to write the functions R, P and Q as

$$p(x) = \frac{1}{x} \frac{P_0 + P_1 x + P_2 x^2 + \cdots}{R_0 + R_1 x + R_2 x^2 + \cdots}$$
$$q(x) = \frac{1}{x^2} \frac{Q_0 + Q_1 x + Q_2 x^2 + \cdots}{R_0 + R_1 x + R_2 x^2 + \cdots} \quad (5.24)$$

$R(x)$ is, by hypothesis, nonvanishing at the origin, so $R_0 \neq 0$. Thus, if $P_0 = 0$ and $Q_0 = Q_1 = 0$, $p(x)$ and $q(x)$ can be written as

$$p(x) = \frac{P_1 + P_2 x + P_3 x^2 + \cdots}{R_0 + R_1 x + R_2 x^2 + \cdots} = \frac{P_1}{R_0} + \left(\frac{P_2}{R_0} - \frac{P_1 R_1}{R_0^2}\right) x + \cdots$$
$$q(x) = \frac{Q_2 + Q_3 x + Q_4 x^2 + \cdots}{R_0 + R_1 x + R_2 x^2 + \cdots} = \frac{Q_2}{R_0} + \left(\frac{Q_3}{R_0} - \frac{Q_2 R_1}{R_0^2}\right) x + \cdots \quad (5.25)$$

Since R, P and Q are presumed to have convergent power series expansions near the origin, the quotients in (5.25) also represent convergent power series (Knopp 1956). We conclude therefore that the origin is a regular point if $P_0 = 0$ and $Q_0 = Q_1 = 0$, and a regular singular point otherwise. With these restrictions on R, P and Q, the indicial equation (5.19) for a regular point becomes

$$a_0 R_0 s(s-1) = 0 \quad (5.26)$$

which yields the two roots $s=0$ and $s=1$, as expected from the discussion following Equation (5.5), with the quantity a_0 being an arbitrary constant.

Turning our attention to the recursion relations (5.21), we have for $n=1$ and with $Q_1 = 0$,

$$a_1 R_0 s(s+1) = -a_0 s[(s-1)R_1 + P_1] \quad (5.27)$$

By choosing $s=0$ as the solution of the indicial equation, both sides of this equation are made to vanish, so this recursion relation is satisfied for *any* values of a_0 and a_1. Thus, a_1 is a second arbitrary constant of the solution. For $n>1$ and $s=0$, the recursion relations (5.22) become

$$a_n n(n-1) R_0 = -\sum_{k=0}^{n-1} a_k [k(k-1) R_{n-k} + k P_{n-k} + Q_{n-k}] \quad (5.28)$$

Since the left-hand side of this equation does not vanish for $n > 1$, we can solve for each of the remaining a_n to obtain expressions that involve the arbitrary constants a_0 and a_1. Thus, the application of the method of Frobenius at a regular point of a differential equation produces a series solution with two arbitrary constants a_0 and a_1 which, within the radius of convergence of the series, is the general solution to (5.11).

Example 5.2. Consider the problem of applying the method of Frobenius to construct the general solution near the origin of the following differential equation:

$$(1 - x^2)\frac{d^2y}{dx^2} - 2x\frac{dy}{dx} + 2y = 0 \qquad (5.29)$$

The origin is readily seen to be a regular point of this differential equation, while the points $x = \pm 1$ are regular singular points. In the notation of Equation (5.11), the functions P, Q and R are given by

$$R(x) = 1 - x^2, \qquad P(x) = -2x^2, \qquad Q(x) = 2x^2 \qquad (5.30)$$

so the nonvanishing expansion coefficients of these functions are $R_0 = 1$, $R_2 = -1$, $P_2 = -2$ and $Q_2 = 2$. By substituting the series expansion (5.14) into (5.18), the indicial equation for (5.29) is found to be

$$s(s - 1) = 0 \qquad (5.31)$$

The roots of this equation are $s = 0$ and $s = 1$, as expected from (5.26), with a_0 being arbitrary. The recursion relation for $n = 1$ is

$$a_1 s(s + 1) = 0 \qquad (5.32)$$

If $s = 0$ is chosen as the root of the indicial equation, this equation can be satisfied with any value of a_1. Thus, the solution of the recursion relations with $s = 0$ produces the general solution of the differential equation within the radius of convergence of the series, again as expected from the discussion following (5.24). For $n > 1$, the recursion relations are given by

$$a_n(n + s)(n + s - 1)$$
$$= -a_{n-2}[(n + s - 2)(n + s - 3)R_2 + (n + s - 2)P_2 + Q_2]$$
$$= a_{n-2}(n + s)(n + s - 3) \qquad (5.33)$$

With $s = 0$, this equation may be solved to obtain an expression for a_n in terms of a_{n-2}:

$$a_n = \frac{n - 3}{n - 1} a_{n-2} \qquad (5.34)$$

This recursion relation reveals several characteristic features of the method of Frobenius. The simplicity of the recursion relations—a given a_n is given *only* in terms a_{n-2}—is due entirely to the fact that only P_2, Q_2 and R_2 contribute to the right-hand side of (5.22). As is evident from the general form of the recursion relations in (5.22) and (5.23), the greater the number of nonvanishing coefficients in the series expansions of P, Q and R, the greater the number of terms in the recursion relations. With simple recursion relations, closed-form expressions for the expansion coefficients are much easier to obtain than from recursion relations with many terms and the convergence of the series can be assessed much more readily. In the case of (5.34) an application of the ratio test yields

$$\lim_{n\to\infty} \left|\frac{a_{n-2}}{a_n}\right| = \lim_{n\to\infty} \frac{n-3}{n-1} = 1 \qquad (5.35)$$

The solution obtained by solving the recursion relations (5.34) is seen to converge for values of x that lie in the open interval $-1 < x < 1$ and diverge if $|x| > 1$. The ratio test fails at the end-points $x = \pm 1$. In fact, the Gauss test shows that this series diverges at $x = \pm 1$ (Problem 3).

The solution of the recursion relations proceeds by calculating successively higher-order expansion coefficients. The form of (5.34), which couples even-order coefficients only to other even-order coefficients and odd-order coefficients only to other odd-order coefficients, indicates that the solution is composed of the sum of a series of even terms proportional to a_0, and a series of odd terms proportional to a_1. The solution for the odd-order terms turns out for this equation to be very simple. We can deduce immediately from the recursion relations that $a_3 = 0$, so all of the odd-order coefficients beyond a_3 also vanish. Thus, the part of the solution composed of odd-order terms is simply $a_1 x$.

To solve the recursion relations for the even-order terms, we first set $2k = n$ in (5.34). An expression for the general even-order term can then be obtained by recursive substitution:

$$\begin{aligned}
a_{2k} &= \frac{2k-3}{2k-1} a_{2k-2} \\
&= \frac{2k-5}{2k-3}\frac{2k-3}{2k-1} a_{2k-4} \\
&\;\;\vdots \\
&= \frac{1}{3}\frac{3}{5}\frac{5}{7}\cdots \frac{2k-7}{2k-5}\frac{2k-5}{2k-3}\frac{2k-3}{2k-1} a_2 \\
&= -\frac{1}{2k-1} a_0
\end{aligned} \qquad (5.36)$$

Thus, by combining the expressions for the even-order and odd-order coefficients, we obtain the series expansion for the general solution to (5.29), which is valid for $-1 < x < 1$:

$$y(x) = a_1 x - a_0 \sum_{k=0}^{\infty} \frac{x^{2k}}{2k-1} \tag{5.37}$$

where a_0 and a_1 are arbitrary constants. In Problem 5, a closed-form expression for the series in (5.37) is obtained, with the result that the general solution to (5.29) can be written as

$$y(x) = a_1 x + a_0 - \tfrac{1}{2} a_0 x \ln\left(\frac{1+x}{1-x}\right) \tag{5.38}$$

The argument of the logarithm shows explicitly why this solution diverges at $x = \pm 1$ and shows that for $|x| \geq 1$, the series in (5.37) is not a valid solution to (5.29). To obtain the general solution in this region, the transformation $x \to 1/x$ is made, and an expansion is carried out about the point at infinity, which is a regular singular point of the equation. This is left as an exercise (Problem 9).

The steps used earlier in *Mathematica* for the general case of the method of Frobenius can be applied to solving the differential equation in (5.29). The expression Frobenius is set up as before by defining the coefficient functions R, P and Q and the solution y:

```
R[x_]:=1-x^2

P[x_]:=-2x^2

Q[x_]:=2x^2

y[x_,s_,n_]:=x^s Sum[a[k]x^k,{k,0,n}]

Frobenius[n_]:=Collect[R[x] D[y[x,s,n],{x,2}] +
    x^(-1)P[x] D[y[x,s,n],x] + x^(-2)Q[x] y[x,s,n],x]
```

Frobenius can now be used either to generate the indicial equation and the recursion relations for any given n, as in (5.31) and (5.32). Coefficient is invoked to obtain individual terms in Frobenius, e.g.

```
Factor[Coefficient[Frobenius[0],x^(s-2)]]

(-1 + s) s a[0]

Factor[Coefficient[Frobenius[1],x^(s-1)]]

s (1 + s) a[1]
```

The solution to the differential equation can be obtained by first constructing the recursion relations,

```
a[0]=a0;
a[1]=a1;
a[n_]:=If[n>=2,((n-3)/(n-1))a[n-2]]
```

The constants a0 and a1 have been left as arbitrary. The series for the solution can now be displayed to any desired order. For example,

```
y[x,0,10]
```

$$a0 + a1\, x - a0\, x^2 - \frac{a0\, x^4}{3} - \frac{a0\, x^6}{5} - \frac{a0\, x^8}{7} - \frac{a0\, x^{10}}{9}$$

which is the same as (5.37).

∎

5.3 Constructing Complementary Solutions

The results of the preceding section show that at a regular point of a differential equation, the method of Frobenius yields directly the general solution in the form of Taylor series expansions about that point. In this section we study two special cases that occur in applying the method of Frobenius near a regular singular point of a differential equation: the case where the indicial equation yields equal roots, and the case where the roots of the indicial equation differ by an integer. We will find for these situations that a solution of the form (5.6) is obtained, but that only one arbitrary constant is involved, so we must resort to auxiliary methods for constructing the complementary solution.

At a regular singular point, either or both of the coefficients P_0 and Q_0 are nonvanishing. We consider first the situation where the roots of the indicial equation are equal, $s_1 = s_2$. In this case, the recursion relations (5.22) for $n \geq 1$ are given by

$$a_n n^2 = -\sum_{k=0}^{n-1} a_k \left[(k+s_1)(k+s_1-1)R_{n-k} + (k+s_1)P_{n-k} + Q_{n-k} \right] \quad (5.39)$$

Since the coefficient of a_n on the left-hand side of (5.39) does not vanish for $n \geq 1$, these equations, with $n = 1$, can be solved for a_1 in terms of a_0, then with $n = 2$ solved for a_2 in terms of a_1 and a_0, and so on. The series generated by solving the recursion relations for successively increasing values of n therefore yields a solution of the differential equation containing only a single arbitrary constant, a_0. Thus, only one solution of the form (5.14) is obtained by direct application of the method of Frobenius. This solution must be supplemented by an independent solution to obtain the general solution of the differential equation.

To construct the second solution of the differential equation, we first solve the recursion relations for the a_n in terms of a_0, but we retain the dependence upon s. These solutions of the recursion relations are then written in the notation $a_n(s)$ to emphasize the dependence upon the exponent s. Since the $a_n(s)$ have been chosen to satisfy the recursion relations (5.21) for $n>0$ for *any* value of s, the only term in the series representation (5.16) that does not vanish if $s \neq s_1$ is that corresponding to $n=0$, i.e. the indicial equation. In other words, the recursion relations can be solved for the a_n as a function of s, but only a root of the indicial equation yields an equality in (5.16). We now define the function $y(x;s)$ generated by the s-dependent expansion coefficients by

$$y(x;s) = x^s \sum_{n=0}^{\infty} a_n(s) x^n \qquad (5.40)$$

If the root s_1 of the indicial equation is substituted for s in (5.40) then this expression reduces to the solution obtained by the method of Frobenius. The equation satisfied by (5.40) for general s can be obtained by substituting (5.40) into (5.11), using (5.16), and the fact that the $a_n(s)$ solve the recursion relations:

$$R(x)\frac{d^2 y(x;s)}{dx^2} + \frac{1}{x}P(x)\frac{dy(x;s)}{dx} + \frac{1}{x^2}Q(x)y(x;s) = x^{s-2}\bigg\{ a_0(s-s_1)^2$$
$$+ \sum_{n=1}^{\infty}\bigg\{\sum_{k=0}^{n} a_k(s)\Big[(k+s)(k+s-1)R_{n-k} + (k+s)P_{n-k} + Q_{n-k}\Big]\bigg\}x^n\bigg\}$$
(5.41)

Since the $a_k(s)$ are chosen to satisfy the recursion relations (5.17) for $n>1$, the summation on the right-hand side of (5.41) vanishes for any choice of s. The equation satisfied by (5.40) becomes simply

$$R(x)\frac{d^2 y(x;s)}{dx^2} + \frac{1}{x}P(x)\frac{dy(x;s)}{dx} + \frac{1}{x^2}Q(x)y(x;s) = a_0(s-s_1)^2 x^{s-2}$$
(5.42)

If $s=s_1$, the right-hand side of (5.41) vanishes, in which case (5.40) becomes a solution of (5.11).

To obtain a second solution that is independent of that obtained by the method of Frobenius, we observe that if we differentiate both sides of (5.42) with respect to s and then set $s = s_1$, then the right-hand side still vanishes. Then, noting from the form of (5.40) that the order of the differentiations of $y(x; s)$ with respect to x and s can be exchanged (Problem 6), and using a prime to signify the partial derivatives with respect to s, we obtain

$$R(x)\frac{d^2}{dx^2}y'(x; s) + \frac{1}{x}P(x)\frac{d}{dx}y'(x; s) + \frac{1}{x^2}Q(x)y'(x; s)$$
$$= a_0[2(s - s_1) + (s - s_1)^2 \ln x]x^{s-2} \qquad (5.43)$$

Upon setting $s = s_1$, the right-hand side of (5.43) is seen to vanish, which shows that if the function $y(x; s)$ is a solution of (5.11), then the function y' evaluated with $s = s_1$ is also a solution of this differential equation. Thus, in the case of a doubly degenerate root of the indicial equation, a second independent solution $y_2(x)$ is given by

$$y_2(x) = \left.\frac{\partial y(x; s)}{\partial s}\right|_{s=s_1} = x^{s_1} \ln x \sum_{n=0}^{\infty} a_n x^n + x^{s_1} \sum_{n=0}^{\infty} a'_n(s_1) x^n \qquad (5.44)$$

The presence of the logarithm, which arises from differentiating the factor x^s in (5.40) with respect to s, means that this solution is singular at the origin and so cannot be expressed in the form of a Taylor series expansion about this point. In fact, the solution obtained from (5.40) by s setting equal to s_1 is finite at the origin only if $s_1 \geq 0$.

Example 5.3. Consider the problem of finding the general solution of the following differential equation by the applying the method of Frobenius at the origin:

$$x\frac{d^2y}{dx^2} + \frac{dy}{dx} - y = 0 \qquad (5.45)$$

Dividing this equation by x, we see that the origin is a regular singular point of this equation since, in the notation of Equation (5.1), we have that $p = 1/x$ and $q = -1/x$. To apply the method of Frobenius, Equation (5.45) is written in the standard form in (5.11), which allows us to make the identifications

$$R(x) = 1, \qquad P(x) = 1, \qquad Q(x) = -x \qquad (5.46)$$

Thus, in terms of the series expansions of these functions we have $R_0 = 1$, $P_0 = 1$ and $Q_1 = -1$, with all of the other expansion coefficients vanishing. The indicial equation is

$$s(s - 1) + s = s^2 = 0 \qquad (5.47)$$

which yields the double root $s=0$. The recursion relations take the simple form

$$a_n(n+s)^2 = -a_{n-1}Q_1 = a_{n-1} \tag{5.48}$$

which can be solved for a_n in terms of a_{n-1} as

$$a_n = \frac{a_{n-1}}{(n+s)^2} \tag{5.49}$$

The simple form of this recursion relation again permits solutions to be readily obtained for each of the a_n and for the radius of convergence to be determined using the ratio test. Setting $s=0$ in (5.49), the ratio test yields

$$\lim_{n\to\infty} \left|\frac{a_n}{a_{n-1}}\right| = \lim_{n\to\infty} \frac{1}{n^2} = 0 \tag{5.50}$$

so the series solution of (5.45) corresponding to $s=0$ converges over the entire real line. The recursion relations can be solved for every a_n in terms of a_0 by following the steps of recursive substitution in Example 5.2:

$$a_n = \frac{1}{(n+s)^2} a_{n-1}$$

$$= \frac{1}{(n+s)^2} \frac{1}{(n+s-1)^2} a_{n-2}$$

$$\vdots$$

$$= \frac{1}{(s+1)^2} \cdots \frac{1}{(n+s-1)^2} \frac{1}{(n+s)^2} a_0 \tag{5.51}$$

Thus, setting $s=0$ in (5.51), one solution to (5.45) is

$$y_1(x) = a_0 \sum_{n=0}^{\infty} \frac{x^n}{(n!)^2} \tag{5.52}$$

According to (5.44), the form of a second solution to (5.45) involves differentiating $a_n(s)$ with respect to s and then evaluating the resulting expression with $s=0$. To perform these operations, we first rewrite $a_n(s)$ as

$$a_n(s) = a_0 \left(\frac{1}{s+1} \cdots \frac{1}{n+s-1} \frac{1}{n+s}\right)^2$$

$$= a_0 \exp\left[2\ln\left(\frac{1}{s+1} \cdots \frac{1}{n+s-1} \frac{1}{n+s}\right)\right] \tag{5.53}$$

The derivative can now be carried out, and we obtain

$$a'_n(s) = a_0 \frac{d}{ds}\left\{\exp\left[-2\ln(n+s) - 2\ln(n+s-1) - \cdots - 2\ln(s+1)\right]\right\}$$

$$= -2a_0\left(\frac{1}{n+s} + \frac{1}{n+s-1} + \cdots + \frac{1}{s+1}\right)$$

$$\times \left(\frac{1}{s+1} \cdots \frac{1}{n+s-1}\frac{1}{n+s}\right)^2 \qquad (5.54)$$

Thus, substitution of this expression into (5.44) yields the second solution of (5.45):

$$y_2(x) = a_0 \ln x \sum_{n=0}^{\infty} \frac{x^n}{(n!)^2} - 2a_0 \sum_{n=1}^{\infty} \frac{1}{(n!)^2}\left(1 + \frac{1}{2} + \cdots + \frac{1}{n}\right)x^n \qquad (5.55)$$

so the general solution of (5.45) can be written as a linear combination of (5.52) and (5.55). Notice that the solution (5.51) is regular at the origin as is the second summation in (5.55). However, because of the presence of the logarithm, the first term in (5.55) has a singularity at the origin. The radius of convergence of each of the series in (5.55) is readily seen to be the entire real line.

The modifications to the *Mathematica* procedure used in Example 5.2 to account for the s-dependence in the a_n are quite minimal. We begin as before by constructing the series representation of the solution directly from the recursion relations, but now we retain the s-dependence in the recursion relations and declare s as an argument of the expansion coefficients:

```
a[0,s]=a0;
a[n_,s_]:=If[n>0,a[n-1,s]/(n+s)^2]

y[x_,s_,n_]:=x^s Sum[a[k,s]x^k,{k,0,n}]
```

The solution (5.52) produced by the direct application of the method of Frobenius is then obtained by evaluating y[x,s,n] with s set equal to zero. Thus, up to terms of 10th order in x, we obtain the solution directly from the recursion relations as in Example 5.2:

```
y[x,s,10]/.s->0
```

$$a0 + a0\,x + \frac{a0\,x^2}{4} + \frac{a0\,x^3}{36} + \frac{a0\,x^4}{576} + \frac{a0\,x^5}{14400} + \frac{a0\,x^6}{518400} +$$

$$\frac{a0\,x^7}{25401600} + \frac{a0\,x^8}{1625702400} + \frac{a0\,x^9}{131681894400} + \frac{a0\,x^{10}}{13168189440000}$$

The denominators corresponding to x^n in this series are readily verified to be equal to $(n!)^2$. The complementary solution (5.55) is obtained by performing the construction in (5.44). Again retaining terms to 10th order in y[x,s,n], we obtain

D[y[x,s,10],s]/.s->0

$$-2\,a0\,x - \frac{3\,a0\,x^2}{4} - \frac{11\,a0\,x^3}{108} - \frac{25\,a0\,x^4}{3456} - \frac{137\,a0\,x^5}{432000} -$$

$$\frac{49\,a0\,x^6}{5184000} - \frac{121\,a0\,x^7}{592704000} - \frac{761\,a0\,x^8}{227598336000} - \frac{7129\,a0\,x^9}{165919186944000} -$$

$$\frac{7381\,a0\,x^{10}}{16591918694400000} + \Big(a0 + a0\,x + \frac{a0\,x^2}{4} + \frac{a0\,x^3}{36} + \frac{a0\,x^4}{576} +$$

$$\frac{a0\,x^5}{14400} + \frac{a0\,x^6}{518400} + \frac{a0\,x^7}{25401600} + \frac{a0\,x^8}{1625702400} + \frac{a0\,x^9}{131681894400} +$$

$$\frac{a0\,x^{10}}{13168189440000}\Big)\,\text{Log}[x]$$

The verification that (5.55) yields the same expression as that shown above is left as an exercise.

Consider now the case where there are two distinct roots s_1 and s_2 of the indicial equation. We take $s_2 > s_1$ and define the quantity Δs as the difference between the two roots $\Delta s = s_2 - s_1 > 0$. For the larger of the two roots, s_2, the left-hand side of (5.22) becomes $a_n n(n+\Delta s)$, so the recursion relations can be solved to obtain a series with one arbitrary constant, a_0. However, for the smaller root s_1, the left-hand side of (5.22) becomes $a_n n(n-\Delta s)$ which, if Δs equals an integer N, vanishes for $n=N$. If the corresponding right-hand side of (5.22) does not also vanish, then the recursion relations do not have a solution for a_n. On the other hand, if the right-hand side does vanish for $n=N$, then the series obtained by setting $s=s_1$ is given in terms of two arbitrary constants, a_0 and a_N, and so yields the general solution.

The case where the indicial roots differ by an integer, $s_2 - s_1 = N$, and where the recursion relations cannot be solved for a_N, is the second special case that we will consider. We begin by following a similar approach to that used in dealing with double roots of the indicial equation, namely, solving the recursion relations for the expansion coefficients a_n as a function of s. Since the right-hand side of (5.22) is presumed not to vanish for $n=N$, we can solve the equation for $a_N(s)$. The recursion relation for $n=N$ can be written as

$$a_N(N+s-s_1)(N+s-s_2) = a_N(N+s-s_1)(s-s_1)$$

$$= -\sum_{k=0}^{n-1} a_k \left[(k+s)(k+s-1)R_{n-k} + (k+s)P_{n-k} + Q_{n-k}\right] \quad (5.56)$$

so the solution for $a_N(s)$ will be proportional to $(s-s_1)^{-1}$. This factor of $(s-s_1)^{-1}$ will propagate through the recursion relations and appear in the higher-order expansion coefficients but, of course, there will be no such factor in the lower-order coefficients. Therefore, if we form the product $(s-s_1)a_k(s)$ and take the limit as $s \to s_1$, the coefficients $a_k(s_1)$ for $k<N$ will vanish, while the coefficients $a_k(s_1)$ for $k \geq N$ will take nonzero values. The consequences of this for obtaining a series solution can be seen by multiplying the recursion relations by $s-s_1$ and defining

$$\tilde{a}_k(s) \equiv (s-s_1)a_k(s) \quad (5.57)$$

Then, as $s \to s_1$, the recursion relations for $n < N$ are clearly satisfied by $\tilde{a}_n(s_1) = 0$. For $n=N$, we can use (5.22) and (5.56) to obtain an expression for $\tilde{a}_N(s_1)$:

$$\tilde{a}_N(s_1)$$
$$= -\frac{1}{N} \sum_{k=0}^{N-1} a_k(s_1) \left[(k+s_1)(k+s_1-1)R_{N-k} + (k+s_1)P_{N-k} + Q_{N-k}\right]$$
$$(5.58)$$

Notice that the factor $s - s_1$ in (5.56) cancels against the corresponding factor in each of the $\tilde{a}_k(s)$, so the solution for $a_N(s)$ is given in terms of $a_k(s_1)$ rather than $\tilde{a}_k(s_1)$.

For $k > N$, the recursion relations can be written by introducing a new index j given by $n = N + j$. Then, after taking the limit $s \to s_1$, the recursion relations become (Problem 12)

$$\tilde{a}_{N+j}(s_1) j(N+j)$$
$$= - \sum_{i=0}^{j-1} \tilde{a}_{N+i}(s_1) \left[(i+s_2)(i+s_2-1) R_{j-i} + (i+s_2) P_{j-i} + Q_{j-i} \right] \tag{5.59}$$

The reason for writing the recursion relations in this way is seen by setting $s = s_2$ in the recursion relations (5.56):

$$a_n(s_2) n(n+N)$$
$$= - \sum_{k=0}^{n-1} a_k(s_2) \left[(k+s_2)(k+s_2-1) R_{N-k} + (k+s_2) P_{N-k} + Q_{N-k} \right] \tag{5.60}$$

We see that the recursion relations in (5.59) produce a series whose first term is proportional to $x^{s_1+N} = x^{s_2}$. Furthermore, comparing (5.59) with (5.60) shows that the recursion relations for $\tilde{a}_{N+n}$ in (5.59) are identical to those for a_n in (5.60). Therefore, the two series solutions are proportional to one another, and we are confronted with a situation that is analogous to that encountered earlier in this section where the roots of the indicial equation are equal, only in this case with the solutions $y(x; s_2)$ and $\lim_{s \to s_1} (s-s_1) y(x; s)$ being proportional. In fact, an independent solution can be obtained by following a similar strategy to that used for the case of degenerate indicial roots by first substituting $(s-s_1) y(x; s)$ into the differential equation (5.11). Then, by carrying out steps analogous to those leading to (5.41), we obtain

$$\left[R(x) \frac{d^2}{dx^2} + \frac{1}{x} P(x) \frac{d}{dx} + \frac{1}{x^2} Q(x) \right] (s-s_1) y(x; s)$$
$$= a_0 (s-s_1)^2 (s-s_2) x^{s-2} \tag{5.61}$$

We can see immediately that by taking the limit $s \to s_1$ of this equation a solution to the differential equation is obtained. However, as Equations (5.59) and (5.60) show, this solution is proportional to that obtained by setting $s = s_2$. To generate a linearly independent solution, we proceed as in (5.43) by differentiating (5.61) with respect to s and then setting $s = s_1$

to obtain a second independent solution $y_2(x)$ given by:

$$y_2(x) = \frac{d}{ds}\left[(s-s_1)y(x;s)\right]\bigg|_{s=s_1}$$

$$= \ln x \left[\sum_{n=0}^{\infty}(s-s_1)a_n(s)x^{n+s}\right]\bigg|_{s=s_1}$$

$$+ x^{s_1}\sum_{n=0}^{\infty}\frac{d}{ds}\left[(s-s_1)a_n(s)\right]\bigg|_{s=s_1} x^n \qquad (5.62)$$

Example 5.4. Consider the differential equation

$$x\frac{d^2y}{dx^2} + (x-1)\frac{dy}{dx} + y = 0 \qquad (5.63)$$

To apply the method of Frobenius at the origin, we divide the equation by x to obtain the standard form (5.11). This allows us to identify the origin as a regular singular point with

$$R(x) = 1, \qquad P(x) = x - 1, \qquad Q(x) = 1 \qquad (5.64)$$

Thus, the nonvanishing expansion coefficients of these functions are $R_0 = 1$, $P_0 = -1$, $P_1 = 1$ and $Q_1 = 1$. The indicial equation is

$$s(s-1) - s = s(s-2) = 0 \qquad (5.65)$$

whose roots are $s_1 = 0$ and $s_2 = 2$. The recursion relation (5.21) is

$$a_n(n+s)(n+s-2) = -a_{n-1}(n+s) \qquad (5.66)$$

which, upon cancelling the common factor from both sides of the equation, becomes simply

$$a_n(n+s-2) = -a_{n-1} \qquad (5.67)$$

By substituting the larger root $s_1 = 2$ for s into this equation, an expression for the expansion coefficients is readily obtained:

$$a_n = -\frac{1}{n}a_{n-1}$$

$$= (-1)^2 \frac{1}{n(n-1)}a_{n-2}$$

$$\vdots$$

$$= (-1)^n \frac{1}{n!}a_0 \qquad (5.68)$$

The series with these coefficients can be summed to obtain a closed-form expression for the solution:

$$y(x) = a_0 x^2 \sum_{n=0}^{\infty} (-1)^n \frac{x^n}{n!} = a_0 x^2 e^{-x} \tag{5.69}$$

We turn our attention now to finding the series solution corresponding to the smaller root $s=0$. The recursion relation (5.67) becomes

$$a_n(n-2) = -a_{n-1} \tag{5.70}$$

For $n=2$, the left-hand side of this equation vanishes, but the right-hand side does not, so we must construct the second independent solution by following the steps outlined in (5.62). The recursion relation (5.67) is first solved as a function of s to obtain

$$a_n(s) = (-1)^n \frac{a_0}{(s+n-2)(s+n-3)\cdots(s+1)s(s-1)} \tag{5.71}$$

This expression is now used to construct the two parts of the complementary solution in (5.62). The first, which is proportional to $\ln x$, is obtained by evaluating the quantity $s a_n(s)$ in the limit that $s \to 0$. Applying this to (5.71) we obtain

$$\lim_{s \to 0} s a_1(s) = -a_0 \lim_{s \to 0} \frac{s}{s-1} = 0 \tag{5.72a}$$

$$\lim_{s \to 0} s a_n(s) = (-1)^n \lim_{s \to 0} \frac{a_0}{(s+n-2)(s+n-3)\cdots(s+1)(s-1)}$$

$$= (-1)^n \frac{a_0}{(n-2)(n-3)\cdots 2 \cdot 1 \cdot (-1)}$$

$$= (-1)^{n+1} \frac{1}{(n-2)!} a_0 \tag{5.72b}$$

Thus, the first term on the right-hand side of (5.62) yields

$$\ln x \left[a_0 \sum_{n=2}^{\infty} (-1)^{n+1} \frac{x^n}{(n-2)!} \right] = -a_0 \ln x \left[x^2 \sum_{n=0}^{\infty} (-1)^n \frac{x^n}{n!} \right]$$

$$= -a_0 x^2 e^{-x} \ln x \tag{5.73}$$

As expected from the discussion following Equation (5.60), this term is proportional to the solution (5.69) corresponding to the larger root of the indicial equation.

To construct the second term on the right-hand side of (5.62), we follow steps analogous to those in (5.53) and (5.54):

$$\frac{d}{ds}[sa_n(s)] = (-1)^{n+1} a_0 \left(\frac{1}{s+n-2} + \cdots + \frac{1}{s+1} + \frac{1}{s-1} \right)$$

$$\times \frac{1}{(s+n-2)\cdots(s+1)(s-1)} \quad (5.74)$$

Thus, by taking the limit as $s \to 0$ in (5.74), we obtain for a_1, a_2 and a_3 the expressions

$$\lim_{s \to 0} \left\{ \frac{d}{ds}[sa_1(s)] \right\} = a_0 \quad (5.75a)$$

$$\lim_{s \to 0} \left\{ \frac{d}{ds}[sa_2(s)] \right\} = -a_0 \quad (5.75b)$$

$$\lim_{s \to 0} \left\{ \frac{d}{ds}[sa_3(s)] \right\} = 0 \quad (5.75c)$$

For $n \geq 4$, a general expression can be obtained for a_n:

$$\lim_{s \to 0} \left\{ \frac{d}{ds}[sa_n(s)] \right\} = (-1)^n a_0 \left(\frac{1}{2} + \cdots + \frac{1}{n-2} \right) \frac{1}{(n-2)!} \quad (5.76)$$

Combining (5.73) with (5.75) and (5.76), the complementary solution for Equation (5.63) is

$$y_2(x) = -a_0 x^2 e^{-x} \ln x + a_0(1 + x - x^2)$$

$$+ a_0 x^2 \sum_{n=2}^{\infty} (-1)^n \left(\frac{1}{2} + \cdots + \frac{1}{n} \right) \frac{x^n}{n!} \quad (5.77)$$

The procedures used in *Mathematica* to construct the complementary solution in Example 5.3 can be used here with little modification. Here, we will use *Mathematica* to check that the complementary solution is indeed a solution of (5.63). We first construct (5.77):

```
S[n_]:=Sum[1/i,{i,2,n}]

y[x_,n_]:=-x^2 Exp[-x] Log[x] + (1 + x - x^2) +
    x^2 Sum[(-1)^k S[k] (x^k/k!),{k,2,n}]
```

This solution is now substituted into the differential equation and the command Series is used to expand each term in the equation to the specified order of the expansion of the equation as a whole. In the example below, the solution is expanded to 10th order.

```
Series[x D[y[x,10],{x,2}]+(x-1)D[y[x,10],{x,1}]+y[x,10],
    {x,0,10}]
```

$$O[x]^{11}$$

The resulting expression has the value zero with the leading-order corrections being proportional to x^{11}. Thus, the differential equation is satisfied up to the order to which the solution is expanded.

In contrast to the method just used to find the general solution of Equation (5.63), the equation

$$x\frac{d^2y}{dx^2} + (x-1)\frac{dy}{dx} - y = 0 \qquad (5.78)$$

also has the indicial roots $s_1 = 0$ and $s_2 = 2$, but the general solution can be obtained directly from the recursion relations for the root $s_1 = -1$, with the result

$$y(x) = A e^{-x} + B(1-x) \qquad (5.79)$$

where A and B are arbitrary constants. This shows that the coefficients of a differential equation sometimes conspire in a such a way as to bypass the need for applying (5.62) to obtain a second solution of a differential equation near a regular singular point. The solution of this equation is the subject of Problem 23. ∎

5.4 Standard Forms of Equations with Singular Points

Most of the richness of solutions to differential equations in physical applications is due to the existence of singular points. In fact, the influence of singular points is so profound that the number of such points, particularly regular singular points, can be used to classify differential equations and to derive relatively simple standard forms of equations and solutions by placing the singular points at convenient positions. In addition to having a great deal of aesthetic appeal, this approach can be used to identify interrelationships among solutions of different equations based simply on the number and type of the singular points of the governing differential equations. This will be taken up in the next chapter. In this section we will illustrate the procedure for obtaining the standard form of a differential equation using as an example an equation that is regular everywhere except at two regular singular points. Other cases are treated in the problems at the end of the chapter.

We return to the general form of a second-order differential equation in (5.1) and we suppose that there are regular singular points at $x=a$ and $x=b$. The coefficient functions $p(x)$ and $q(x)$ must then have the form

$$p(x) = \frac{f(x)}{(x-a)(x-b)}, \qquad q(x) = \frac{g(x)}{(x-a)^2(x-b)^2} \qquad (5.80)$$

The functions $f(x)$ and $g(x)$ can have no singularities except possibly at infinity. The corresponding differential equation is thus given by

$$\frac{d^2 y}{dx^2} + \frac{f(x)}{(x-a)(x-b)} \frac{dy}{dx} + \frac{g(x)}{(x-a)^2(x-b)^2} y = 0 \qquad (5.81)$$

Under the stated assumptions for $f(x)$ and $g(x)$, these functions have Taylor series expansions that converge over the real line:

$$f(x) = \sum_{k=0}^{\infty} f_k x^k, \qquad g(x) = \sum_{k=0}^{\infty} g_k x^k \qquad (5.82)$$

To insure that the points at $x=a$ and $x=b$ are the *only* singular points of this differential equation, we must check the analytic behavior of the equation at infinity. This is done as before by introducing a new variable ξ through the transformation $\xi = 1/x$, which brings the point at infinity to the origin and vice versa. Under this transformation, the differential equation (5.81) becomes

$$\frac{d^2 y}{d\xi^2} + \left[\frac{2}{\xi} - \frac{f(\xi^{-1})}{(1-a\xi)(1-b\xi)}\right] \frac{dy}{d\xi} + \frac{g(\xi^{-1})}{(1-a\xi)^2(1-b\xi)^2} y = 0 \qquad (5.83)$$

For the origin of the transformed independent variable, i.e. the point at infinity in the origin variable, to be a regular point, the coefficient functions of y and $dy/d\xi$ must be analytic about $\xi = 0$. Thus, taking the limit of these functions as ξ approaches the origin yields the expansions

$$\lim_{\xi \to 0} \left[\frac{2}{\xi} - \frac{f(\xi^{-1})}{(1-a\xi)(1-b\xi)}\right] = \frac{2}{\xi} - f_0 - \frac{f_1}{\xi} - \frac{f_2}{\xi^2} - \cdots \qquad (5.84a)$$

$$\lim_{\xi \to 0} \frac{g(\xi^{-1})}{(1-a\xi)^2(1-b\xi)^2} = g_0 + \frac{g_1}{\xi} + \frac{g_2}{\xi^2} + \cdots \qquad (5.84b)$$

For the expression on the right-hand side of (5.84a) to be regular at $\xi = 0$, the terms involving powers of ξ^{-1} must be eliminated. This can be achieved by setting $f_1 = 2$ and $f_k = 0$ for $k > 2$. The remaining quantity f_0 can take any finite value. For similar reasons, the right-hand side of (5.84b) is regular provided $g_k = 0$ for $k \geq 1$, with g_0 also restricted only in being a finite quantity. Thus, the requirement that the Equation (5.83) be regular at infinity means that $f(x)$ and $g(x)$ must be chosen as

$$f(x) = f_0 + 2x, \qquad g(x) = g_0 \tag{5.85}$$

The general form of a differential equation with regular singular points at $x = a$ and $x = b$ and regular elsewhere is therefore given by

$$\frac{d^2 y}{dx^2} + \frac{f_0 + 2x}{(x-a)(x-b)} \frac{dy}{dx} + \frac{g_0}{(x-a)^2(x-b)^2} y = 0 \tag{5.86}$$

Equation (5.86) can be written in a simpler form by examining the roots of the indicial equation about each of the singular points. Consider first the point at $x = a$. Performing the change of variables $x \to w = x - a$ to bring the point at $x = a$ to the origin, $w = 0$, Equation (5.86) becomes

$$\frac{d^2 y}{dw^2} + \frac{f_0 + 2(w+a)}{w(w+a-b)} \frac{dy}{dw} + \frac{g_0}{w^2(w+a-b)^2} y = 0 \tag{5.87}$$

To determine the roots of the indicial equation (5.19) about $w = 0$, only the quantities R_0, P_0 and Q_0 are required. Comparing (5.87) with our standard form in Equation (5.11), we can make the identifications

$$R_0 = 1, \qquad P_0 = \frac{f_0 + 2a}{a - b}, \qquad Q_0 = \frac{g_0}{(a-b)^2} \tag{5.88}$$

Using the notation α_1 and α_2 for the roots of the indicial equation, we obtain

$$s^2 + \frac{f_0 + a + b}{a - b} s + \frac{g_0}{(a-b)^2} = (s - \alpha_1)(s - \alpha_2) \tag{5.89}$$

which implies

$$\alpha_1 + \alpha_2 = -\frac{f_0 + a + b}{a - b}, \qquad \alpha_1 \alpha_2 = \frac{g_0}{(a-b)^2} \tag{5.90}$$

The same procedure can be applied to determine the indices β_1 and β_2 about the point at $x = b$, with the result

$$\beta_1 + \beta_2 = -\frac{f_0 + a + b}{b - a}, \qquad \beta_1 \beta_2 = \frac{g_0}{(b-a)^2} \tag{5.91}$$

Comparing (5.90) and (5.91), we deduce the following relationships between the two sets of exponents:

$$\alpha_1 + \alpha_2 = -(\beta_1 + \beta_2), \qquad \alpha_1 \alpha_2 = \beta_1 \beta_2 \tag{5.92}$$

Using these equations, we can determine two of the indices in terms of the other two. In particular, solving for β_1 and β_2 in terms of α_1 and α_2 we obtain

$$\beta_1 = -\alpha_1, \qquad \beta_2 = -\alpha_2 \tag{5.93}$$

Thus, the indices of the points $x=a$ and $x=b$ are simply the negatives of one another. Equations (5.90) or (5.91) can now be used to express f_0 and g_0 in terms of the indices of the two regular singular points of the equation. Doing so, (5.87) becomes

$$\frac{d^2 y}{dx^2} + \frac{2x - (\alpha_1 + \alpha_2)(a-b) - (a+b)}{(x-a)(x-b)} \frac{dy}{dx} + \frac{\alpha_1 \alpha_2 (a-b)^2}{(x-a)^2 (x-b)^2} y = 0 \tag{5.94}$$

The final step is to place the points at $x=a$ and $x=b$ at 'convenient' positions to simplify as much as possible the form of the equation. For this equation, these points are at the origin and at infinity. Thus, setting $a=0$ and taking the limit $b \to \infty$, we find that Equation (5.94) reduces to

$$\frac{d^2 y}{dx^2} + \frac{1 - \alpha_1 - \alpha_2}{x} \frac{dy}{dx} + \frac{\alpha_1 \alpha_2}{x^2} y = 0 \tag{5.95}$$

This equation is regular everywhere except at the origin and at infinity, which are regular singular points, and is the standard form for an equation with two regular singular points. The indices of the origin are α_1 and α_2 and those of infinity are $-\alpha_1$ and $-\alpha_2$. The general solution of (5.95) is easily obtained as

$$y(x) = c_1 x^{\alpha_1} + c_2 x^{\alpha_2} \tag{5.96}$$

Notice that this solution is expressed in terms of elementary functions. For more complicated differential equations, such as those involving a greater number of regular singular points, or possibly even an irregular singular point, this is no longer true, and the solutions are given in terms of infinite series. Several well-known examples of such equations are considered in the problems at the end of this chapter and in the next chapter.

Further Reading

The discussion in this chapter has provided a general introduction to the basic principles of the series solution of ordinary differential equations. An introductory treatment is given by Boyce and DiPrima (1977). More advanced treatments may be found in the books by Morse and Feshbach (1953), Ince (1956), Hildebrand (1962) and Whittaker and Watson (1963). Morse and Feshbach (1953) and Whittaker and Watson (1963) also provide comprehensive discussions on the classification of equations with singular points. Discussions of the method of Frobenius from the standpoint of complex analysis, including behavior near singular points and analytic continuation, are given by Ahlfors (1966), Birkhoff and Rota (1989) and Hille (1976).

References

Ahlfors L. V. (1966). *Complex Analysis* 2nd edn. New York: McGraw-Hill.
Birkhoff G. and Rota G.-C. (1989). *Ordinary Differential Equations* 4th edn. New York: Wiley.
Boyce W. E. and DiPrima R. C. (1977). *Elementary Differential Equations and Boundary Value Problems* 3rd edn. New York: Wiley.
Hildebrand F. B. (1962). *Advanced Calculus for Applications*. Englewood Cliffs NJ: Prentice-Hall.
Hille E. (1976). *Ordinary Differential Equations in the Complex Domain*. New York: Wiley.
Ince E. L. (1956). *Ordinary Differential Equations*. New York: Dover.
Knopp K. (1956). *Infinite Sequences and Series*. New York: Dover.
Morse P. M. and Feshbach H. (1953). *Methods of Theoretical Physics* Vol. 1. New York: McGraw-Hill.
Whittaker E. T. and Watson G .N. (1963). *A Course of Modern Analysis* 4th edn. Cambridge: Cambridge University Press.

Problems

1. Apply the Frobenius about the origin to obtain formally the power series solution of equation (5.7):

$$\frac{d^2y}{dx^2} - \frac{2}{x}\frac{dy}{dx} - \frac{y}{x^4} = 0$$

Expand the solution as $y(x) = x^s \sum a_n x^n$, substitute into (5.7), to obtain

$$x^{s-2}\left\{a_0 x^{-2} + a_1 x^{-1} + \sum_{n=0}^{\infty}\left[a_{n+2} - (n+s)(n+s-3)a_n\right]x^n\right\} = 0$$

Show that the first two terms in this equation require that $a_0 = 0$ and $a_1 = 0$. For the remaining expansion coefficients, obtain the recursion relation

$$a_{n+2} = (n+s)(n+s-3)a_n$$

Deduce that the only power series solution satisfying the differential order by order is the trivial solution. Compare this result with that obtained by attempting to expand the function $\exp(-1/x)$ in a Taylor series about the origin.

2. From the identifications in (5.13),

$$xp(x) = \frac{P(x)}{R(x)}, \qquad x^2 q(x) = \frac{Q(x)}{R(x)}$$

and the expansions for R, P and Q in (5.12), with $R_0 \neq 0$, show explicitly that the origin is either a regular point or a regular singular point of the equation. Identify the conditions that must be satisfied by these series for the origin to be a regular point.

3. By introducing the quantity $c_n = a_{2n}$ write the recursion relations (5.33) as

$$\frac{c_n}{c_{n-1}} = \frac{2n-3}{2n-1}$$

By writing the right-hand side of this expression as

$$1 - \frac{1}{n} - \frac{2}{n(2n-1)}$$

deduce from Gauss' test that the series $y(x) = \sum a_n x^n$, with a_n obtained by solving (5.33), diverges at $x = \pm 1$.

4. Suppose that $u_1(x)$ is a solution of

$$R(x)\frac{d^2y}{dx^2} + \frac{1}{x}P(x)\frac{dy}{dx} + \frac{1}{x^2}Q(x)y = 0$$

Show that $u_2 = u_1 v$ is also a solution of this equation provided v is chosen as

$$v(x) = \int^x \frac{1}{u_1^2(s)} \exp\left[-\int^s \frac{P(t)}{tR(t)} dt\right] ds$$

By calculating the Wronskian between u_1 and u_2, deduce that the general solution of the differential equation is

$$y(x) = Au_1(x) + Bu_2(x)$$

where A and B are constants.

5. Beginning with the solution $u_1(x) = x$ of Equation (5.29),

$$(1-x^2)\frac{d^2y}{dx^2} - 2x\frac{dy}{dx} + 2y = 0$$

use the result of Problem 4 to obtain a second independent solution:

$$u_2(x) = \tfrac{1}{2}x \ln\left(\frac{1+x}{1-x}\right) - 1$$

Using *Mathematica*, or otherwise, verify that this expression does indeed solve the differential equation. Then, by using the command Series, expand $u_2(x)$ in a Taylor series about the origin and deduce that

$$\sum_{k=0}^{\infty} \frac{x^{2k}}{2k-1} = 1 - \tfrac{1}{2}x \ln\left(\frac{1+x}{1-x}\right)$$

The argument of the logarithm shows clearly that this solution diverges at $x = \pm 1$ and that this solution is not valid for $|x| \geq 1$.

6. By differentiating the following function term by term,

$$y(x; s) = x^s \sum_{n=0}^{\infty} a_n(s) x^n$$

show that the operations of differentiation with respect to x and with respect to s in the expressions which are required for (5.42) commute:

$$\frac{\partial^3 y}{\partial x^2 \partial s} = \frac{\partial^3 y}{\partial s \partial x^2}, \quad \frac{\partial^2 y}{\partial x \partial s} = \frac{\partial^2 y}{\partial s \partial x}$$

7. Identify the singular points of the following differential equations and determine the general solutions of each of the equations in a power series

expansion about the origin. In each case determine the radius of convergence of the series.

(a) $\dfrac{d^2y}{dx^2} + y = 0$

(b) $\dfrac{d^2y}{dx^2} + xy = 0$

(c) $\dfrac{d^2y}{dx^2} + x^2 y = 0$

(d) $\dfrac{d^2y}{dx^2} - x\dfrac{dy}{dx} + y = 0$

(e) $\dfrac{d^2y}{dx^2} - x\dfrac{dy}{dx} + x^2 y = 0$

(f) $(1+x^2)\dfrac{d^2y}{dx^2} - 2x\dfrac{dy}{dx} + 2y = 0$

8. For each of the equations in Problem 7, use *Mathematica* to (i) show that the solutions are independent by calculating their Wronskian, as in Section 1.4, (ii) verify that the series obtained do satisfy the differential equation (to some prescribed order) and (iii) plot the two solutions.

9. Obtain the general solution of (5.29) in the region $|x|>1$ by first transforming the point at infinity to the origin by introducing the variable $\xi = 1/x$. Show that in terms of ξ (5.29) becomes

$$(\xi^2 - 1)\dfrac{d^2y}{d\xi^2} + 2\xi\dfrac{dy}{d\xi} + \dfrac{2}{\xi^2}y = 0$$

Apply the method of Frobenius to obtain the general solution of this equation in an expansion about $\xi = 0$. In deriving the general solution explain why the auxiliary methods discussed in Section 5.3 are not required, although the roots of the indicial equation differ by an integer. By comparing equations (5.37) and (5.38), show that the general solution for $|x|>1$ can be expressed in the original variable x as

$$y(x) = Ax - B + \tfrac{1}{2} Bx \ln\left(\dfrac{x+1}{x-1}\right)$$

where A and B are constants.

10. Identify and characterize the singular points of the differential equation

$$x^2 \dfrac{d^2y}{dx^2} + (3x - 1)\dfrac{dy}{dx} + y = 0$$

In particular, show that the origin is an *irregular* singular point and the point at infinity is a regular singular point. By applying the method of Frobenius to obtain a solution in an expansion about the origin, obtain the formal expansion

$$y(x) = \sum_{k=0}^{\infty} k! x^k$$

Show that this series has a radius of convergence of zero. This shows that the application of the Frobenius method at an irregular singular point may yield an expression that satisfies the recursion relations, but the radius of convergence of the series may vanish, due to the presence of the essential singularity.

11. Consider again the differential equation in Problem 10. By transforming the point at infinity to the origin, apply the method of Frobenius to obtain the general solution. Express your solution in terms of the original variable and show that the solution to the original differential equation is

$$y(x) = A x^{-1} e^{-1/x} + B\left[x^{-1} e^{-1/x} \ln(1/x) - x^{-1} \sum_{k=1}^{\infty} \frac{(-1)^k}{k!}\left(1 + \frac{1}{2} + \cdots + \frac{1}{k}\right)\frac{1}{x^k}\right]$$

where A and B are constants.

12. Beginning with the recursion relations (5.21), multiply both sides by $s - s_1$ to obtain

$$\tilde{a}_n(n + s - s_1)(n + s - s_2)$$
$$= -\sum_{k=0}^{n-1} \tilde{a}_k\left[(k+s)(k+s-1)R_{n-k} + (k+s)P_{n-k} + Q_{n-k}\right]$$

where $\tilde{a}_k = (s - s_1)a_k$. By introducing the index j by $n = N + j$, observing that $\tilde{a}_n = 0$ for $n < N$, show that this recursion relation can be written as

$$\tilde{a}_{N+j}(N + j + s - s_1)(N + j + s - s_2)$$
$$= -\sum_{k=N}^{N+j-1} \tilde{a}_k\left[(k+s)(k+s-1)R_{N+j-k} + (k+s)P_{N+j-k} + Q_{N+j-k}\right]$$

Then, by making the appropriate change in the summation index in this equation and then taking the limit $s \to s_1$, obtain

$$\tilde{a}_{N+j} j(N+j)$$
$$= -\sum_{i=0}^{j-1} \tilde{a}_{N+i}\left[(i+s_2)(i+s_2-1)R_{j-i} + (i+s_2)P_{j-i} + Q_{j-i}\right]$$

which is the same as the recursion relations in (5.59).

13. The procedure used in Section 5.4 to derive the standard form of a differential equation with two regular singular points can be extended to the case of an equation with three regular singular points. Beginning with the general equation (5.1) with $p(x)$ and $q(x)$ given by (5.80), show that requiring the points at $x=0$, $x=1$, and $x=\infty$ to be regular singular points means that the equation must be of the form

$$\frac{d^2y}{dx^2} + \frac{f_0 + f_1 x}{x(x-1)}\frac{dy}{dx} + \frac{g_0 + g_1 x + g_2 x^2}{x^2(x-1)^2}y = 0$$

where the quantities f_i and g_i are the expansion coefficients for the functions $f(x)$ and $g(x)$, as in Equation (5.82).

Let the indices of $x=0$ be signified by α_1 and α_2, those of $x=1$ by β_1 and β_2, and those of $x=\infty$ by γ_1 and γ_2. Show that this differential equation can be expressed in terms of these indices as

$$\frac{d^2y}{dx^2} + \left(\frac{1-\alpha_1-\alpha_2}{x} + \frac{1-\beta_1-\beta_2}{x-1}\right)\frac{dy}{dx} - \left(\frac{\alpha_1\alpha_2}{x} - \frac{\beta_1\beta_2}{x-1} - \gamma_1\gamma_2\right)\frac{y}{x(x-1)} = 0$$

and that the indices are related by the 'sum rule'

$$\alpha_1 + \alpha_2 + \beta_1 + \beta_2 + \gamma_1 + \gamma_2 = 1$$

14. The equation derived in Problem 13 can be simplified by making the transformation

$$y(x) = x^{\alpha_1}(x-1)^{\beta_1}F(x)$$

Determine the behavior of F near the three regular singular points of the differential equation. Hence, deduce that the differential equation satisfied by F is

$$\frac{d^2F}{dx^2} + \left(\frac{1+\alpha_1-\alpha_2}{x} + \frac{1+\beta_1-\beta_2}{x-1}\right)\frac{dF}{dx} + \frac{(\alpha_1+\beta_1+\gamma_1)(\alpha_1+\beta_1+\gamma_2)}{x(x-1)}F = 0$$

and obtain the modified sum rule that relates these indices. By direct substitution of the transformation relating y and F into the equation for y, use *Mathematica* to verify that F satisfies this differential equation.

There are four unknown quantities in this differential equation: the differences $\alpha_1-\alpha_2$ and $\beta_1-\beta_2$, and the indices at infinity, γ_1 and γ_2. Since these quantities are constrained by a sum rule, only three of them are independent, so only three quantities are required to determine the indices. By convention, the three independent quantities are taken as the indices at infinity, and assigned the values α and β,

$$\gamma_1 = \alpha, \qquad \gamma_2 = \beta$$

and the value of the index at the origin, which is assigned the value $1-\gamma$:

$$\alpha_1 - \alpha_2 = 1 - \gamma$$

Series Solutions of Ordinary Differential Equations 201

The value of the index at $x=1$ is therefore $\beta_1-\beta_2=\gamma-\alpha-\beta$. By substituting these quantities into the differential equation for F and using the sum rule for the indices, obtain

$$x(x-1)\frac{d^2 F}{dx^2} + [(\alpha+\beta+1)x - \gamma]\frac{dF}{dx} + \alpha\beta F = 0$$

This is the **hypergeometric equation**, which is the standard form of a differential equation with three regular singular points.

15. By using the notation

$$(\alpha)_k = \alpha(\alpha+1)\cdots(\alpha+k-1)$$

together with $(\alpha)_0 = 1$, show that a regular solution of the hypergeometric equation derived in Problem 14 can be obtained by applying the method of Frobenius in an expansion about the origin, with the result given by

$$F(\alpha,\beta;\gamma;x) \equiv \sum_{k=0}^{\infty} \frac{(\alpha)_k (\beta)_k}{(\gamma)_k} \frac{x^k}{k!}$$

where the solution is standardized by the normalization $F(\alpha,\beta;\gamma;0)=1$.

Show that if γ is not an integer, then the second independent solution obtained from the Frobenius method with the second root $m=1-\gamma$ is

$$x^{1-\gamma} F(\alpha+1-\gamma;\beta+1-\gamma;2-\gamma;x)$$

The notation $_2F_1(\alpha,\beta;\gamma;x)$ is also frequently used for this hypergeometric function.

16. Use the series representation of the hypergeometric function derived in Problem 15 to obtain the following special cases:

$$F(\alpha,\beta;\beta;x) = (1-x)^\alpha$$

$$F(1,1;2;-x) = x^{-1}\ln(1+x)$$

$$\lim_{\beta\to\infty} F(\alpha,\beta;\alpha;x/\beta) = e^x$$

17. The behavior of solutions to a differential equation in response to the merging, or *confluence*, of two singular points is an important feature for several types of special functions. Consider the general form (5.86) for an equation with two regular singular points at $x=a$ and $x=b$. If the two singular points are merged by setting $b=a$, then the equation becomes

$$\frac{d^2 y}{dx^2} + \frac{f_0+2x}{(x-a)^2}\frac{dy}{dx} + \frac{g_0}{(x-a)^4} y = 0$$

Show that the solution to this equation is

$$y(x) = A e^{k^+/(x-a)} + B e^{k^-/(x-a)}$$

where A and B are constants and the quantities $k^\pm$ are the two solutions of the quadratic equation

$$k^{\pm 2} - (f_0 + 2a)k^\pm + g_0 = 0$$

Notice that in the solution of the 'confluent' equation the regular singular point at $x=a$ has become an irregular singular point, where the solution now has an essential singularity. Thus, this solution has no power series expansion about $x=a$.

18. The transformation

$$\xi = \frac{x-a}{x-b}$$

moves the point $x = a$ to the origin and the point $x = b$ to infinity. Use *Mathematica* to show that introducing ξ as the independent variable in (5.94) transforms this equation to

$$\frac{d^2 y}{d\xi^2} + \frac{1-\alpha_1-\alpha_2}{\xi}\frac{dy}{d\xi} + \frac{\alpha_1 \alpha_2}{\xi^2} y = 0$$

which is precisely Equation (5.95). Hence, deduce that the general solution to (5.94) is

$$y(x) = A\left(\frac{x-a}{x-b}\right)^{\alpha_1} + B\left(\frac{x-a}{x-b}\right)^{\alpha_2}$$

19. An alternative way of deriving the result of Problem 17 is to work directly with the solution obtained in Problem 18. First set $b = a + \epsilon$ in this solution, where ϵ is a small quantity. Then substitute this expression into Equation (5.86) to obtain the solution in the form

$$y(x) = A\left(\frac{x-a}{x-a-\epsilon}\right)^{k^+(\epsilon)/\epsilon} + B\left(\frac{x-a}{x-a-\epsilon}\right)^{k^-(\epsilon)/\epsilon}$$

where

$$k^\pm(\epsilon) = \tfrac{1}{2}(f_0 + 2a + \epsilon) \pm \tfrac{1}{2}\sqrt{(f_0 + 2a + \epsilon)^2 - 4g_0}$$

Then, by using

$$e^x = \lim_{\epsilon \to 0}(1+\epsilon x)^{1/\epsilon}$$

take the limit $\epsilon \to 0$ in the solution to obtain

$$y(x) = A e^{k^+/(x-a)} + B e^{k^-/(x-a)}$$

where $k^\pm$ are defined as in Problem 17.

20. Beginning with the hypergeometric equation derived in Problem 14, make the change of variable $\xi = x\beta$ to obtain

$$\xi\left(\frac{\xi}{\beta} - 1\right)\frac{d^2y}{d\xi^2} + \left(\xi + \frac{\alpha+1}{\beta}\xi - \gamma\right)\frac{dy}{d\xi} + \alpha y = 0$$

Identify and characterize the singular points of this equation. By taking the limit $\beta \to \infty$ show that this equation becomes

$$\xi\frac{d^2y}{d\xi^2} + (\gamma - \xi)\frac{dy}{d\xi} - \alpha y = 0$$

This is the **confluent hypergeometric function**. Show that this equation has a regular singular point at $\xi = 0$ but $\xi = \infty$ has become an *irregular* singular point.

21. Obtain a series expansion about the origin for the regular solution of the confluent hypergeometric function derived in Problem 20. This solution is often denoted by $_1F_1(\alpha; \beta; x)$ or by $M(\alpha; \beta; x)$.

If β is not an integer, show that a second independent solution is given by

$$x^{1-\beta}M(\alpha - \beta + 1; 2 - \beta; x)$$

22. Deduce from the limits taken in Problem 20 that the hypergeometric function $F(\alpha, \beta; \gamma; x)$ and the confluent hypergeometric function $M(\alpha; \gamma; x)$ are related by

$$M(\alpha; \gamma; x) = \lim_{\beta \to \infty} F(\alpha, \beta; \gamma; x/\beta)$$

Verify this relationship from the series solutions for the hypergeometric function in Problem 14 and that for the confluent hypergeometric function in Problem 20.

Either from the series representation of $M(\alpha; \gamma; x)$ or by combining the results of this problem with that of Problem 15 show that

$$M(\alpha; \alpha; x) = e^x$$

23. Identify and classify the singular points of the differential equation

$$x\frac{d^2y}{dx^2} + (x-1)\frac{dy}{dx} - y = 0$$

Obtain the general solution of this equation in an expansion around the origin directly from the method of Frobenius. In particular, show that the roots of the indicial equation are $s_1 = 0$ and $s_2 = 2$ and that the recursion relations are

$$a_n(n+s)(n+s-2) = -(n+s-2)a_{n-1}$$

Thus, since the problematic factor $n+s-2$ cancels from both sides of the equation, deduce that the general solution can be obtained by solving these recursion relations by substituting in turn the two roots of the indicial equation. Solve the recursion relations and sum the two series to obtain the general solution in (5.79).

24. Use the method of Frobenius to obtain the general solutions of the following differential equations and compare your results with the solution in (5.96) to the equation in (5.95).

(a) $\dfrac{d^2y}{dx^2} + \dfrac{1-2\alpha}{x}\dfrac{dy}{dx} + \dfrac{\alpha^2}{x^2}y = 0$

(b) $\dfrac{d^2y}{dx^2} + \dfrac{1}{x}\dfrac{dy}{dx} - \dfrac{1}{x^2}y = 0$

(c) $\dfrac{d^2y}{dx^2} - \dfrac{4}{x}\dfrac{dy}{dx} - \dfrac{4}{x^2}y = 0$

25. Show that if $y(x)$ is a solution of

$$\dfrac{d^2y}{dx^2} + p(x)\dfrac{dy}{dx} + q(x)y = 0$$

then the function $u(x) = y'(x)/y(x)$ (where the prime signifies differentiation with respect to x) satisfies the **Ricatti equation**

$$\dfrac{du}{dx} + p(x)u + u^2 + q(x) = 0$$

Thus, deduce that if $y_1(x)$ and $y_2(x)$ are linearly independent solutions of the first equation, then the general solution to Ricatti's equation is given by

$$u(x) = \dfrac{Ay_1'(x) + By_2'(x)}{Ay_1(x) + By_2(x)}$$

where A and B are arbitrary constants. Notice that there is only a single arbitrary constant in this expression: the ratio A/B or B/A.

26. With the result of Problem 25, use *Mathematica* to construct a series representation of the general solution of the first-order nonlinear equation

$$\dfrac{dy}{dx} + y^2 = x^2$$

by applying the Frobenius method to the appropriate linear second-order equation. Show that up to third order in x, the solution is given by

$$u(x) = \dfrac{a_1}{a_0} - \dfrac{a_1^2}{a_0^2}x + \dfrac{a_1^3}{a_0^3}x^2 + \left(\dfrac{1}{3} - \dfrac{a_1^4}{a_0^4}\right)x^3 + \cdots$$

where a_0 and a_1 are the arbitrary constants in the general solution of the linear equation. This clearly shows the presence of the single arbitrary constant a_1/a_0 in the solution of the Ricatti equation. Verify that the series representation of the general solution does indeed solve the Ricatti equation order by order in x.

27. Identify the singular points of the following differential equations and determine the general solutions of each of the equations in a power series expansion about the origin. In each case determine the radius of convergence of the series. If possible, express the series in terms of elementary functions.

(a) $\dfrac{d^2y}{dx^2} + \dfrac{1}{x}\dfrac{dy}{dx} - y = 0$

(b) $x\dfrac{d^2y}{dx^2} + 2\dfrac{dy}{dx} + xy = 0$

(c) $x\dfrac{d^2y}{dx^2} + \dfrac{dy}{dx} + xy = 0$

(d) $x\dfrac{d^2y}{dx^2} + (1-x)\dfrac{dy}{dx} + 2y = 0$

(e) $x\dfrac{d^2y}{dx^2} + (1-x)\dfrac{dy}{dx} + \tfrac{1}{2}y = 0$

(f) $x(1-x)\dfrac{d^2y}{dx^2} - 3\dfrac{dy}{dx} + 2y = 0$

Chapter 6

Special Functions and Orthogonal Polynomials

Fourier series are the simplest examples of eigenfunction expansions arising from the solution of a Sturm–Liouville problem. However, the form of the Liouville equation admits many more different types of eigenvalue equation. The solutions of these equations can be obtained with the methods of the preceding chapter, and the eigenfunctions of these equations take the form either of an infinite series, a finite series or sometimes a combination of elementary functions. These eigenfunctions are known as special functions and orthogonal polynomials.

In this chapter we will derive some of the orthogonal polynomials and special functions that are commonly encountered in mathematical physics, applied mathematics and engineering. In each case, we will begin with a physical problem whose solution gives rise to the particular eigenfunctions, apply the method of Frobenius to obtain expressions for the eigenfunctions, and discuss the consequences of some of their properties for the physical problem either in the main text or in the problems at the end of the chapter. The equations solved in the main text have been chosen to be representative of those encountered in applying the method of Frobenius to Sturm–Liouville problems. Once these solutions are understood, other equations can be solved with little additional difficulty. The special functions and orthogonal polynomials derived in this chapter are the Hermite polynomials, the Legendre polynomials, Legendre functions and spherical harmonics, and the Bessel functions of the first and second kinds. Other special functions are discussed in the problems at the end of the chapter and in Chapter 8.

As an alternative to the direct solution of differential equations by the method of Frobenius, a concise representation of special functions and orthogonal polynomials can be obtained in terms of generating functions, which is a relatively simple function that can be expressed as an infinite series of a particular set of eigenfunctions. Generating functions are useful for obtaining recursion relations, i.e. linear relationships among eigenfunctions corresponding to neighboring eigenvalues, for normalization constants, and for various other types of interrelationship. Generating functions arise most naturally in terms of integral representations of solutions to differential equations, as will be discussed in Chapter 8, but can sometimes be derived by physical or heuristic arguments. Examples of generating functions obtained from this latter approach, and their uses, will be given in the main text and in the problems at the end of the chapter.

6.1 Hermite Polynomials

We begin our discussion of orthogonal polynomials with the Hermite polynomials. Hermite polynomials are encountered in quantum mechanics, probability theory and statistical mechanics, and in solutions of Laplace's equation in parabolic coordinates. In quantum mechanics, these polynomials appear in the eigenfunctions of the Schrödinger equation for the harmonic oscillator, which is the example we will consider here. An example from statistical mechanics is given in Problem 1.

A harmonic oscillator consists of a particle of mass m attached to a linear spring with stiffness k. The classical expression for the energy of a harmonic oscillator as a function of the velocity v and displacement x is the sum of the kinetic energy, $\frac{1}{2}mv^2$, and the potential energy, $\frac{1}{2}kx^2$. By writing the kinetic energy in terms of the linear momentun of the mass, the time-independent Schrödinger equation for the harmonic oscillator is obtained from the classical energy as (Pauling and Wilson, 1935)

$$-\frac{\hbar^2}{2m}\frac{\mathrm{d}^2\psi}{\mathrm{d}x^2} + \tfrac{1}{2}kx^2\psi = E\psi \qquad (6.1)$$

where $\hbar$ is Planck's constant and E is the energy of the oscillator. This equation can be written in a somewhat simpler form by introducing the quantities

$$a = \frac{2mE}{\hbar^2}, \qquad b = \frac{m\omega_0}{\hbar} \qquad (6.2)$$

where $\omega_0 = \sqrt{k/m}$ is the natural frequency of the oscillator. After some rearrangement, Equation (6.1) becomes

$$\frac{\mathrm{d}^2\psi}{\mathrm{d}x^2} + (a - b^2 x^2)\psi = 0 \qquad (6.3)$$

The origin is readily identified as a regular point of this equation, while the transformation $x \to 1/x$ reveals the point at infinity to be an irregular singular point. Therefore, while the series solution of (6.3) is expected to converge for any finite value of x, the convergence of this series must be tested explicitly in the limit that x approaches infinity to determine if the solution remains finite for all values of x.

Equation (6.3) is already in the standard notation of (5.11), with $R=1$, $P=0$ and $Q=-b^2x^4$, but is not a very convenient starting point for a series solution for the following reason. Since the origin is a regular point of the equation, expanding the solution as $\psi(x) = x^s \sum_n a_n x^n$, substituting into (6.3), and choosing the root $s=0$ of the indicial equation, generates a series with two arbitrary constants, a_0 and a_1. The expansion coefficients a_n are obtained by solving the recursion relation:

$$a_n = -\frac{a a_{n-2} + b a_{n-4}}{n(n-1)} \tag{6.4}$$

Since this equation couples a given a_n to two coefficients with lower indices, a general solution for the a_n is more difficult to obtain, and the convergence of the series is more difficult to assess than if only one term is present on the right-hand side of (6.4). Even using *Mathematica* as in Examples 5.2 and 5.3 does not allow the conditions for convergence of the series to be easily identified, since an expression for large-order terms is required. To overcome this difficulty, we can appeal to the behavior of the solution near the singular points of the equation to motivate a transformation of (6.3) into a more convenient form for applying the Frobenius method. Since Equation (6.3) has irregular singular points at $x = \pm\infty$, we will examine the solutions of this equation as $x \to \pm\infty$. In this limit, the equation becomes

$$\frac{d^2\psi}{dx^2} - b^2 x^2 \psi = 0 \tag{6.5}$$

The solution to (6.5) that remains finite as x becomes large is, to within a proportionality constant, given by (Problem 4)

$$\psi(x) = \exp\left(-\tfrac{1}{2}bx^2\right) \tag{6.6}$$

We now incorporate the asymptotic behavior obtained in (6.6) to introduce a reduced quantity $y(x)$ through the transformation

$$\psi(x) = y(x) \exp\left(-\tfrac{1}{2}bx^2\right) \tag{6.7}$$

By substituting (6.7) into (6.3), the equation for y is found to be

$$\frac{d^2y}{dx^2} - 2xb\frac{dy}{dx} + (a-b)y = 0 \tag{6.8}$$

Upon making the transformation $\xi = x\sqrt{b}$, Equation (6.8) becomes

$$\frac{d^2 y}{d\xi^2} - 2\xi \frac{dy}{d\xi} + 2\lambda y = 0 \qquad (6.9)$$

where $2\lambda + 1 = a/b$ (the factor of 2 is introduced simply for convenience in writing and solving the recursion relations). Equation (6.9) is **Hermite's equation**.

To apply the method of Frobenius to obtain the solutions of Hermite's equation in an expansion about the origin, we again observe that the origin is a regular point of the equation. Thus, we expect to obtain the general solution directly from the Frobenius method without the need for the auxiliary methods described in Section 5.3 In the standard notation of Equation (5.11), we identify

$$R(\xi) = 1, \quad P(\xi) = -2\xi^2, \quad Q(\xi) = 2\lambda\xi^2 \qquad (6.10)$$

from which we have $R_0 = 1$, $P_2 = -2$ and $Q_2 = 2\lambda$, with all other expansion coefficients vanishing. Thus, by expanding the solution as $y(\xi) = \xi^s \sum a_n \xi^n$ and substituting into Hermite's equation, we obtain the indicial equation as

$$a_0 s(s-1) = 0 \qquad (6.11)$$

which yields the roots $s = 0$ and $s = 1$, as expected. The recursion relation for $n = 1$ is

$$a_1 s(s+1) = 0 \qquad (6.12)$$

Thus, by choosing $s = 0$, both (6.11) and (6.12) can be solved with arbitrary a_0 and a_1, and the general solution of Hermite's equation will be obtained by solving the recursion relations with $s = 0$. The recursion relations for $n > 1$ are

$$a_n (n+s)(n+s-1) = -a_{n-2}[-2(n-2+s) + 2\lambda] \qquad (6.13)$$

The form of these equations is seen to be much simpler than that in (6.4). Each even-order coefficient is coupled only to one other even-order coefficient and each odd-order coefficient is coupled only to one other odd-order coefficient. Thus, the general solution will take the form of two series, one comprised only of even-order terms and proportional to a_0, and the other comprised only of odd-order terms and proportional to a_1. This form also allows the convergence of the series solution to be much more easily assessed, which we will see below after we solve the recursion relations.

To solve the recursion relations we set $s = 0$ in (6.13) and solve for a_n in terms of a_{n-2}:

$$a_n = -2 \frac{\lambda - n + 2}{n(n-1)} a_{n-2} \qquad (6.14)$$

This equation can be solved by the method of recursive substitution used in Examples 5.2 and 5.3. We obtain for the general even-order expansion coefficient a_{2k} the expression,

$$a_{2k} = \frac{(-1)^k 2^k}{(2k)!} \lambda(\lambda - 2) \cdots (\lambda - 2k + 2) a_0 \qquad (6.15a)$$

Similarly, for the general odd-order expansion coefficient a_{2k+1}, we obtain

$$a_{2k+1} = \frac{(-1)^k 2^k}{(2k+1)!} (\lambda - 1)(\lambda - 3) \cdots (\lambda - 2k + 1) a_1 \qquad (6.15b)$$

To investigate the convergence of the power series with coefficients given by (6.15), we perform the ratio test using the recursion relation (6.14):

$$\lim_{n \to \infty} \left| \frac{a_n}{a_{n-2}} \right| = \lim_{n \to \infty} \frac{\lambda - 2(n-2)}{n(n-1)} = \lim_{n \to \infty} \frac{2}{n} = 0 \qquad (6.16)$$

The radius of convergence of the general solution is thus seen to be the entire real line. However, we must look more closely to see whether the series converges quickly enough to insure that (6.7) remains finite as $x \to \pm\infty$, where both the original equation and Hermite's equation have irregular singular points. To address this question, we first observe that the limit in (6.16) shows that for large order the recursion relations (6.14) for even-order and odd-order terms become

$$a_{2k} = \frac{1}{k} a_{2k-2}, \qquad a_{2k+1} = \frac{1}{k} a_{2k-1} \qquad (6.17)$$

Solving these recursion relations for the expansion coefficients, the asymptotic form of the solution of Hermite's equation (6.9) for large values of ξ is found to be

$$\lim_{\xi \to \infty} y(\xi) = a_0 \sum_{k=0}^{\infty} \frac{\xi^{2k}}{k!} + a_1 \sum_{k=0}^{\infty} \frac{\xi^{2k+1}}{k!}$$

$$= a_0 \exp(\xi^2) + a_1 \xi \exp(\xi^2) \qquad (6.18)$$

By substituting these solutions into (6.7), we see that the asymptotic form of the solutions generated by (6.14) is

$$\lim_{\xi \to \infty} \psi(\xi) = \lim_{n \to \infty} y(\xi) \exp\left(-\tfrac{1}{2}\xi^2\right) = (a_0 + a_1 x) \exp\left(\tfrac{1}{2}\xi^2\right) \qquad (6.19)$$

so the eigenfunctions ψ of (6.3) become infinite as $\xi \to \infty$. Thus, even though the solutions of Hermite's equation obtained from the Frobenius method converge over the entire real line, normalization cannot be carried out, since (6.7) indicates that the weight function for Equation (6.3) is $\exp(-\xi^2)$.

This situation can be remedied by observing that

$$\lim_{\xi \to \infty} \xi^m e^{-\xi^2} = 0 \tag{6.20}$$

for any integer m, which can be derived by repeated application of L'Hospital's rule. Equation (6.20) suggests that a set of finite and normalizable eigenfunctions of (6.3) can be obtained by forcing the infinite series of terms generated by the recursion relations to truncate, producing *polynomial* solutions of Hermite's equation. The solutions of the recursion relations in (6.15) indicate that either the even series of terms or the odd series of terms can be truncated by requiring λ to be a nonnegative even integer, $\lambda = N$. If N is an even integer, $N = 2k$, then the even series of terms is truncated after the kth term, while the odd series of terms is unaffected. Alternatively, if n is an odd integer, $N = 2k+1$, then the odd series of terms is truncated after the kth term, but the even series is unaffected. Thus, to insure that a normalizable solution of (6.3) is obtained, we must set $a_1 = 0$ if $\lambda = 2k$ and we must set $a_0 = 0$ if $\lambda = 2k+1$:

$$\lambda = \begin{cases} 0, 2, 4, \ldots & \text{if } a_1 = 0 \\ 1, 3, 5, \ldots & \text{if } a_0 = 0 \end{cases} \tag{6.21}$$

The resulting polynomial solutions are called the **Hermite polynomials** and are denoted as H_n, where n indicates the order of the polynomial. The 'standard' form of the Hermite polynomials is chosen with the coefficient of ξ^n in $H_n(\xi)$ set equal to 2^n. Combining this with (6.15), the general forms of the even and odd-order Hermite polynomials are obtained as (Problem 5)

$$H_{2k}(\xi) = \sum_{n=0}^{k} (-1)^{n-k} \frac{(2k)!}{(2n)!(k-n)!} (2\xi)^{2n}$$

$$H_{2k+1}(\xi) = \sum_{n=0}^{k} (-1)^{n-k} \frac{(2k+1)!}{(2n+1)!(k-n)!} (2\xi)^{2n+1} \tag{6.22}$$

Expressions for the first few Hermite polynomials are given below:

$$H_0(\xi) = 1 \qquad\qquad H_1(\xi) = 2\xi$$
$$H_2(\xi) = 4\xi^2 - 2 \qquad\qquad H_3(\xi) = 8\xi^3 - 12\xi \tag{6.23}$$
$$H_4(\xi) = 16\xi^4 - 48\xi^2 + 12 \quad H_5(\xi) = 32\xi^5 - 160\xi^3 + 120\xi$$

Having found the solutions of Hermite's equation (6.9), we can now construct the solution of the original Equation (6.3) from (6.7) and (6.22).

These functions, which are called **Weber–Hermite** functions, are then obtained as

$$\psi_n(x) = \exp\left(-\tfrac{1}{2}bx^2\right) H_n(\sqrt{b}x) \tag{6.24}$$

With the definitions of a and b in (6.2), 2λ can be written as $2\lambda = a/b - 1 = 2E/\hbar\omega_0 - 1$. Setting λ equal to an integer N and solving for the energy yields the expression $E = \hbar\omega_0(N + \tfrac{1}{2})$, which are the familiar energy eigenvalues for the harmonic oscillator.

The Weber–Hermite functions in (6.24) are plotted in Figure 6.1 for $n = 0, \ldots, 9$. There are several features to be noticed. Since H_n is an nth-order polynomial, the function ψ_n has n zeros. Thus, with increasing n, as more zeros need to be accommodated, the ψ_n become increasingly oscillatory. For sufficiently large $|x|$, the exponential factor dominates because of (6.20), so the solution decays toward zero as $|x| \to \infty$. The normalization and other properties of these functions are covered in the problems at the end of the chapter.

The Hermite polynomials can be obtained in a number of ways within *Mathematica*, e.g. by calling the in-built function HermiteH. We can also solve the recursion relations (6.14) to generate the Hermite polynomials directly by using the ability of *Mathematica* to recognize and solve such equations. We first construct (6.14), retaining both the λ-dependence and the n-dependence in the expansion coefficients:

```
a[k_,n_]:=-2((k-n+2)/(n(n-1)))a[k,n-2]
```

The initial values for these equations are determined by stipulating that a_0 must vanish if k is an odd integer, and that a_1 must vanish if k is an even integer. Furthermore, to obtain the standard form of the H_n in (6.22), whereby the coefficient of x^n is set equal to 2^n, the coefficients a_0 and a_1 must be assigned the values that are calculated in Problem 5:

```
a[k_,0]:=If[EvenQ[k],(-1)^(k/2) k!/(k/2)!,0]
a[k_,1]:=If[EvenQ[k],0,2(-1)^((k-1)/2) k!/((k-1)/2)!]
```

The Hermite polynomials are now defined by using the series solutions in (6.22). Both series can be combined into a single expression by using conditional statements to distinguish between even-order and odd-order polynomials:

```
One[k_]  :=If[EvenQ[k],0,1]
kMax[k_]:=If[EvenQ[k],k/2,(k-1)/2]
H[k_,x_]:=Sum[a[k,2j+One[k]]x^(2j+One[k]),{j,0,kMax[k]}]
```

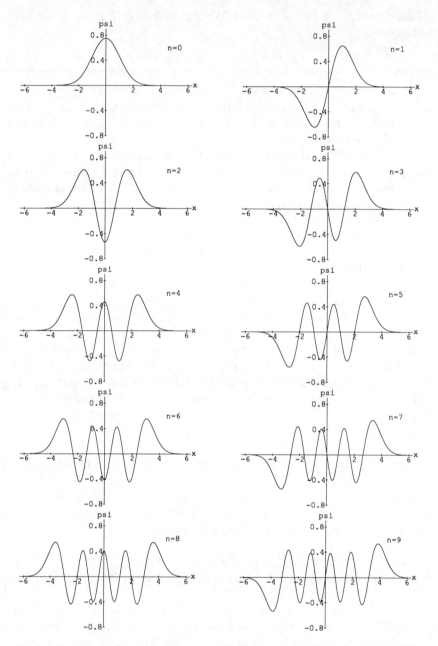

Figure 6.1 The Weber–Hermite functions ψ_n (6.24) for $n = 0, \ldots, 9$. The nth-order function has n zeros, because of the Hermite polynomial factor, and the region within which the ψ_n have appreciable magnitude expands slightly with increasing n. The parity of the ψ_n is seen to alternate between even and odd because the order of the Hermite polynomial factor alternates between even and odd accordingly.

Expressions for the Hermite polynomial of order k can now be generated by calling the function H[k,x]:

```
Do[Print[H[k,x]],{k,0,8}]
```

```
1

2 x

         2
-2 + 4 x

           3
-12 x + 8 x

         2        4
12 - 48 x  + 16 x

        3        5
120 x - 160 x + 32 x

            2        4       6
-120 + 720 x  - 480 x  + 64 x

              2          4         6        8
1680 - 13440 x  + 13440 x  - 3584 x  + 256 x

          3          5         7        9
30240 x - 80640 x + 48384 x - 9216 x + 512 x
```

6.2 Legendre Polynomials

Legendre's equation and Legendre polynomials are encountered in several contexts of mathematical physics, usually involving the eigenfunctions of the angular part of the Laplacian operator, ∇^2, in spherical polar coordinates. Applications include quantum mechanics, where central field problems in atomic physics require solving Schrödinger's equation and determining angular momentum eigenfunctions, and electrostatics and steady state heat conduction in the presence of spherical boundaries, where Laplace's equation must be solved.

We begin by considering Laplace's equation, $\nabla^2 V = 0$, for a function $V(\mathbf{r}) = V(x, y, z)$ in spherical polar coordinates

$$x = r \sin\theta \cos\phi, \quad y = r \sin\theta \sin\phi, \quad z = r \cos\theta \quad (6.25)$$

where $0 \leq r < \infty$, $0 \leq \phi < 2\pi$, and $0 \leq \theta \leq \pi$. According to Equation (1.75), Laplace's equation in spherical polar coordinates is

$$\nabla^2 V = \frac{1}{r}\frac{\partial^2}{\partial r^2}(rV) + \frac{1}{r^2 \sin^2\theta}\frac{\partial^2 V}{\partial \phi^2} + \frac{1}{r^2 \sin\theta}\frac{\partial}{\partial \theta}\left(\sin\theta \frac{\partial V}{\partial \theta}\right) \quad (6.26)$$

216 Partial Differential Equations with Mathematica

Equation (6.26) can be solved with the method of separation of variables. We factor the solution by writing

$$V(r,\theta,\phi) = \frac{R(r)}{r}\Theta(\theta)\Phi(\phi) \qquad (6.27)$$

and substitute into (6.36):

$$\Theta\Phi\frac{d^2R}{dr^2} + \frac{R\Theta}{r^2\sin^2\theta}\frac{d^2\Phi}{d\phi^2} + \frac{R\Phi}{r^2\sin\theta}\frac{d}{d\theta}\left(\sin\theta\frac{d\Theta}{d\theta}\right) = 0 \qquad (6.28)$$

By multiplying (6.28) by $r^2\sin^2\theta$ and dividing by $V = RP\Phi$, the equation becomes

$$r^2\sin\theta\left[\frac{1}{R}\frac{d^2R}{dr^2} + \frac{1}{\Theta r^2\sin\theta}\frac{d}{d\theta}\left(\sin\theta\frac{d\Theta}{d\theta}\right)\right] + \frac{1}{\Phi}\frac{d^2\Phi}{d\phi^2} = 0 \qquad (6.29)$$

The last term on the left-hand side of (6.29) is a function of ϕ only, while the first two terms are functions of r and θ only. Since these two quantities can be varied independently, the last term and the sum of the first two terms must therefore be equal to the same constant, which we signify by μ^2. This produces the two equations

$$\frac{1}{\Phi}\frac{d^2\Phi}{d\phi^2} + \mu^2 = 0 \qquad (6.30a)$$

$$\frac{1}{R}\frac{d^2R}{dr^2} + \frac{1}{\Theta r^2\sin\theta}\frac{d}{d\theta}\left(\sin\theta\frac{d\Theta}{d\theta}\right) - \frac{\mu^2}{r^2\sin\theta} = 0 \qquad (6.30b)$$

Equation (6.30b) can simplified by multiplying by r^2 to separate the r and θ dependencies:

$$\frac{r^2}{R}\frac{d^2R}{dr^2} + \frac{1}{\Theta\sin\theta}\frac{d}{d\theta}\left(\sin\theta\frac{d\Theta}{d\theta}\right) - \frac{\mu^2}{\sin\theta} = 0 \qquad (6.31)$$

The first term on the left-hand side of (6.31) is a function of r only, and the second and third terms are functions of θ only. Therefore, the first term and the sum of the second and third terms each must be equal to a constant, which we call κ. Combining this result with that in (6.30), the solution of Laplace's equation (6.26) has been reduced to solving the following three ordinary differential equations:

$$\frac{d^2\Phi}{d\phi^2} + \mu^2\Phi = 0 \qquad (6.32a)$$

$$\frac{d^2R}{dr^2} - \frac{\kappa}{r^2}R = 0 \qquad (6.32b)$$

$$\frac{1}{\sin\theta}\frac{d}{d\theta}\left(\sin\theta\frac{d\Theta}{d\theta}\right) + \left(\kappa - \frac{\mu^2}{\sin^2\theta}\right)\Theta = 0 \qquad (6.32c)$$

Equations (6.32a) and (6.32b) can be solved easily. The general solution of the azimuthal Equation (6.32a) can be written in several ways. One convenient form is

$$\Phi_\mu(\phi) = \begin{cases} A_0 + B_0\phi, & \text{if } \mu = 0 \\ A_\mu e^{i\mu\phi} + B_\mu e^{-i\mu\phi}, & \text{if } \mu \neq 0 \end{cases} \quad (6.33)$$

where A_0, B_0, A_μ and B_μ are constants. If there is no restriction on the range of ϕ, i.e. if ϕ can take any value between 0 and 2π, then a rotation by 2π cannot cause the value of Φ_μ to change. Thus, Φ_μ must be **single-valued**: $\Phi_\mu(\phi + 2\pi) = \Phi_\mu(\phi)$. Thus, if $\mu \neq 0$, the property of single-valuedness requires

$$\begin{aligned}\Phi_\mu(\phi + 2\pi) &= A_\mu e^{i\mu(\phi+2\pi)} + B_\mu e^{-i\mu(\phi+2\pi)} \\ &= A_\mu e^{2\mu\pi i} e^{i\mu\phi} + B_\mu e^{-2\mu\pi i} e^{-i\mu\phi} \\ &= A_\mu e^{i\mu\phi} + B_\mu e^{-i\mu\phi}\end{aligned} \quad (6.34)$$

Since the functions $\exp(i\mu\phi)$ and $\exp(-i\mu\phi)$ are orthogonal over the interval $(0, 2\pi)$, Equation (6.34) can be satisfied only if $\exp(2\mu i\phi) = 1$ and $\exp(-2\mu i\phi) = 1$, i.e. μ must be an integer, which we signify by m. Referring to (6.33), we see that we must also require that $B_0 = 0$. Thus, the single-valued solution to (6.32a) can be written as

$$\Phi_m(\phi) = A_m e^{im\phi} \quad (6.35)$$

where m can be any positive or negative integer.

The radial equation in (6.32b) can be solved with a trial solution of the form $R(r) = r^\alpha$. This expression yields a solution provided α is a root of the quadratic equation $\alpha(\alpha - 1) - \kappa = 0$. By writing the separation constant as $\kappa = \ell(\ell+1)$, with ℓ as yet undetermined, the roots of this quadratic equation can be expressed as $\alpha = \ell+1$ and $\alpha = -\ell$. The general solution to (6.32b) then may be written as

$$R_\ell(r) = C_\ell r^{\ell+1} + D_\ell r^{-\ell} \quad (6.36)$$

where C_ℓ and D_ℓ are constants.

If solutions are required for the inhomogeneous Laplace equations of the form $\nabla^2 V = f(r)$, where the right-hand side depends only on the radial coordinate, the only modification to the discussion in this section is in the solution of the radial equation. The function $f(r)$ contributes an additional term to (6.32b), which changes the radial functions $R_\ell(r)$, but leaves the polar and azimuthal solutions of Equations (6.32a) and (6.32c) unaffected. This situation arises in applications to quantum mechanical problems with spherically symmetric potentials. An example is Laguerre's equation, which is discussed in Problems 9 and 10.

We now turn our attention to the equation for the polar variable (6.32c). This equation can be simplified somewhat by introducing a new variable ξ defined by $\xi = \cos\theta$. Since the range of θ is $0 \leq \theta \leq \pi$, the range of ξ is $-1 \leq \xi \leq 1$. With this transformation, Equation (6.32c) becomes the **generalized Legendre equation**

$$\frac{d}{d\xi}\left[(1-\xi^2)\frac{d\Theta}{d\xi}\right] + \left[\ell(\ell+1) - \frac{m^2}{1-\xi^2}\right]\Theta = 0 \qquad (6.37)$$

This equation is regular in the *open* interval $(-1,1)$, but has regular singular points at the end-points of this interval, $\xi = \pm 1$, so we must be mindful of the convergence of the series solution near these points. The singular points correspond to the values $\theta = 0$ and $\theta = \pi$ in the original variables. These are the accumulation points of the spherical polar coordinate system, i.e. where the lines of constant r and ϕ intersect, and are also singular points of the coordinate transformation between rectangular and spherical polar coordinates. This is a general feature of applying the separation of variables method in coordinate systems with accumulation points, and illustrates the profound influence the chosen coordinate system for a differential equation has on the analytic behavior of the solutions obtained in that coordinate system. A comprehensive discussion may be found in Morse and Feshbach (1953).

To apply the method of Frobenius to solve the generalized Legendre equation as an expansion about the origin, we carry out the derivative in the first term of (6.37) to place this equation in the standard form of Equation (5.11):

$$(1-\xi^2)\frac{d^2\Theta}{d\xi^2} - 2\xi\frac{d\Theta}{d\xi} + \left[\ell(\ell+1) - \frac{m^2}{1-\xi^2}\right]\Theta = 0 \qquad (6.38)$$

The functions P, Q and R are then readily identified as

$$R(\xi) = 1 - \xi^2, \quad P(\xi) = -2\xi^2, \quad Q(\xi) = \ell(\ell+1)\xi^2 - \frac{m^2\xi^2}{1-\xi^2} \qquad (6.39)$$

While P and R are given by quite simple expressions, Q is seen to have an infinite number of terms when written in the form (5.12):

$$Q(\xi) = \ell(\ell+1)\xi^2 - m^2\xi^2(1 + \xi^2 + \xi^4 + \cdots) \qquad (6.40)$$

Thus, the application of the Frobenius method directly to (6.37) leads to recursion relations that have an expanding number of terms as the order increases. In addition to obscuring the convergence of the series, the solution obtained from such recursion relations is difficult to express in a simple closed form. Therefore, motivated by the approach taken in obtaining Hermite's equation (6.9) from Equation (6.3), where we examined the behavior of the solutions near the singular points, we seek a transformation that converts (6.37) into a more manageable form. This will be done in the next section. In this section, to gain some familiarity with the generalized Legendre equation, we will obtain the solutions of the equation obtained by setting $m=0$. This choice of m, which is appropriate to problems involving azimuthal symmetry, eliminates the problematic term in (6.38). In this case, the generalized Legendre equation reduces to the simplified equation

$$\frac{d}{d\xi}\left[(1-\xi^2)\frac{d\Theta}{d\xi}\right] + \ell(\ell+1)\Theta = 0 \tag{6.41}$$

which is called the **ordinary Legendre equation**. This equation is also readily seen to have regular singular points at $\xi=\pm 1$.

To apply the method of Frobenius to solve (6.41) in an expansion about the origin, we first identify the nonvanishing expansion coefficients of the functions R, P and Q from (6.39):

$$R_0 = 1, \quad R_2 = -1, \quad P_2 = -2, \quad Q_2 = \ell(\ell+1) \tag{6.42}$$

Expanding the solution as $\Theta(\xi) = \xi^s \sum a_n \xi^n$ and substituting into (6.41), the indicial equation is found to be $s(s-1) = 0$. The two roots of this equation are $s=0$ and $s=1$, as expected. The recursion relation for $n=1$ is

$$a_1 s(s+1) = 0 \tag{6.43}$$

and for $n > 1$ takes the form

$$a_n(n+s)(n+s-1) = a_{n-2}[(n-2+s)(n-1+s) - \ell(\ell+1)] \tag{6.44}$$

Thus, by choosing the indicial root $s=0$, both a_0 and a_1 are arbitrary and, according to the form of the recursion relation (6.44), the general solution of (6.41) takes the form of a sum of an even-order series proportional to a_0 and an odd-order series proportional to a_1. The required recursion relations are obtained by setting $s=0$ in (6.44):

$$a_n = \frac{(n-2)(n-1) - \ell(\ell+1)}{n(n-1)} a_{n-2}$$

$$= \frac{(n+\ell-1)(n-\ell-2)}{n(n-1)} a_{n-2} \tag{6.45}$$

By applying the steps used in Examples 5.2 and 5.3, this equation can be solved readily for the even-order coefficients

$$a_{2n} = \frac{(-1)^n}{(2n)!}(\ell - 2n + 2)\cdots\ell(\ell+1)\cdots(\ell+2n-1)a_0 \tag{6.46a}$$

and for the odd-order coefficients

$$a_{2n+1} = \frac{(-1)^{n+1}}{(2n+1)!}(\ell - 2n + 1)\cdots(\ell - 1)(\ell + 2)\cdots(\ell + 2n)a_1 \quad (6.46b)$$

These expansion coefficients generate the formal general solution of the ordinary Legendre equation, though, as always, we must determine the radius of convergence of the series. In particular, we must insure that the solution converges everywhere within the *closed* interval $[0, 1]$. Applying the ratio test to (6.46) yields

$$\lim_{n\to\infty}\left|\frac{a_n}{a_{n-2}}\right| = 1 \quad (6.47)$$

Thus, the series solution generated by the recursion relations (6.45) is guaranteed to converge if $|\xi| < 1$, i.e. within the *open* interval $(-1, 1)$, and to diverge if $|\xi| > 1$. At the end-points $\xi = \pm 1$, where (6.41) has regular singular points, the ratio test fails.

To examine the behavior of solutions determined by (6.45) at $\xi = 1$, we apply the ratio test of Gauss discussed in Section 1.4. We consider the series of even terms in (6.46a). For $n > \ell - 2$, all of the a_n have the same sign, so we can apply the Gauss test for a series of positive terms. At $\xi = 1$, the pertinent series is given by $\sum c_k$, where $c_k = a_{2k}$. Equation (6.45) can be written in terms of c_k as

$$\frac{c_k}{c_{k-1}} = \frac{(2k-2)(2k-1) - \ell(\ell+1)}{2k(2k-1)}$$

$$= 1 - \frac{1}{k} - \frac{\ell(\ell+1)}{2k(2k-1)} \quad (6.48)$$

In the limit $k \to \infty$, the leading-order terms on the right-hand side of (6.48) are

$$\lim_{k\to\infty}\frac{c_k}{c_{k-1}} = 1 - \frac{1}{k} - \frac{\ell(\ell+1)}{4k^2} + \cdots \quad (6.49)$$

According to Gauss' test, since the coefficient of k^{-1} is unity, the infinite series generated by the even-order coefficients in (6.46) *diverges* at $\xi = 1$. Similar reasoning shows that the series of odd-order terms in (6.46) is divergent at these points as well.

To obtain regular solutions of (6.37) for all values of ξ within the closed interval $[-1, 1]$, we proceed as in Section 6.1 and require one of the infinite series in (6.46) to truncate and choose the multiplicative constant of the other series to be zero. Referring to (6.45), the truncation of either of two series of terms in (6.46) can be achieved by choosing ℓ such that the numerator vanishes for some integer n: $(n-2)(n-1) = \ell(\ell+1)$. There are two roots of this quadratic equation: $\ell_1 = n-2$ and $\ell_2 = -n+1$, where $\ell_1 \geq 0$ and $\ell_2 < 0$. In fact, we can restrict ourselves to the first solution and consider only positive values of ℓ, since the two solutions are related by $\ell_1 + \ell_2 = -1$, so any solution obtained for a value of ℓ_2 has a unique counterpart corresponding to $\ell_1 = -(\ell_2+1)$. Thus, in keeping with standard usage, we choose $\ell = n-2$. Combining this result with (6.46), we obtain the condition for obtaining a finite solution of the ordinary Legendre equation as

$$\ell = \begin{cases} 0, 2, 4, \ldots & \text{if } a_1 = 0 \\ 1, 3, 5, \ldots & \text{if } a_0 = 0 \end{cases} \qquad (6.50)$$

The resulting polynomials, written as P_ℓ, are the **Legendre polynomials** of order ℓ. These polynomials are standardized by the condition that $P_\ell(1) = 1$. Thus, by making appropriate choices for a_0 and a_1 to observe this standardization, general expressions for the even-order and odd-order Legendre polynomials can be obtained from (6.46):

$$P_{2k}(\xi) = \sum_{n=0}^{k} (-1)^{k-n} \frac{(2k+2n)!}{2^{2k}(2n)!(k-n)!(k+n)!} \xi^{2n}$$

$$P_{2k+1}(\xi) = \sum_{n=0}^{k} (-1)^{k-n} \frac{(2k+2n+2)!}{2^{2k+1}(2n+1)!(k-n)!(k+n+1)!} \xi^{2n+1}$$

(6.51)

The first few Legendre polynomials are given by

$$P_0(\xi) = 1 \qquad P_1(\xi) = \xi$$
$$P_2(\xi) = \tfrac{1}{2}(3\xi^2 - 1) \qquad P_3(\xi) = \tfrac{1}{2}(5\xi^3 - 3\xi) \qquad (6.52)$$
$$P_4(\xi) = \tfrac{1}{8}(35\xi^4 - 30\xi^2 + 3) \qquad P_5(\xi) = \tfrac{1}{8}(63\xi^5 - 70\xi^3 + 15\xi)$$

Since (6.32c) is already in the form of a Liouville equation, the weight function for the Legendre polynomials is unity.

Plots of $P_0(\cos\theta)$, $P_1(\cos\theta)$, $P_2(\cos\theta)$ and $P_3(\cos\theta)$ are shown in Figure 6.2. These diagrams clearly show the azimuthal symmetry of the Legendre polynomials. With increasing ℓ these plots also reveal that the P_ℓ acquire more lobes, which result from the increasing number of zeros of the P_ℓ as a function of ξ. Furthermore, as ℓ increases, P_ℓ becomes more localized along the z axis. Thus, the representation of an axially-symmetric function that is directed along the z axis as an expansion in Legendre polynomials requires more higher-order polynomials than the expansion of a comparatively isotropic function.

Having determined the solution of the ordinary Legendre equation, the general solution to Laplace's equation for the case of azimuthal symmetry ($m = 0$) can be written as a sum of products of solutions to each of the equations in (6.32). Since the solution to (6.32a) is simply a constant, the

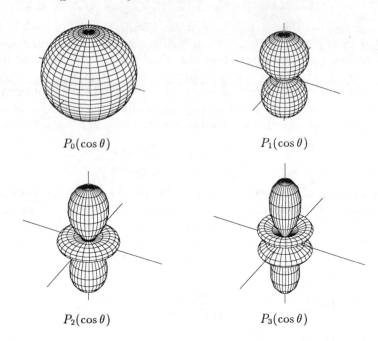

Figure 6.2 Plots generated by `ParametricPlot3D` for the *absolute values* of the Legendre polynomials $P_0(\cos\theta)$, $P_1(\cos\theta)$, $P_2(\cos\theta)$ and $P_3(\cos\theta)$. With increasing ℓ, the Legendre polynomials become more localized along the z direction and acquire more lobes along the polar direction.

solution to Laplace's equation is given, in terms of the original variables, by the series

$$V(r,\theta) = \sum_{\ell=0}^{\infty} \left(A_\ell r^\ell + B_\ell r^{-\ell+1} \right) P_\ell(\cos\theta) \qquad (6.53)$$

Notice that the term r^ℓ becomes infinite as $r \to \infty$, and that the term $r^{-\ell+1}$ is infinite at the origin. Thus, depending upon the boundary conditions, either or both of these terms can contribute. For example, the solution of Laplace's equation within a bounded region that contains the origin necessitates setting $B_\ell = 0$, while in an unbounded region that excludes the origin, a finite solution is obtained by setting $A_\ell = 0$.

6.3 Legendre Functions and Spherical Harmonics

We now return to the generalized Legendre equation (6.37). As we have already mentioned, this equation is not in a convenient form for applying directly the method of Frobenius. Therefore, in the spirit of the approach taken in deriving Hermite's equation, we seek a transformation that will reduce the generalized Legendre equation into a more suitable form by examining the behavior of the solutions near the regular singular points at $\xi = \pm 1$.

To extract the dominant behavior of P near $\xi = \pm 1$, we need only solve the indicial equation for the expansions of the solution about these points. The roots of the indicial equations for both $\xi = 1$ and $\xi = -1$ are $s = \pm\frac{1}{2}|m|$. Thus, only the positive root, $s = \frac{1}{2}|m|$, produces behavior that is finite at the end-points, and yields the following asymptotic forms of the solution as $\xi \to \pm 1$:

$$\begin{aligned} \Theta(\xi) &\longrightarrow (1-\xi)^{|m|/2} \quad \text{as} \quad \xi \to 1 \\ \Theta(\xi) &\longrightarrow (1+\xi)^{|m|/2} \quad \text{as} \quad \xi \to -1 \end{aligned} \quad (6.54)$$

We now use these results to define a reduced function F from which the dominant behavior near the singular points has been 'factored out':

$$\Theta(\xi) = (1+\xi)^{|m|/2}(1-\xi)^{|m|/2} F(\xi) = (1-\xi^2)^{|m|/2} F(\xi) \qquad (6.55)$$

The differential equation for F is obtained by substituting (6.55) into the generalized Legendre equation, with the result:

$$(1-\xi^2)\frac{d^2 F}{d\xi^2} - 2(|m|+1)\xi\frac{dF}{d\xi} + \left[\ell(\ell+1) - |m|(|m|+1)\right] F = 0 \quad (6.56)$$

Thus, introducing the function F is again seen to transform the original differential equation in (6.37) into a form that is more suitable for applying the method of Frobenius. The analytic character of the generalized Legendre equation has not been changed by (6.55), namely the points $\xi = \pm 1$ remain regular singular points and the equation is regular elsewhere in the interval $[-1, 1]$. Therefore, we must again take care to insure the series solution converges at the end-points.

In the standard notation of Section 5.2, the only nonvanishing expansion coefficients of the functions R, P and Q are

$$\begin{aligned} R_0 &= 1, \quad R_2 = -1, \quad P_2 = -2(|m|+1), \\ Q_2 &= \ell(\ell+1) - |m|(|m|+1) \end{aligned} \qquad (6.57)$$

Thus, expanding the solution as $F(\xi) = \xi^s \sum a_n \xi^n$ and substituting into Equation (6.56), the indicial equation is obtained as $s(s-1) = 0$, which again yields the roots $s = 0$ and $s = 1$. The recursion relation for $n = 1$ yields

$$a_1 s(s+1) = 0 \qquad (6.58)$$

Thus, by choosing $s = 0$ in the recursion relations, both a_0 and a_1 are arbitrary, and the solution of the recursion relations again produces the general solution. The recursion relations for $n \geq 2$ are

$$a_n(n+s)(n+s-1) = a_{n-2}\Big[(n-2+s)(n-1+2|m|+s) + (|m|+\ell+1)(|m|-\ell)\Big] \quad (6.59)$$

The general solution of (6.56) is again seen to comprise the sum of an even-powered series proportional to a_0 and an odd-powered series proportional to a_1. Setting $s = 0$ in (6.58), and performing manipulations analogous to those used in obtaining the simplified form (6.45) for the recursion relations of the ordinary Legendre equation, the recursion relations for the reduced function F are obtained as

$$a_n = \frac{(n+\ell+|m|-1)(n-\ell+|m|-2)}{n(n-1)} a_{n-2} \quad (6.60)$$

These recursion relations are similar in form to those in (6.45), and we will see that their solutions have several common properties. For example, given the solutions (6.46) of the recursion relations, we can immediately obtain the corresponding solutions of (6.60). Introducing the notation $\lambda_{\pm} = \ell \pm |m|$, the expansion coefficients are given by

$$a_{2n} = \frac{(-1)^n}{(2n)!}(\lambda_- - 2n + 2)\cdots\lambda_-(\lambda_+ + 1)\cdots(\lambda_+ + 2n - 1)a_0 \quad (6.61a)$$

$$a_{2n+1} = \frac{(-1)^{n+1}}{(2n+1)!}(\lambda_- - 2n + 1)\cdots(\lambda_- - 1)(\lambda_+ + 2)\cdots(\lambda_+ + 2n)a_1 \quad (6.61b)$$

This provides a solution of Equation (6.56), though we must again check the convergence of the series, particularly to insure that the condition (6.54) is met. In fact, applying the ratio test to (6.60), we again see that

$$\lim_{n\to\infty}\left|\frac{a_{n-2}}{a_n}\right| = 1 \quad (6.62)$$

which guarantees convergence only on the open interval $(-1, 1)$. The same procedure we used in (6.48) and (6.49) to show that the infinite series solution for the ordinary solution diverges at the end-points also applies for the infinite series solution with the coefficients in (6.61). Thus, obtaining a finite solution of Equation (6.56) everywhere in the closed interval $[-1, 1]$ again requires that the infinite series generated by (6.61) be truncated.

A glance at the recursion relations (6.60) reveals that this truncation can be achieved for *one* of the series by choosing λ_- such that

$\lambda_- = n - 2 \geq 0$ for some integer n, and the multiplicative constant of the other series must then be set equal to zero, i.e.

$$\lambda_- = \ell - |m| = \begin{cases} 0, 2, 4, \ldots & \text{if } a_1 = 0 \\ 1, 3, 5, \ldots & \text{if } a_0 = 0 \end{cases} \quad (6.63)$$

In particular, this condition states that ℓ and $|m|$ must be chosen to satisfy the inequality $\ell - |m| \geq 0$. We have already established in (6.34) that the single-valuedness of the solution in the azimuthal angle ϕ requires that m be an integer. Therefore, ℓ must also be an integer and the allowed values of m are seen from (6.63) to be

$$m = -\ell, -\ell+1, \ldots, \ell-1, \ell \quad (6.64)$$

In other words, there are $2\ell + 1$ allowed values of m for a given ℓ. The solution of the recursion relations to obtain the polynomials $F_{\ell,m}$ is now straightforward, and the solutions of the generalized Legendre equation are obtained from (6.55) as

$$P_\ell^m(\xi) = (1 - \xi^2)^{|m|/2} F_{\ell,m}(\xi) \quad (6.65)$$

The functions P_ℓ^m are the **associated Legendre functions** (sometimes called simply the **Legendre functions**).

An alternative form of the Legendre functions can be obtained by differentiating the ordinary Legendre equation (6.41) $|m|$ times with respect to ξ (Problem 13). This shows that the function $F_{\ell,m}$ is proportional to $d^{|m|} P_\ell / d\xi^{|m|}$. By convention, this proportionality constant is chosen to be unity, so (6.65) can be expressed as

$$P_\ell^{|m|}(\xi) = (1 - \xi^2)^{|m|/2} \frac{d^{|m|} P_\ell}{d\xi^{|m|}} \quad (6.66)$$

The first few Legendre functions calculated by substituting the expressions in (6.52) for the Legendre polynomials into this formula are given below:

$$\begin{aligned}
P_0^0(\xi) &= 1 \\
P_1^0(\xi) &= \xi \qquad\qquad P_1^1(\xi) = (1 - \xi^2)^{1/2} \\
P_2^0(\xi) &= \tfrac{1}{2}(3\xi^2 - 1) \quad P_2^1(\xi) = 3\xi(1 - \xi^2)^{1/2} \\
P_2^2(\xi) &= 3(1 - \xi^2)
\end{aligned} \quad (6.67)$$

Since the generalized Legendre equation (6.37) is effectively a function only of $|m|$, and not of m, as (6.56) shows, the solution $P_\ell^{-m}(\xi)$ must be proportional to $P_\ell^m(\xi)$. The usual way of obtaining this proportionality constant is the subject of Problem 44, where we find

$$P_\ell^{-m}(\xi) = (-1)^m \frac{(\ell - m)!}{(\ell + m)!} P_\ell^m(\xi) \quad (6.68)$$

The angular part of Laplace's equation, and related equations such as Schrödinger's equation and Helmholtz's equation, obtained after a separation of variables in spherical polar coordinates, occurs in many applications. Since the difference between the solutions of these problems is confined to the radial part of the equation, it is convenient for many purposes to combine the solutions of the angular variables θ and ϕ into orthogonal functions defined over the 4π steradians of the unit sphere. These functions, which are signified by $Y_{\ell,m}(\theta,\phi) = \Theta_{\ell,m}(\theta)\Phi_m(\phi)$, are called **spherical harmonics**. The normalization for these functions can be determined from that for the Legendre functions (Problem 45), with the result that the normalized spherical harmonics are given by

$$Y_{\ell,m}(\theta,\phi) = \sqrt{\frac{2\ell+1}{4\pi}\frac{(\ell-m)!}{(\ell+m)!}} P_\ell^m(\cos\theta) \, e^{im\phi} \qquad (6.69)$$

Equations (6.68) and (6.69) imply that

$$Y_{\ell,-m}(\theta,\phi) = (-1)^m Y_{\ell,m}^*(\theta,\phi) \qquad (6.70)$$

where the asterisk signifies complex conjugation. The unnormalized spherical harmonics for $\ell \leq 2$ and $m \geq 0$ obtained from (6.52), (6.68) and (6.69) are:

$$\begin{aligned}
Y_{0,0}(\theta,\phi) &= 1 \\
Y_{1,1}(\theta,\phi) &= \sin\theta \, e^{-i\phi} & Y_{1,0}(\theta,\phi) &= \cos\theta \\
Y_{2,2}(\theta,\phi) &= 3\sin^2\theta \, e^{2i\phi} & Y_{2,1}(\theta,\phi) &= 3\sin\theta\cos\theta \, e^{i\phi} \\
Y_{2,0}(\theta,\phi) &= \tfrac{3}{2}\cos^2\theta - \tfrac{1}{2}
\end{aligned} \qquad (6.71)$$

The real parts of these functions are shown in Figure 6.3. With increasing ℓ and m these functions show an increasing number of angular nodes, both in the polar and azimuthal directions. The structure of these functions has important implications in several areas of atomic physics, quantum scattering and other areas where the governing equation is of the Laplace or Helmholtz form.

6.4 Bessel Functions

Bessel's equation and Bessel functions are encountered in applications such as electrostatics and other problems involving cylindrical or circular boundaries, and a related equation is solved in quantum mechanical scattering in spherical polar coordinates. In cylindrical coordinates,

$$x = r\cos\phi, \qquad y = r\sin\phi, \qquad z = z \qquad (6.72)$$

Special Functions and Orthogonal Polynomials   227

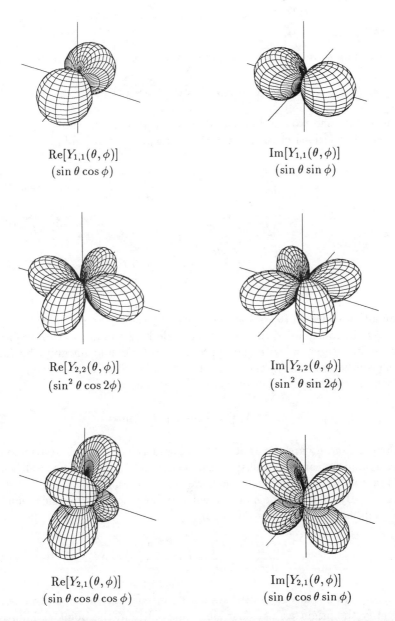

Re[$Y_{1,1}(\theta,\phi)$]
($\sin\theta\cos\phi$)

Im[$Y_{1,1}(\theta,\phi)$]
($\sin\theta\sin\phi$)

Re[$Y_{2,2}(\theta,\phi)$]
($\sin^2\theta\cos 2\phi$)

Im[$Y_{2,2}(\theta,\phi)$]
($\sin^2\theta\sin 2\phi$)

Re[$Y_{2,1}(\theta,\phi)$]
($\sin\theta\cos\theta\cos\phi$)

Im[$Y_{2,1}(\theta,\phi)$]
($\sin\theta\cos\theta\sin\phi$)

Figure 6.3 The *absolute values* of the real and imaginary parts of the spherical harmonics (designated by 'Re' and 'Im', respectively) in (6.71) generated by `ParametricPlot3D`. For each function, the expression in spherical polar coordinates is also given. The functions $Y_{\ell,0}(\theta,\phi)$ are proportional to $P_\ell(\cos\theta)$, which are shown in Figure 6.2.

Laplace's equation becomes

$$\nabla^2 V = \frac{\partial^2 V}{\partial r^2} + \frac{1}{r}\frac{\partial V}{\partial r} + \frac{1}{r^2}\frac{\partial^2 V}{\partial \phi^2} + \frac{\partial^2 V}{\partial z^2} = 0 \qquad (6.73)$$

Proceeding as in (6.27) et seq., we separate the variables of V by writing

$$V(r,\phi,z) = R(r)\Phi(\phi)Z(z) \qquad (6.74)$$

and reduce the partial differential equation (6.73) into three ordinary differential equations in each of the three independent variables:

$$\frac{d^2\Phi}{d\phi^2} + \nu^2 \Phi = 0 \qquad (6.75a)$$

$$\frac{d^2 Z}{dz^2} - \kappa^2 Z = 0 \qquad (6.75b)$$

$$\frac{d^2 R}{dr^2} + \frac{1}{r}\frac{dR}{dr} + \left(\kappa^2 - \frac{\nu^2}{r^2}\right) R = 0 \qquad (6.75c)$$

The quantities ν and κ are the separation constants. The solution to the azimuthal equation (6.75a) is again given by (6.33). Single-valuedness with respect to rotations of the azimuthal angle by 2π radians again requires ν to be an integer, n, with the appropriate solutions again given by (6.35). The solution of (6.75b) can be expressed in terms of elementary functions either as real exponentials or as hyperbolic functions, e.g.

$$Z(z) = c_\kappa \cosh\kappa z + d_\kappa \sinh\kappa z \qquad (6.76)$$

where c_κ and d_κ are constants. The allowed values of κ are restricted only if a boundary condition must be satisfied along the z direction. Otherwise κ takes an unrestricted continuous range of values.

By changing the independent variable to $\xi = \kappa r$, the radial equation (6.75c) becomes **Bessel's equation**:

$$\frac{d^2 R}{d\xi^2} + \frac{1}{\xi}\frac{dR}{d\xi} + \left(1 - \frac{\nu^2}{\xi^2}\right) R = 0 \qquad (6.77)$$

This equation is already in the standard form of Section 5.3, and we readily identify the origin as a regular singular point and the point at infinity as an irregular singular point. The nonvanishing expansion coefficients of the functions R, P and Q are

$$R_0 = 1, \quad P_0 = 1, \quad Q_0 = -\nu^2, \quad Q_2 = 1 \qquad (6.78)$$

Expanding the solution as $R(\xi) = x^2 \sum a_n \xi^n$ and substituting into Bessel's equation, we obtain the indicial equation as $s^2 - \nu^2 = 0$, which yields the two roots $s = \pm \nu$. Thus, if ν is neither an integer nor a half-integer, the general solution to Bessel's equation will be obtained directly from the method of Frobenius. Otherwise, we will need to apply the methods of Section 5.3. The recursion relations are given by the expression

$$a_n \left[(n+s)^2 - \nu^2\right] = a_{n-2} \tag{6.79}$$

For $n=1$, this yields $a_1[(s+1)^2 - \nu^2] = 0$ which, combined with the roots of the indicial equation, implies that $a_1 = 0$. Applying the ratio test to (6.79),

$$\lim_{n \to \infty} \left| \frac{a_{n-2}}{a_n} \right| = 0 \tag{6.80}$$

we see that the radius of convergence of the series solution generated by the recursion relations is the entire real line.

To solve the recursion relations, we substitute the root $s = \nu$ into (6.79) and solve for a_n to obtain

$$a_n = -\frac{1}{(n+\nu)^2 - \nu^2} a_{n-2} = -\frac{1}{n(n+2\nu)} a_{n-2} \tag{6.81}$$

Since $a_1 = 0$, (6.81) shows that all odd-order expansion coefficients of the solution vanish. The expression for the even-order coefficients can be simplified somewhat by writing $n = 2k$, whereupon (6.81) becomes

$$a_{2k} = -\frac{1}{2k(k+\nu)} a_{2k-2} \tag{6.82}$$

This equation can be readily solved for the a_{2k} again by using recursive substitution:

$$a_{2k} = (-1)^k \frac{1}{2^{2k} k! (\nu+k)(\nu+k-1) \cdots (\nu+1)} a_0 \tag{6.83}$$

If ν is an integer then (6.83) can be written in a more compact form by making use of factorials as

$$a_{2k} = (-1)^k \frac{\nu!}{2^{2k} k! (\nu+k)!} a_0 \tag{6.84}$$

An expression for the a_{2k} that combines the case when ν is an integer, Equation (6.84), with that when ν is not an integer can be obtained by using the Gamma function (Problem 18):

$$a_{2k} = (-1)^k \frac{\Gamma(\nu+1)}{2^{2k} k! \Gamma(k+\nu+1)} a_0 \tag{6.85}$$

These expansion coefficients can now be used to construct a solution of Bessel's equation. By convention, a_0 is chosen to cancel the factors in (6.85) that involve only ν, $a_0 = 1/2^\nu \Gamma(\nu+1)$. Thus, one solution to Bessel's equation is

$$J_\nu(\xi) = \sum_{k=0}^{\infty} \frac{(-1)^k}{k!\Gamma(k+\nu+1)} \left(\frac{\xi}{2}\right)^{2k+\nu} \qquad (6.86)$$

The function J_ν is called the **Bessel function of the first kind** of order ν. If ν is neither an integer nor a half-integer, a second solution that is independent of (6.86) can be obtained by solving the recursion relations with $s = -\nu$. If ν is an integer, then either $\nu = 0$ or $\nu \neq 0$. We consider each of these two cases in turn. The case where ν is a half-integer is treated in the problems at the end of the chapter.

Case I. If $\nu = 0$, then the indicial equation yields the double root $s = 0$ and we must apply Equation (5.44) to obtain a second solution. In fact, this problem has already been solved in Example 5.3, where the change of variables $x = \frac{1}{2}\xi^2$ transforms Equation (5.45) into Bessel's equation (6.77) with $m = 0$. The complementary solution obtained is

$$Y^{(0)}(\xi) = J_0(\xi) \ln \xi + \sum_{k=1}^{\infty} \frac{(-1)^{k+1}}{(k!)^2} \left[1 + \frac{1}{2} + \cdots + \frac{1}{k}\right] \left(\frac{\xi}{2}\right)^{2k} \qquad (6.87)$$

Some of the features of this solution will be discussed below after we perform the corresponding analysis for the case where ν is a non-zero integer.

Case II. If ν is equal to an integer n, then the two roots of the indicial equation are $s = n$ and $s = -n$, the difference between which is clearly an integer. The method of Frobenius yields a solution which is given by (6.85) and we must apply (5.62) to generate a complementary solution. The recursion relations (6.79) can be written as

$$a_{2k}(s) = -\frac{1}{(2k+s+n)(2k+s-n)} a_{2k-2} \qquad (6.88)$$

The solution to this equation can be expressed as

$$a_{2k}(s) = a_0(-1)^k \prod_{i=1}^{k} \frac{1}{(s+2i+n)(s+2i-n)} \qquad (6.89)$$

To construct the first term on the right-hand side of (5.62), we must evaluate the expression $\lim_{s \to -n}[(s+n)a_{2k}(s)]$. From (6.89), we obtain

$$\lim_{s \to -n} [(s+n)a_{2k}(s)] = 0 \qquad (6.90)$$

if $k<n$, and

$$\lim_{s\to -n}[(s+n)a_{2k}(s)] = (-1)^{k+n+1}\frac{a_0}{2^{2k-1}}\frac{1}{k!(n-1)!(k-n)!} \quad (6.91)$$

if $k\geq n$. Using these expansion coefficients, the contribution from the first term on the right-hand side of (5.62) is found to be

$$\ln\xi \lim_{s\to -n}\left[\sum_{k=n}^{\infty}(s+n)a_{2k}(s)\xi^{2k+s}\right]$$

$$= \ln\xi\left[\frac{a_0}{(n-1)!}\sum_{k=n}^{\infty}\frac{(-1)^{k+n+1}}{2^{2k-1}k!(k-n)!}\xi^{2k-n}\right]$$

$$= \frac{a_0}{2^{2n}(n-1)!}\left[-2\ln\xi\sum_{j=0}^{\infty}\frac{(-1)^j}{2^{2j}j!(j+n)!}\xi^{2j+n}\right]$$

$$= \frac{a_0}{2^{2n}(n-1)!}\left[-2\ln\xi\, J_n(\xi)\right] \quad (6.92)$$

For the second term on the right-hand side of (5.62), we again break up the evaluation into the two cases $k<n$ and $k\geq n$. If $k<n$, then

$$\frac{d}{ds}\left[(s+n)a_{2k}(s)\right]\bigg|_{s=-n} = \frac{a_0}{2^{2k}}\frac{(n-k-1)!}{k!(n-1)!} \quad (6.93)$$

If $k\geq n$, applying the method of Example 5.4 produces

$$\frac{d}{ds}\left[(s+n)a_{2k}(s)\right]\bigg|_{s=-n}$$

$$= (-1)^{k+n}\frac{a_0}{2^{2k}}\frac{1}{k!(n-1)!(k-n)!}\left[\sum_{j=1}^{k}\frac{1}{j}+\sum_{j=1}^{k-n}\frac{1}{j}-\sum_{j=1}^{n-1}\frac{1}{j}\right] \quad (6.94)$$

Collecting the terms in (6.91), (6.93) and (6.94), the complementary solution of Bessel's equation when ν is equal to a nonnegative integer n is

$$y^{(n)}(\xi) = \sum_{k=0}^{n-1}\frac{(n-k-1)!}{k!}\left(\frac{\xi}{2}\right)^{2k-n} - J_n(\xi)\left[2\ln\xi + \sum_{k=1}^{n-1}\frac{1}{k}\right]$$

$$+ \sum_{k=0}^{\infty}\frac{(-1)^k}{k!(n+k)!}\left[\sum_{j=1}^{k}\frac{1}{j}+\sum_{j=1}^{k+n}\frac{1}{j}\right]\left(\frac{\xi}{2}\right)^{2k+n} \quad (6.95)$$

The summation in the second term on the right-hand side of (6.95) is excluded if $n\leq 1$.

As the presence of the second term on the right-hand side of (6.95) indicates, any multiple of $J_n(\xi)$ can be added to this solution while retaining the linear independence of $J_n(\xi)$ and $\mathcal{Y}^{(n)}(\xi)$. This freedom in choosing the coefficient of $J_n(\xi)$ (apart from the logarithm) has led to several versions of the complementary solution for integer-order Bessel functions. The standard that has been adopted arises most naturally from the solution of Bessel's equation using complex integrals (Copson, 1935), which will be discussed in Chapter 8:

$$Y_\nu(\xi) = \frac{\cos(\nu\pi)J_\nu(\xi) - J_{-\nu}(\xi)}{\sin(\nu\pi)} \tag{6.96}$$

The function $Y_\nu(\xi)$ is called the **Bessel function of the second kind** of order ν. If ν is not equal to an integer then this expression is well-defined as written, and clearly yields a solution that is independent of J_ν, since J_ν and $J_{-\nu}$ are already linearly independent. Of course, in this case, the construction of Y_ν is not *necessary* for obtaining the general solution of Bessel's equation. If ν is equal to an integer n, (6.96) must be interpreted as the limiting value of the expression on the right-hand side as ν approaches n. Applying L'Hospital's rule, we obtain

$$Y_n(\xi) = \lim_{\nu \to n} Y_\nu(\xi)$$

$$= \frac{1}{\pi}\left[\frac{\partial J_\nu(\xi)}{\partial \nu} - (-1)^n \frac{\partial J_{-\nu}(\xi)}{\partial \nu}\right]\bigg|_{\nu=n} \tag{6.97}$$

The series representation of this function can be obtained by performing the derivatives in (6.97) on the series in (6.86) for $J_\nu(\xi)$ and setting $\nu = n$ in the resulting expression (Problem 27):

$$Y_n(\xi) = -\frac{1}{\pi}\sum_{k=0}^{n-1}\frac{(n-k-1)!}{k!}\left(\frac{\xi}{2}\right)^{2k-n} + \frac{2}{\pi}J_n(\xi)\ln(\tfrac{1}{2}\xi)$$

$$-\frac{1}{\pi}\sum_{k=0}^{\infty}\frac{(-1)^k}{k!(n+k)!}\Big[\psi(k+1) + \psi(k+n+1)\Big]\left(\frac{\xi}{2}\right)^{2k+n} \tag{6.98}$$

The function $\psi(z)$ is called the **digamma function** and is the logarithmic derivative of the gamma function, $\psi(z) = \Gamma'(z)/\Gamma(z)$. The appearance of the digamma function is a natural consequence of taking the derivatives in (6.97), as a glance at the series representation (6.86) of $J_\nu(\xi)$ reveals. For integer values n of the argument, the digamma function is given by

$$\psi(n) = \sum_{k=1}^{n-1}\frac{1}{k} - \gamma \tag{6.99}$$

with $\psi(1)=-\gamma$. The quantity γ is called **Euler's constant** and is defined by

$$\gamma = \lim_{k\to\infty}\left[\sum_{n=1}^{k}\frac{1}{n} - \ln k\right] \tag{6.100}$$

A comparison of (6.95) and (6.98) shows that $Y_n(\xi)$ differs from $\mathcal{Y}^{(n)}(\xi)$ only by terms proportional to $J_n(\xi)$:

$$Y_n(\xi) = -\frac{1}{\pi}\left\{\mathcal{Y}^{(n)}(\xi) + J_n(\xi)\left[2\ln 2 - 2\gamma + \sum_{k=1}^{n-1}\frac{1}{k}\right]\right\} \tag{6.101}$$

The Bessel functions $J_n(x)$ and $Y_n(x)$ for $n=0,1$ and 2 are plotted in Figure 6.4.

Having obtained the general solution of Bessel's equation, we can assemble the solutions to the equations in (6.75) to form general solution of Laplace's equation in cylindrical coordinates. If the range of ϕ is unrestricted, then ν must be an integer and the solution can be written as

$$V(r,\phi,z) = \sum_{n=-\infty}^{\infty}\sum_{\kappa} \Phi_n(\phi) R_{n\kappa}(r) Z_\kappa(z) \tag{6.102}$$

The summation over κ can be either a discrete sum or an integral over continuous values, depending on whether or not the solution is required to take particular values on boundaries in the z direction. The individual eigenfunctions are given by the solutions to (6.75a), (6.75b) and (6.75c), respectively:

$$\Phi_n(\phi) = e^{in\phi} \tag{6.103a}$$

$$Z_\kappa(z) = c_\kappa \cosh(\kappa z) + d_\kappa \sinh(\kappa z) \tag{6.103b}$$

$$R_{n\kappa}(r) = a_n J_n(\kappa r) + b_n Y_n(\kappa r) \tag{6.103c}$$

The quantities a_n, b_n, c_κ and d_κ are constants to be determined by the boundary conditions.

Although the Bessel functions of the first and second kinds are in-built functions in *Mathematica* (`BesselJ` and `BesselY`, respectively), we can use the series representations (6.86) and (6.98) to investigate their convergence. Before doing so, we first examine the convergence of the limit in (6.100), which is Euler's constant. We first construct the partial sum in (6.100) and then take the limit $k\to\infty$:

```
S[k_]:=Sum[1/n,{n,1,k}]
```

```
Do[Print[N[S[k]-Log[k],10]],{k,1,2000,200}]
```

1.
0.5797011644
0.5784620295
0.5780473809
0.5778397547
0.5777150822
0.5776319269
0.5775725104
0.5775279372
0.5774932628

The convergence is seen to be very slow, as even after 2000 terms, the sequence has converged only to the first three decimal places. This result is to be compared with the 'exact' value obtained from EulerGamma to ten decimal places:

```
N[EulerGamma,10]
```

0.5772156649

This slow convergence of our construction makes it much more convenient to use the in-built function PolyGamma to calculate the logarithmic derivative of the Gamma function instead of performing explicitly the operations in (6.99).

To examine the convergence of the series representation for $Y_n(\xi)$, we first construct the partial sums of (6.98):

```
Y[n_,x_,k_]:=(2/Pi)BesselJ[n,x]Log[x/2]-
    (1/Pi)Sum[((n-j-1)!/j!)(x/2)^(2j-n),{j,0,n-1}]-
    (1/Pi)Sum[((-1)^j/(j!(n+j)!))(PolyGamma[j+1]+
        PolyGamma[j+n+1])(x/2)^(2j+n),{j,0,k}]
```

This series is now evaluated as a function of the cutoff p in the summation of the last term on the right-hand side of (6.98) for values of the argument of $\xi=1$ and $\xi=5$ for Y_2 and Y_8:

```
Do[Print[N[Y[2,1,i],10]," ",N[Y[8,1,i],10]],{i,1,5}]
```

−1.650435988 −425674.6185
−1.650687662 −425674.6185
−1.650682547 −425674.6185
−1.650682607 −425674.6185
−1.650682607 −425674.6185

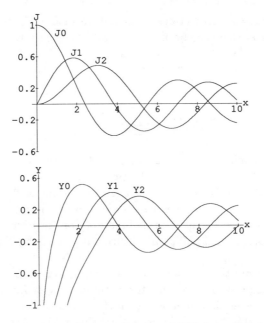

Figure 6.4 The Bessel functions $J_n(x)$ and $Y_n(x)$ for $n = 0, 1$ and 2. The Bessel functions $J_n(x)$ are regular at the origin, while the functions $Y_n(x)$ have a logarithmic singularity resulting from the origin being a regular singular point of Bessel's equation. Notice the damped oscillatory form of these functions.

```
Do[Print[N[Y[2,5,i],10]," ",N[Y[8,5,i],10]],{i,1,5}]
```

```
2.793418762   -2.813905226
-1.138992535  -2.822465489
0.8592916835  -2.820634179
0.2657078391  -2.82089412
0.3824604801  -2.82086741
```

The convergence for $Y_2(1)$ and $Y_8(1)$ is seen to be very rapid. This is because the first two terms on the right-hand side of (6.98) dominate, with the contribution from the series providing a relatively small correction. However, with increasing n, the convergence is faster for larger arguments, as can be seen by comparing the sequence of partial sums $Y_2(5)$ and $Y_8(5)$.

6.5 Generating Functions

The series solutions of the equations in the preceding sections do not suggest in any obvious way that there might be interrelationships within a given set of eigenfunctions. In this sections we introduce a closed-form representation of eigenfunctions that can be used to establish such interrelationships as well as having many other important uses. For a given set of eigenfunctions $u_n(x)$, consider a function $\mathcal{G}$ of two variables x and t which has an expansion of the form

$$\mathcal{G}(x,t) = \sum_{n=0}^{\infty} g_n t^n u_n(x) \qquad (6.104)$$

The expansion coefficients g_n are constants that are independent of x and t, though they may depend on n. The right-hand side of (6.104) shows that by expanding $\mathcal{G}$ in a Taylor series in t the coefficient of t^n is proportional to $u_n(x)$. For this reason, $\mathcal{G}$ is called the **generating function** for the eigenfunctions $u_n(x)$. The generating functions for the most common special functions and orthogonal polynomials turn out to be fairly simply functions of x and t.

Aside from the obvious aesthetic appeal of such a simple and concise representation of a set of functions, there are a number of important practical advantages that generating functions offer. Generating functions are useful for exhibiting the interrelationships within a particular set of functions, such as recursion relations, which are expressions for a particular function, $u_n(x)$, or a derivative of that function in terms of functions with adjacent indices, e.g. $u_{n\pm1}(x)$. Another use of generating functions is the calculation of integrals involving either the $u_n(x)$ and products of the $u_n(x)$ and other functions. Although the method of Frobenius provides explicit series expansions for these eigenfunctions, it is much easier use generating functions to evaluate a particular type of integral involving the u_n and express the result as a closed-form expression as a function of n.

We will see in Chapter 8 that the generating function for a particular set of eigenfunctions are closely related to integral representations of the solutions of the governing equation. The methods used to derive integral representations will provide a systematic way of obtaining generating functions for those functions. However, it is also sometimes possible to obtain a generating function by a simple physical argument. In this section, we will follow this latter course and derive the generating function for Legendre polynomials from a particular solution of Laplace's equation in spherical polar coordinates. The use of the generating function will then be illustrated by obtaining some useful properties of Legendre polynomials. An example involving the generating function of Hermite polynomials is provided in the problems at the end of the chapter.

Consider the potential at a point **r** due to a unit point charge at the point **r**$'$. From elementary electrostatics, this is given by the Coulomb potential (see Section 7.4 for a derivation). By choosing **r**$'$ to lie along the z axis, the potential is seen to be independent of the azimuthal angle, and the appropriate solution of Laplace's equation for this problem is given in terms of Legendre polynomials by (6.53):

$$\frac{1}{|\mathbf{r} - \mathbf{r}'|} = \sum_{n=0}^{\infty} (a_n r^n + b_n r^{-n-1}) P_n(\cos \varphi) \qquad (6.105)$$

where $r = |\mathbf{r}|$, $r' = |\mathbf{r}'|$ and φ is the angle between $\mathbf{r}$ and $\mathbf{r}'$. The quantities a_n and b_n are expansion coefficients that will be determined below. We let $\varphi \to 0$, so the positions of both the charge and the observation point are located on the z axis, whereupon the solution simplifies to

$$\frac{1}{|r - r'|} = \sum_{n=0}^{\infty} (a_n r^n + b_n r^{-n-1}) \qquad (6.106)$$

since the Legendre polynomials are standardized by $P_n(1) = 1$. Suppose now that $r > r'$. Then the allowed values of r include points out to infinity, so the coefficients a_n must be set equal to zero to avoid an unphysical singularity in the potential at infinity. In this case, (6.106) reduces to

$$\frac{1}{r - r'} = \sum_{n=0}^{\infty} b_n r^{-n-1} \qquad (6.107)$$

Since $r > r'$, the left-hand side of (6.107) can be expanded as

$$\frac{1}{r - r'} = \frac{1}{r} \sum_{n=0}^{\infty} \left(\frac{r'}{r}\right)^n \qquad (6.108)$$

Equating the series in (6.107) and (6.108) term by term shows that $b_n = r'^n$. Thus, since the coefficients b_n do not depend on φ, the solution (6.105) becomes

$$\frac{1}{r - r'} = \sum_{n=0}^{\infty} \frac{r'^n}{r^{n+1}} P_n(\cos \varphi) \qquad (6.109)$$

for $r > r'$.

Suppose now that $r < r'$. Then the domain of r includes the origin, and we must set $b_n = 0$ to avoid the unphysical singularity obtained if $\mathbf{r} = 0$. In this case, Equation (6.106) becomes

$$\frac{1}{r' - r} = \sum_{n=0}^{\infty} a_n r^n \qquad (6.110)$$

The a_n are obtained by a procedure similar to that used to obtain the b_n. The left hand side of (6.110) is expanded as

$$\frac{1}{r' - r} = \frac{1}{r'} \sum_{n=0}^{\infty} \left(\frac{r}{r'}\right)^n \qquad (6.111)$$

and a comparison of this series with that in (6.110) yields $a_n = r'^{-n-1}$. Since the a_n are also independent of φ, Equation (6.105) takes the form

$$\frac{1}{r'-r} = \sum_{n=0}^{\infty} \frac{r^n}{r'^{n+1}} P_n(\cos\varphi) \tag{6.112}$$

for $r < r'$. The solutions (6.109) and (6.112) can be consolidated into a single expression by introducing the notation $r_>$ and $r_<$ for the larger and smaller of r and r', respectively:

$$\frac{1}{|\mathbf{r}-\mathbf{r}'|} = \sum_{n=0}^{\infty} \frac{r_<^n}{r_>^{n+1}} P_n(\cos\varphi) \tag{6.113}$$

The final step in the derivation of the generating function is carried out by observing that the left-hand side of (6.113) is symmetric with respect to the interchange of $r_<$ and $r_>$ and can therefore be written as

$$\frac{1}{|\mathbf{r}-\mathbf{r}'|} = \frac{1}{\sqrt{r^2+r'^2-2\mathbf{r}\cdot\mathbf{r}'}} = \frac{1}{\sqrt{r_>^2+r_<^2-2r_>r_<\cos\varphi}}$$

$$= \frac{1}{r_>}\left[1+\left(\frac{r_<}{r_>}\right)^2 - 2\frac{r_<}{r_>}\cos\varphi\right]^{1/2} \tag{6.114}$$

Combining (6.113) and (6.114) and introducing the variables $t = r_</r_>$ and $x = \cos\varphi$, we obtain

$$\frac{1}{\sqrt{1+t^2-2tx}} = \sum_{n=0}^{\infty} t^n P_n(x) \tag{6.115}$$

The left-hand side of (6.115) is the generating function for Legendre polynomials. In applying (6.115), we must observe the restrictions on the allowed values of t and x: $0 \leq t < 1$ and $-1 \leq x \leq 1$. This generating function is typical of those obtained for the most common special functions (Abramowitz and Stegun, 1965) in that it is a fairly simple function of x and t.

Special functions and orthogonal polynomials can be obtained easily from their generating function in *Mathematica*. For the Legendre polynomials, we first construct the right-hand side of (6.115):

```
G[x,t]=1/Sqrt[1+t^2-2 t x];
```

The Legendre polynomial of nth order is obtained by differentiating this function n times with respect to t, dividing by $n!$ and then setting $t=0$:

```
P[n_,x_]:=Together[D[G[x,t],{t,n}]/n!/.t->0];
```

The Legendre polynomials can now be generated directly from P[n,x]. For example, to obtain $P_5(x)$, we simply enter

```
P[5,x]

       3       5
15 x - 70 x  + 63 x
-------------------
         8
```

The generating function can also be used to obtain relationships among Legendre polynomials for neighboring orders n. By differentiating (6.115) with respect to t, we obtain

$$\frac{\partial \mathcal{G}}{\partial t} = \frac{x-t}{(1+t^2-2tx)^{3/2}} = \frac{x-t}{1+t^2-2tx}\mathcal{G} \tag{6.116}$$

This equation can be rearranged as

$$(1+t^2-2tx)\frac{\partial \mathcal{G}}{\partial t} + (t-x)\mathcal{G} = 0 \tag{6.117}$$

The left-hand side of this equation can be written in terms of Legendre polynomials by using (6.115) and performing the indicated operations:

$$(1+t^2-2tx)\frac{\partial \mathcal{G}}{\partial t} + (t-x)\mathcal{G}$$

$$= \sum_{n=0}^{\infty} nt^{n-1}P_n + \sum_{n=0}^{\infty}(n+1)t^{n+1}P_n - \sum_{n=0}^{\infty}(2n+1)xt^n P_n$$

$$= \sum_{k=0}^{\infty}[kP_{k-1} + (k+1)P_{k+1} - (2k+1)xP_k]t^k \tag{6.118}$$

Since, according to (6.117), this expression must vanish, we can equate each power of t separately to zero to obtain a recurrence relation for the Legendre polynomials:

$$(2k+1)xP_k(x) = kP_{k-1}(x) + (k+1)P_{k+1}(x) \tag{6.119}$$

Thus, this recursion relation expresses the Legendre polynomial $P_{k+1}(x)$ in terms of simple algebraic terms involving $P_k(x)$ and $P_{k-1}(x)$.

Mathematica can be used to verify the recursion relations explicitly. Equation (6.119) is first entered by writing the kth recursion relation in terms of the in-built function LegendreP as

```
Recursion[k_]:=Simplify[k LegendreP[k-1,x]+
    (k+1)LegendreP[k+1,x]-(2k+1) x LegendreP[k,x]==0]
```

We can now test this for any value of k:

```
Recursion[10]
```

```
True
```

The recurrence relation (6.119) can also be used to generate the hierarchy of Legendre polynomials by specifying only the 'initial conditions' $P_0(x)=1$ and $P_1(x)=x$.

```
P[0,x_]:=1
```

```
P[1,x_]:=x
```

```
P[n_,x_]:=Together[((2n-1)x P[n-1,x]-(n-1)P[n-2,x])/n]
```

Any of the Legendre polynomials can now be generated directly from the recursion relations. For example,

```
P[5,x]
```

```
           3       5
15 x - 70 x  + 63 x
---------------------
          8
```

Further Reading

There are many textbooks that cover various aspects of special functions. Morse and Feshbach (1953) and Whittaker and Watson (1963) provide a comprehensive discussion of different methods of deriving special functions, with Morse and Feshbach (1953) giving a particularly clear treatment of generating functions and related topics. One of the standard references for all aspects of Bessel functions is Watson (1952). A compilation of properties, integral representations, interrelationships and numerical tables of special functions may be found in Abramowitz and Stegun (1965) and Erdelyi et al. (1953). Both tables and illustrations of special functions are provided by Jahnke and Emde (1945). Discussions similar in scope but covering a wider variety of special functions than the discussion in this chapter are given by Bell (1968), Hochstadt (1971) and Andrews (1985). Hochstadt (1971) also addresses the completeness of orthogonal polynomials in some detail. A treatment of special functions with examples of applications may be found in Pauling and Wilson (1935), Hildebrand (1962), Lebedev (1972) and Jackson (1975).

References

Abramowitz M. and Stegun I. A. (1965). *Handbook of Mathematical Functions.* New York: Dover.

Andrews L. C. (1985). *Special Functions for Engineers and Applied Mathematicians.* New York: Macmillan.

Bell W. W. (1968). *Special Functions for Scientists and Engineers.* London: Van Nostrand.

Copson E. T. (1935). *An Introduction to the Theory of Functions of a Complex Variable.* Oxford: Oxford University Press.

Erdelyi A., Magnus W., Oberhettinger F. and Tricomi F. G. (1953). *Higher Transcendental Functions* (Bateman Manuscript Project) Vols. 1–3. New York: McGraw-Hill.

Hildebrand F. B. (1962). *Advanced Calculus for Applications.* Englewood Cliffs NJ: Prentice-Hall.

Hochstadt H. (1971). *The Functions of Mathematical Physics.* New York: Wiley.

Jackson J. D. (1975). *Classical Electrodynamics* 2nd edn. New York: Wiley.

Jahnke E. and Emde F. (1945). *Tables of Functions.* New York: Dover.

Lebedev N. N. (1972). *Special Functions and their Applications* (translated and edited by R. A. Silverman). New York: Dover.

Morse P. M. and Feshbach H. (1953). *Methods of Theoretical Physics* Vol. 1. New York: McGraw-Hill.

Pauling L. and Wilson E. B. (1935). *Introduction to Quantum Mechanics.* New York: McGraw-Hill.

Reif F. (1965). *Fundamentals of Statistical and Thermal Physics.* New York: McGraw-Hill.

Watson G. N. (1952). *A Treatise on the Theory of Bessel Functions* 2nd edn. Cambridge: Cambridge University Press.

Whittaker E. T. and Watson G. N. (1963). *A Course of Modern Analysis* 4th edn. Cambridge: Cambridge University Press.

Problems

1. Another way to arrive at Hermite's equation is through the **Fokker–Planck equation** of nonequilibrium statistical mechanics (Reif, 1965):

$$\frac{\partial u}{\partial t} = \frac{\partial^2 u}{\partial x^2} + \frac{\partial}{\partial x}(xu)$$

Write the solution to this equation as $u(x,t) = X(x)T(t)$ and apply the separation of variables method to obtain the two equations

$$\frac{dT}{dt} + kT = 0, \qquad \frac{d^2X}{dx^2} + x\frac{dX}{dx} + (1+k)X = 0$$

where k is the separation constant. Then, with $X(x)$ taken to be of the form

$$X(x) = \exp(-\tfrac{1}{2}x^2)f(x)$$

show that after a rescaling of the independent variable, the equation for f becomes Hermite's equation:

$$\frac{d^2 f}{d\xi^2} - 2\xi\frac{df}{d\xi} + 2kf = 0$$

Hence, obtain the solution to the Fokker–Planck equation in the form

$$u(x,t) = \sum_{k=0}^{\infty} a_n \exp(-kt - \tfrac{1}{2}x^2) H_k(\tfrac{1}{2}\sqrt{2}x)$$

where the a_n are expansion coefficients (see Problem 2). What is the limiting form of this solution as $t \to \infty$?

2. The expansion coefficients a_k in Problem 1 can be determined once the initial value of the solution of the Fokker–Planck equation is specified, $u(x,0) = F(x)$. Thus, by writing the ordinary differential equation for X in the Liouville form shown in Equation (4.86), show that the coefficients a_k of the eigenfunction expansion are given by

$$a_k = \frac{1}{\mathcal{N}_k}\int_{-\infty}^{\infty} F(x) H_k(\tfrac{1}{2}\sqrt{2}x)\, dx$$

where $\mathcal{N}_k$ is the normalization integral of the $X_k(x)$. Show that this integral can be written as

$$\mathcal{N}_k = \sqrt{2}\int_{-\infty}^{\infty} \exp(-x^2) H_k(x) H_k(x)\, dx$$

The integral in this expression, which is the normalization integral for the Hermite polynomials, is evaluated in Problem 39.

3. Use *Mathematica* to construct the two series solutions of Equation (6.3) about the origin by choosing a case where the quantities a and b satisfy the requirement $2\lambda = a/b - 1$ and a case where they do not. By comparing the first several terms in each series, deduce that the two solutions are the same as those obtained from the series with the coefficients in (6.15) multiplied by $\exp(-\frac{1}{2}bx^2)$.

4. The solution to Equation (6.5) appropriate for large values of x can be obtained by first introducing a new independent variable, $\xi = x^2$. Then by writing (6.5) in terms of ξ and applying the method of Frobenius, deduce that the solution for large values of ξ satisfies the differential equation

$$\frac{d^2\psi}{d\xi^2} - \frac{1}{4}b^2\psi = 0$$

Hence, show that the expression shown in Equation (6.6) is the solution to this equation that remains finite as $x \to \infty$.

5. By choosing $\lambda = 2k$ in (6.15a) and $\lambda = 2k+1$ in (6.15b) obtain the solutions of the even-order and odd-order Hermite polynomials in the form

$$H_{2k}(x) = a_0 \sum_{n=0}^{k} (-1)^k \frac{k!}{(2n)!(k-n)!} (2x)^{2n}$$

$$H_{2k+1}(x) = a_1 \sum_{n=0}^{k} (-1)^n \frac{k!}{(2n+1)!(k-n)!} (2x)^{2n+1}$$

By requiring the coefficient of x^n in $H_n(x)$ to be equal to 2^{2n} determine a_0 and a_1, and obtain the standard Hermite polynomials in (6.22). Show that these two forms can be combined into a single expression as a *descending* series:

$$H_n(x) = \sum_{k=0}^{[n/2]} (-1)^k \frac{n!}{k!(n-2k)!} (2x)^{n-2k}$$

where

$$[\tfrac{1}{2}n] = \begin{cases} \tfrac{1}{2}n, & \text{if } n \text{ is even} \\ \tfrac{1}{2}(n-1), & \text{if } n \text{ is odd} \end{cases}$$

6. Use *Mathematica* to construct and display solutions of Hermite's equation from the solutions of the recursion relations in (6.15) for the case where λ is *not* an integer. The infinite series solutions of Hermite's equation with

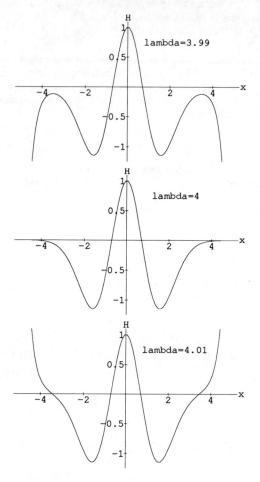

Figure 6.5 The solutions to Hermite's equation (6.9) corresponding to three values of λ: $\lambda=3.99$ (top), $\lambda=4.00$ (center) and $\lambda=4.01$ (bottom).

the coefficients in (6.15) are called **Hermite functions**. The solutions for $\lambda=3.99$, $\lambda=4.00$ and $\lambda=4.01$ are shown in Figure 6.5.

7. The confluent hypergeometric equation

$$x\frac{d^2y}{dx^2} + (\gamma - x)\frac{dy}{dx} - \alpha y = 0$$

was solved in Problem 5.21. In doing so, the roots of the indicial equation were found to be $s_1 = 0$ and $s_2 = 1-\gamma$. Use this result to observe that the equation obtained with the choice $\gamma = \frac{1}{2}$ together with the change of variables to $\xi = \sqrt{x}$ has *regular* point at the origin while infinity remains an irregular singular point.

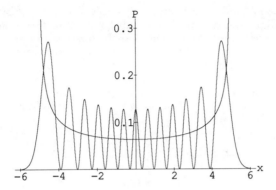

Figure 6.6 Comparison of the normalized probability calculated from the Weber–Hermite function $\psi_{12}(\xi)$ compared with the probability distribution for a classical harmonic oscillator with the same total energy.

Motivated by this observation, show that the Hermite polynomials of even order can be expressed in terms of the solution $M(\alpha; \frac{1}{2}; x)$ of the confluent hypergeometric function for an appropriate choice of α. By comparing the series representations in (6.22) for $H_{2n}(\xi)$ with that obtained in Problem 5.21 for $M(\alpha; \gamma; x)$ show that

$$H_{2n}(\xi) = (-1)^n \frac{(2n)!}{n!} M(-n; \tfrac{1}{2}; \xi^2)$$

8. Show that the trajectory of a classical harmonic oscillator with energy $E = \hbar\omega_0(n+\frac{1}{2})$ is given by

$$\xi = \sqrt{2n+1}\cos(\omega_0 t)$$

where ξ is the dimensionless position, $\xi = bx$, $b = m\omega_0/\hbar$ and $\omega_0 = \sqrt{k/m}$ is the natural frequency of the oscillator. Obtain the classical probability that the position of the particle is between ξ and $\xi + d\xi$:

$$P(\xi)\,d\xi = \frac{1}{\pi}\frac{d\xi}{\sqrt{2n+1-\xi^2}}$$

Compare this expression with that obtained for the *normalized* Weber–Hermite functions with increasing order n. An example is shown in Figure 6.6. This example illustrates some important similarities and differences between the classical trajectory and the wavefunction at large order. A discussion has been given by Pauling and Wilson (1935).

9. Determine the two solutions of **Laguerre's equation**,

$$x\frac{d^2y}{dx^2} + (1-x)\frac{dy}{dx} + ny = 0$$

by applying the method of Frobenius. Show that the requirement of regularity at the origin eliminates one of the two solutions. Show furthermore that normalizability imposes the requirement that for large values of x the solutions must grow more slowly than $\exp(\frac{1}{2}x)$. By examining the recursion relations at large order, deduce that the asymptotic form of the solutions obtained with the Frobenius method grows like e^x for large x. Hence, for reasons analogous to those used in Section 6.2 for Hermite's equation, deduce that the series obtained with the Frobenius method must truncate, and obtain the **Laguerre polynomials** of order n:

$$L_n(x) = \sum_{k=0}^{n} (-1)^k \frac{n!}{(n-k)!(k!)^2} x^k$$

where the standard choice for the constant value in the expansion is $a_0 = 1$.

10. In solving Schrödinger's equation for the hydrogen atom (Pauling and Wilson, 1935), the separation of variables in spherical polar coordinates produces the **associated Laguerre equation** for the radial function:

$$x \frac{d^2 y}{dx^2} + (m + 1 - x) \frac{dy}{dx} + ny = 0$$

Show that solutions of this equation are related to the solutions of the ordinary Laguerre equation of order $m+n$,

$$x \frac{d^2 y}{dx^2} + (1 - x) \frac{dy}{dx} + (m + n) y = 0$$

by differentiating the latter equation m times with respect to x. In particular, deduce that the associated Laguerre polynomials $L_n^m(x)$ are proportional to the mth derivative with respect to x of the Laguerre polynomials, $L_n(x)$, derived in Problem 9. By choosing the proportionality constant between $L_n(x)$ and $L_n^m(x)$ to be $(-1)^m$ and using the polynomial solutions obtained in Problem 9, obtain the corresponding polynomial solutions of the associated Laguerre equation as

$$L_n^m(x) = \sum_{k=0}^{n} (-1)^k \frac{(n+m)!}{(n-k)!(m-k)!k!} x^k$$

11. The associated Laguerre equation in Problem 10 is a particular case of the confluent hypergeometric equation. Thus, by comparing the series for the associated Laguerre polynomials determined in Problem 10 with that for the function $M(\alpha; \gamma; x)$ in Problem 5.21, show that these two functions are related by

$$L_n^m(x) = \frac{(n+m)!}{n!m!} M(-n; m+1; x)$$

Use this result and that obtained in Problem 7 to deduce that

$$L_n^{-1/2}(x) = \frac{(-1)^n}{2^n n!} H_{2n}(\sqrt{x})$$

12. Show that the even-order and odd-order Legendre polynomials in (6.51) can be combined into a single expression in terms of a descending series as

$$P_\ell(x) = \sum_{k=0}^{[\ell/2]} (-1)^k \frac{(2\ell - 2k)!}{2^\ell k!(\ell - k)!(\ell - 2k)!} x^{\ell-2k}$$

The quantity $[\frac{1}{2}\ell]$ has the same meaning as in Problem 5.

13. Use induction to show that differentiating the ordinary Legendre equation $|m|$ times with respect to ξ yields

$$(1-\xi^2)\frac{d^2}{dx^2}\frac{d^{|m|}P_\ell}{dx^{|m|}} - 2(|m|+1)\frac{d}{dx}\frac{d^{|m|}P_\ell}{dx^{|m|}} + [\ell(\ell+1) - |m|(|m|+1)]\frac{d^{|m|}P_\ell}{dx^{|m|}} = 0$$

Hence, deduce that

$$F_{\ell,m}(\xi) \propto \frac{d^{|m|}P_\ell(\xi)}{dx^{|m|}}$$

and so obtain the representation of the Legendre functions in Equation (6.66):

$$P_\ell^{|m|}(\xi) = (1 - \xi^2)^{|m|/2} \frac{d^{|m|}P_\ell(\xi)}{d\xi^{|m|}}$$

14. Show that changing the independent variable to $\xi = 1/x$ transforms the ordinary Legendre equation,

$$(1-x^2)\frac{d^2y}{dx^2} - 2x\frac{dy}{dx} + \ell(\ell+1)y = 0$$

into the following equation:

$$(\xi^2 - 1)\frac{d^2y}{d\xi^2} + 2\xi\frac{dy}{d\xi} + \frac{1}{\xi^2}\ell(\ell+1)y = 0$$

Identify and characterize the singular points of this equation. Obtain the *regular* solution about $\xi = 0$ (i.e. $x = \infty$) by using the method of Frobenius. Show that the recursion relations with the appropriate root of the indicial equation require that $a_1 = 0$ and, for $n \geq 2$, can be written as

$$a_n = \frac{(n+\ell)(n+\ell-1)}{n(n+2\ell+1)} a_{n-2}$$

Determine the radius of convergence of this series. By choosing the value of a_0 to be

$$a_0 = 2^\ell \frac{\ell!\ell!}{(2\ell+1)!}$$

(the factor of 2^ℓ is a convention), show that this solution of Legendre's equation as a function of the original variable x, which is signified as $Q_\ell(x)$, is

$$Q_\ell(x) = 2^\ell x^{-\ell-1} \sum_{k=0}^{\infty} \frac{(\ell+k)!(\ell+2k)!}{k!(2\ell+2k+1)!} x^{-2k}$$

This solution is called the **Legendre function of the second kind**. Although our derivation of the Q_ℓ is valid only for $|x|>1$, corresponding functions can also be derived for $|x|<1$.

15. Beginning with Equation (6.56), make the change of variables $\zeta = \frac{1}{2}(1-\xi)$ to obtain

$$\zeta(1-\zeta)\frac{d^2 F}{d\zeta^2} - (|m|+1)(1-2\zeta)\frac{dF}{d\zeta} + \left[\ell(\ell+1) - |m|(|m|+1)\right] F = 0$$

By identifying this equation as a particular case of the hypergeometric equation, derive the relationship between the Legendre functions $P_\ell^m(\xi)$ and the hypergeometric function. Do this in two steps. First, set $m = 0$ and deduce that the Legendre polynomials are proportional to $F[-n, n+1; 1; \frac{1}{2}(1-\xi)]$. By examining the normalization of the Legendre polynomials at, say, $\xi = -1$ and computing the corresponding value of the hypergeometric function, show that

$$P_n(\xi) = F[-n, n+1; 1; \frac{1}{2}(1-\xi)]$$

The Legendre functions can now be obtained from the relation obtained in Problem 13:

$$P_\ell^{|m|}(\xi) = (1-\xi^2)^{|m|/2} \frac{d^{|m|} P_\ell(\xi)}{d\xi^{|m|}}$$

Show that the derivatives of the hypergeometric function are given by

$$\frac{d^k}{dx^k} F(\alpha, \beta; \gamma; x) = \frac{(\alpha)_k (\beta)_k}{(\gamma)_k} F(\alpha+k, \beta+k; \gamma+k; x)$$

for any nonnegative integer k. The notation $(\alpha)_k$, $(\beta)_k$ and $(\gamma)_k$ is defined in Problem 5.15. Hence, obtain

$$P_n^m(\xi) = \frac{1}{2^m m!} \frac{(n+m)!}{(n-m)!} (1-\xi^2)^{m/2} F[m-n, m+n+1; m+1; \frac{1}{2}(1-\xi)]$$

where m is a nonnegative integer.

16. By comparing the series expansion for $Q_\ell(x)$ in Problem 14 with that for the hypergeometric function in Problem 5.15, deduce

$$Q_\ell(x) = \frac{1}{2x^{\ell+1}} \frac{\Gamma(\ell+1)\Gamma(\frac{1}{2}+1)}{\Gamma(\ell+\frac{3}{2})} F(\ell+1, \frac{1}{2}\ell+1; \ell+\frac{3}{2}; x^{-2})$$

Motivated by this observation, perform the substitution

$$F(\xi) = \xi^{-\ell-1} u(\xi^{-2})$$

into Equation (6.56), and obtain the solution in the form

$$Q_\ell^m(x) = (-1)^m \frac{1}{2x^{\ell+1}} \frac{\Gamma(\ell+1)\Gamma(\frac{1}{2}+1)}{\Gamma(\ell+\frac{3}{2})} (\xi^2 - 1)^{m/2}$$
$$\times F[\ell+m+1, \tfrac{1}{2}(\ell+m)+1; \ell+\tfrac{3}{2}; x^{-2}]$$

where $m = 0, 1, 2, \ldots$. The functions $Q_\ell^m(x)$ are **associated Legendre functions of the second kind** of order ℓ.

17. A sum rule for the spherical harmonics, known as **Unsöld's theorem**, states that

$$\sum_{m=-\ell}^{\ell} |Y_{\ell,m}(\theta,\phi)|^2 = \frac{2\ell+1}{4\pi}$$

This relation can be obtained as a special case of what is known as an addition theorem for the spherical harmonics (Jackson, 1975). Using *Mathematica*, or otherwise, verify this result for several values of ℓ. If using *Mathematica*, it is most convenient to call the in-built function SphericalHarmonicY[l,m,theta,phi].

18. The **Gamma function**, signified by $\Gamma(x)$, was introduced by Euler as a way of generalizing the factorial to noninteger numbers x. Two common ways of defining the gamma function are in the form of a product

$$\Gamma(x) = \lim_{k \to \infty} \frac{k! k^x}{x(x+1) \cdots (x+k)}$$

and through the integral representation

$$\Gamma(x) = \int_0^\infty e^{-t} t^{x-1} \, dt$$

The product representation is valid provided x is neither a negative integer nor zero, while the integral representation is restricted to positive values of x. The equivalence of these two representations is discussed by Andrews (1985).

For both definitions, show that

$$\Gamma(1) = 1$$

and obtain the recursion relation

$$\Gamma(x+1) = x\Gamma(x)$$

Hence, deduce by induction that for nonnegative integers n,

$$\Gamma(n) = n!$$

19. Use *Mathematica* to evaluate the two definitions of $\Gamma(x)$ in Problem 18 for a range of positive values of x to demonstrate their equivalence. Use the command Product to construct the product representation as

```
GammaProduct[x_,n_]:=k!k^x Product[1/(x+k),{k,0,n}]}
```

and use NIntegrate to evaluate the integral representation numerically with the command

```
GammaIntegral[x_]:=NIntegrate[Exp[-t] t^(x-1),{t,0,Infinity}]
```

20. Use the integral representation of the gamma function to obtain

$$\Gamma(\tfrac{1}{2}) = \sqrt{\pi}$$

From this result, deduce

$$\Gamma(n + \tfrac{1}{2}) = \frac{1}{2^n} 1 \cdot 3 \cdot 5 \cdots (2n-1)\sqrt{\pi} = \frac{(2n)!}{2^{2n} n!}\sqrt{\pi}$$

21. The Gamma function can be used to evaluate integrals of the form

$$\int_0^{\pi/2} \cos^{2x-1}\theta \sin^{2y-1}\theta \, d\theta$$

by following a procedure commonly applied to evaluating Gaussian integrals. First, show that the integral representation in Problem 18 can be written as

$$\Gamma(x) = 2\int_0^\infty e^{-t^2} t^{2x-1}\, dt$$

by a suitable change of variables. Then, form the product $\Gamma(x)\Gamma(y)$, change integration variables to polar coordinates, and deduce

$$\int_0^{\pi/2} \cos^{2x-1}\theta \sin^{2y-1}\theta \, d\theta = \frac{\Gamma(x)\Gamma(y)}{2\Gamma(x+y)}$$

As a particular case, show that

$$\Gamma\left(\tfrac{1}{2}\right) = \sqrt{\pi}$$

22. By making the substitution $t = \cos^2\theta$, show that the integral in Problem 21 becomes

$$\int_0^{\pi/2} \cos^{2x-1}\theta \sin^{2y-1}\theta \, d\theta = 2\int_0^1 t^x(1-t)^y \, dt$$

Hence, deduce

$$\int_0^1 t^x(1-t)^y \, dt = \frac{\Gamma(x)\Gamma(y)}{\Gamma(x+y)}$$

This integral is called the **beta function**, and is signified by $B(x,y)$ (Andrews, 1985).

23. A useful identity between the gamma function and the sine function is

$$\Gamma(x)\Gamma(1-x) = \frac{\pi}{\sin \pi x}$$

Derive this result in two steps. First, use the integral representation of the gamma function in Problem 18 and an appropriate change of variables to write

$$\Gamma(x)\Gamma(1-x) = \int_0^\infty \int_0^\infty e^{-(s+t)} s^{-x} t^{x-1} \, ds \, dt = \int_0^\infty \frac{u^{x-1}}{1+u} \, du$$

The last integral can be evaluated (Andrews, 1985) by breaking the domain of integration into the region $(0,1)$ and the region $(1,\infty)$. Then, in the integral over the latter region make the change of variables $\xi = u^{-1}$, expand the denominators in both integrals as geometric series, and use the result of Problem 4.4 to obtain

$$\int_0^\infty \frac{u^{x-1}}{1+u} \, du = \pi \sin \pi x$$

24. The reliance on the integral representation of the gamma function appears to restrict the validity of the identity derived in Problem 23 to the range $0 < x < 1$. However, we can use analytic continuation to extend this formula to complex arguments z except where $\sin \pi z$ vanishes ($z = 0, \pm 1, \pm 2, \ldots$):

$$\Gamma(z)\Gamma(1-z) = \frac{\pi}{\sin \pi z}$$

Use *Mathematica* to verify this identity for various values of the argument both within the interval $0 < x < 1$ and for allowed real and complex values outside of this interval. A derivation of this identity beginning with the product representation of the gamma function has been given by Andrews (1985).

25. The **digamma function**, $\psi(x)$, is defined as the logarithmic derivative of the gamma function:

$$\psi(x) = \frac{d}{dx}\left[\log \Gamma(x)\right] = \frac{\Gamma'(x)}{\Gamma(x)}$$

By taking the logarithmic derivative of the recursion relation $\Gamma(x+1)=x\Gamma(x)$, obtain the following recursion relation for $\psi(x)$:

$$\psi(x+1) = \frac{1}{x} + \psi(x)$$

Show that if x is a positive integer n, then

$$\psi(n) = \sum_{k=1}^{n-1} \frac{1}{k} + \psi(1), \qquad n = 1, 2, \ldots$$

where the sum on the right-hand side of this equation vanishes if $n=1$.

The quantity $\psi(1) = -\gamma$, where γ is the Euler constant, can be evaluated using either the integral or product representation of the gamma function. Show from the integral representation that

$$\psi(1) = \int_0^\infty e^{-t} \ln t \, dt$$

and from the product representation that

$$\psi(1) = \lim_{k \to \infty} \left[\ln k - \sum_{n=1}^{k} \frac{1}{n} \right]$$

Use *Mathematica* to evaluate this quantity and compare with the value obtained by calculating directly the quantity $\psi(1)$ from the in-built function PolyGamma.

26. Taking the partial derivatives in Equation (6.97) to produce the Bessel function of the second kind (see Problem 27) produces the quotient

$$\frac{\psi(k-\nu+1)}{\Gamma(k-\nu+1)}$$

where ψ is the digamma function defined in Problem 25. This quotient is well-defined for positive values of the arguments, but is indeterminate if the argument takes negative integer values, since both the Gamma and digamma functions vanish there. To evaluate these quantities for negative integer values of the argument use the product representation of the Gamma function in Problem 18 to write

$$\frac{\psi(x)}{\Gamma(x)} = -\frac{d}{dx}\frac{1}{\Gamma(x)}$$

$$= \lim_{k \to \infty} \frac{x(x+1)\cdots(x+k)}{k!k^x} \left[\ln k - \frac{1}{x} - \frac{1}{x+1} - \cdots - \frac{1}{x+k} \right]$$

Then, show that

$$\lim_{k \to \infty} \frac{(k+x)!}{k!k^x} = 1$$

to obtain
$$\lim_{\nu \to n} \frac{\psi(k-\nu+1)}{\Gamma(k-\nu+1)} = (-1)^{n-k}(n-k-1)!$$
for $k=0,1,\ldots,n-1$. An alternative derivation of this result may be found in Andrews (1985).

27. Carry out the derivatives in Equation (6.97) to obtain the expression
$$\frac{\partial J_{\pm\nu}(\xi)}{\partial \nu} = \pm J_{\pm\nu}(\xi)\ln(\tfrac{1}{2}\xi) \mp \sum_{k=0}^{\infty} \frac{(-1)^k}{k!} \frac{\psi(k\pm\nu+1)}{\Gamma(k\pm\nu+1)} \left(\frac{\xi}{2}\right)^{2k\pm\nu}$$
Then, assemble the terms in (6.97) and use the result obtained in Problem 25 to obtain the series representation of the $Y_n(\xi)$ in Equation (6.98).

28. From the series solution in Equation (6.86) obtain the following recurrence relations for the Bessel functions of the first kind:
$$\frac{d}{dx}\left[x^{\pm k} J_k(x)\right] = \pm x^{\pm k} J_{k\mp 1}(x)$$
From this result deduce the following relations:
$$\frac{2k}{x} J_k(x) = J_{k-1}(x) + J_{k+1}(x)$$
$$2\frac{dJ_k(x)}{dx} = J_{k-1}(x) - J_{k+1}(x)$$
From the definition in Equation (6.96), show that the same recurrence relations apply for the Bessel functions of the second kind.

29. Given that $u(x) = AJ_k(x) + BY_k(x)$ is the general solution of Bessel's equation, where A and B are constants, show that the function
$$y(x) = Ax^\alpha J_k(\beta x^\gamma) + Bx^\alpha Y_k(\beta x^\gamma)$$
is the general solution of the equation
$$x^2\frac{d^2y}{dx^2} + (1-2\alpha)x\frac{dy}{dx} + \left[(\beta\gamma x^\gamma)^2 + \alpha^2 - k^2\gamma^2\right]y = 0$$
In the following two problems, we will use this general form of Bessel's equation to obtain the solution of some special cases.

30. Use the results of Problem 29 to show that the general solution of
$$\frac{d^2y}{dx^2} + x^{2\gamma-2}y = 0$$

is given by
$$y(x) = A\sqrt{x}\, J_{1/2\gamma}(x^\gamma/\gamma) + A\sqrt{x}\, Y_{1/2\gamma}(x^\gamma/\gamma)$$

Use this result to express the solution of
$$\frac{d^2y}{dx^2} + x^n y = 0$$

in terms of Bessel functions of the first and second kinds. Specialize these results to the value $n=0$. In particular, using the results of Problem 29, show that the solution to
$$\frac{d^2y}{dx^2} + y = 0$$

reduces to the familiar form
$$y(x) = A\cos x + B\sin x$$

For $n=1$, obtain the solution of **Airy's equation**:
$$\frac{d^2y}{dx^2} + xy = 0$$

31. Using the result of Problem 29, obtain the general solution of equations of the form
$$\frac{d^2y}{dx^2} + \frac{1}{x}\frac{dy}{dx} + x^{n-2} y = 0$$

32. Using the result of Problem 29, obtain the general solution of
$$x^2 \frac{d^2y}{dx^2} + x\frac{dy}{dx} + (\beta^2 x^2 - \nu^2) y = 0$$

Specialize your result to obtain the general solution of the **modified Bessel equation**,
$$x^2 \frac{d^2y}{dx^2} + x\frac{dy}{dx} - (\nu^2 + \beta^2 x^2) y = 0$$

in the form of Bessel functions with imaginary arguments:
$$y(x) = A J_\nu(ix) + B Y_\nu(ix)$$

Alternatively, show that the *real* solution of this equation that is regular at the origin, which we signify by $I_\nu(x)$, can be obtained directly from Equation (6.86) by multiplying the series representation of $J_\nu(ix)$ by $i^{-\nu}$:
$$I_\nu(x) = i^{-\nu} J_\nu(ix)$$

The series representation of $I_\nu(x)$ is then given by

$$I_\nu(x) = \sum_{k=0}^{\infty} \frac{1}{k!\Gamma(k+\nu+1)} \left(\frac{x}{2}\right)^{2k+\nu}$$

The function $I_\nu(x)$ is the **modified Bessel function of the first kind** of order ν. If ν is not an integer, then a second independent solution is $I_{-\nu}(x)$. If ν is equal to an integer n, the standard form of the second solution of Bessel's modified equation, which is signified by $K_n(x)$ and called the **modified Bessel function of the second kind** of order n, is given by

$$K_n(x) = \tfrac{1}{2}\pi i^{n+1}[iY_n(ix) + i^n I_n(x)] = \tfrac{1}{2}\pi i^{n+1}[iY_n(ix) + J_n(ix)]$$

The modified Bessel functions satisfy recurrence formulae similar to the Bessel functions J_ν and Y_ν (Abramowitz and Stegun, 1965; Bell, 1968; Andrews, 1985).

33. Given that $u(x) = AI_k(x) + BK_k(x)$ is the general solution of the modified Bessel's equation (Problem 32), where A and B are constants, use the method of Problem 29 to show that the expression

$$y(x) = Ax^\alpha I_k(\beta x^\gamma) + Bx^\alpha K_k(\beta x^\gamma)$$

is the general solution of

$$x^2 \frac{d^2 y}{dx^2} + (1 - 2\alpha)x \frac{dy}{dx} - \left[(\beta\gamma x^\gamma)^2 + k^2\gamma^2 - \alpha^2\right] y = 0$$

34. Use the result of Problem 33 to obtain the general solution of

$$\frac{d^2 y}{dx^2} - x^{2\gamma - 2} y = 0$$

Specialize this result to obtain the solution of Equation (6.5). In Chapter 8, the asymptotic form of the Bessel functions and modified Bessel functions will be obtained from the integral representations of these functions and the asymptotic solution derived in Problem 3 will emerge naturally from this general solution.

35. A special case of Bessel's equation that arises in several applications is obtained by solving the **Helmholtz equation**,

$$\nabla^2 \psi + k^2 \psi = 0$$

in spherical polar coordinates. This equation is encountered in solving the wave equation, the heat equation and the Schrödinger equation. By applying the method of separation of variables to the Helmholtz equation, show that the following differential equation is obtained for the radial function:

$$x^2 \frac{d^2 y}{dx^2} + 2x \frac{dy}{dx} + \left[k^2 x^2 - n(n+1)\right] y = 0$$

where n is a nonnegative integer that enters as a separation constant. By introducing the function $z(x) = \sqrt{x} y(x)$, show that the general solution of this equation can be expressed in terms of half-integral order Bessel functions as:

$$y(x) = A x^{-1/2} J_{n+1/2}(kx) + B x^{-1/2} Y_{n+1/2}(kx)$$

The quantities A and B are constants. These solutions are customarily expressed in terms of **spherical Bessel functions**, $j_n(x)$ and $y_n(x)$, of the first and second kind, respectively:

$$j_n(x) = \sqrt{\frac{\pi}{2x}} J_{n+1/2}(x), \qquad y_n(x) = \sqrt{\frac{\pi}{2x}} Y_{n+1/2}(x)$$

36. Using the series solution for the Bessel function J_ν in Equation (6.86), show directly that

$$j_0(x) = \frac{\sin x}{x}$$

Then, using the recurrence formulae for the Bessel functions obtained in Problem 28, show that

$$j_1(x) = \frac{\sin x}{x^2} - \frac{\cos x}{x}$$

$$j_2(x) = \left(\frac{3}{x^3} - \frac{1}{x}\right) - \frac{3}{x^2} \cos x$$

Similarly, using the definition (6.96) for the Bessel function of the second kind, the definition of the corresponding spherical function in Problem 35, and the recurrence relations in Problem 28, show that

$$y_0(x) = -\frac{\cos x}{x}$$

$$y_1(x) = -\frac{\cos x}{x^2} - \frac{\sin x}{x}$$

$$y_2(x) = -\left(\frac{3}{x^3} - \frac{1}{x}\right) \cos x - \frac{3}{x^2} \sin x$$

37. The quantum mechanical Hamiltonian for the harmonic oscillator is, in dimensionless form, given by

$$\mathcal{H} = -\frac{d^2}{dx^2} + x^2$$

The raising and lowering operators, a^+ and a^-, respectively, are given by

$$a^+ = -\frac{\mathrm{d}}{\mathrm{d}x} + x, \qquad a^- = \frac{\mathrm{d}}{\mathrm{d}x} + x$$

Given that the (unnormalized) ground-state wavefunction is $\exp(-\frac{1}{2}x^2)$, show that the wavefunction for the nth level (which is the Hermite polynomial $H_n(x)$) is, in the standard normalization for Hermite polynomials, given by

$$H_n(x) = (-1)^n \exp(\tfrac{1}{2}x^2)\left(\frac{\mathrm{d}}{\mathrm{d}x} - x\right)^n \exp(-\tfrac{1}{2}x^2)$$

Hence, deduce that

$$H_n(x) = (-1)^n \exp(x^2)\frac{\mathrm{d}^n}{\mathrm{d}x^n} \exp(-x^2)$$

38. Use *Mathematica* to construct the two representations of the Hermite polynomials in Problem 37. The second of the two representations is straightforward. The first can done most easily by first defining the function

```
F[f_]:=D[f,x]-x f
```

and then using the command Nest to apply this operation repeatedly to $\exp(-\frac{1}{2}x^2)$:

```
H[x_,n_]:=Simplify[(-1)^nExp[x^2/2]Nest[F,Exp[-x^2/2],n]]
```

39. From the results of Problem 37, derive the generating function for the Hermite polynomials:

$$\exp(-t^2 - 2tx) = \sum_{n=0}^{\infty} \frac{t^n}{n!} H_n(x)$$

Use this generating function to obtain the recursion relation

$$\frac{\mathrm{d}H_n}{\mathrm{d}x} = 2nH_{n-1}$$

and the normalization

$$\int_{-\infty}^{\infty} \exp(-x^2) H_m(x) H_n(x)\, \mathrm{d}x = \begin{cases} 2^n n!\sqrt{\pi} & \text{if } m = n \\ 0 & \text{if } m \neq n \end{cases}$$

To derive the normalization, form the product

$$\exp(-t^2 - 2tx)\exp(-u^2 - 2ux) = \sum_{n=0}^{\infty} \frac{t^n}{n!} H_n(x) \sum_{m=0}^{\infty} \frac{u^m}{m!} H_m(x)$$

Then, multiply both sides by the weight function and integrate over x. Perform the integral on the left-hand side and equate the coefficients of $t^n u^m$ on both sides of the equation to deduce the orthogonality and normalization condition.

40. Expansions in series of special functions are encountered frequently in applications and both integral representations (Chapter 8) and generating functions can be used for evaluating the expansion coefficients. For a function $f(x)$ the expansion in terms of Hermite polynomials take the general form

$$f(x) = \sum_{k=0}^{\infty} a_k H_k(x)$$

Use the orthogonality relation for the Hermite polynomials to show that the expression for the expansion coefficients a_k is given by

$$a_k = \frac{1}{2^k k! \sqrt{\pi}} \int_{-\infty}^{\infty} \exp(-x^2) f(x) H_k(x) \, dx$$

Consider the expansion of the function $f(x) = x^n$, where n is a nonnegative integer, in a series of Hermite polynomials. By using the generating function for the $H_k(x)$ obtained in Problem 39, perform the required number of integrations by parts to obtain the following Hermite polynomial representations of x^n:

$$x^{2n} = \frac{(2n)!}{2^{2n}} \sum_{k=0}^{n} \frac{H_{2k}(x)}{(2k)!(n-k)!}$$

$$x^{2n+1} = \frac{(2n+1)!}{2^{2n+1}} \sum_{k=0}^{n} \frac{H_{2k+1}(x)}{(2k+1)!(n-k)!}$$

Confirm this result for several values of n by constructing the summations on the right-hand side of these equations using *Mathematica*.

41. Apply the procedure in Problem 40 to obtain the corresponding Hermite polynomial expansion for the function $f(x) = e^x$:

$$e^x = e^{1/4} \sum_{k=0}^{\infty} \frac{1}{2^k k!} H_k(x)$$

Use *Mathematica* to construct the partial sums of this expansion to assess the way the series converges. Use the Plot command to display both the function and its partial sums on the same graph.

42. The generating function in (6.115) can be used to obtain the normalization of the Legendre polynomials. First, construct the inner product,

$$\sum_{m=0}^{\infty} \sum_{n=0}^{\infty} \int_{-1}^{1} P_m(x) P_n(x) \, dx = \int_{-1}^{1} \frac{dx}{1 + t^2 - 2tx}$$

Use the orthogonality of the Legendre polynomials (as solutions of a Liouville equation) to eliminate the terms in the summations where $m \neq n$. Then evaluate the integral on the right-hand side, expand the resulting expression as a power series in t, and equate term by term to the left-hand side to obtain

$$\int_{-1}^{1} P_\ell(x) P_{\ell'}(x)\, dx = \frac{2}{2\ell+1} \delta_{\ell,\ell'}$$

43. Beginning with the binomial series expansion of $(x^2-1)^n$,

$$(x^2-1)^n = \sum_{k=0}^{n} (-1)^k \frac{n!}{k!(n-k)!} x^{2n-2k}$$

differentiate both sides n times with respect to x, and compare with the series representation of the Legendre polynomials derived in Problem 12 to deduce **Rodrigues' formula**:

$$P_n(x) = \frac{1}{2^n n!} \frac{d^n}{dx^n}\left[(x^2-1)^n\right]$$

Use this representation to confirm the normalization of the Legendre polynomials obtained in Problem 42.

44. The Rodrigues formula derived in Problem 43 can be used to obtain a generating function for the Legendre functions. By substituting this formula into (6.66), deduce that the resulting expression is valid provided ℓ and m satisfy the inequality $\ell + |m| \geq 0$, so the absolute value signs on m can be removed to obtain a generating function valid for $m = -\ell, \ldots, \ell$:

$$P_\ell^m(x) = \frac{1}{2^\ell \ell!} (1-x^2)^{m/2} \frac{d^{\ell+m}}{dx^{\ell+m}}\left[(x^2-1)^\ell\right]$$

Then, using Leibniz's formula for the derivatives of the product of two functions f and g,

$$\frac{d^n(fg)}{dx^n} = \sum_{k=0}^{n} \frac{n!}{k!(n-k)!} \frac{d^{n-k}f}{dx^{n-k}} \frac{d^k g}{dx^k}$$

show that P_ℓ^m and P_ℓ^{-m} are related by (6.68):

$$P_\ell^{-m}(x) = (-1)^m \frac{(\ell-m)!}{(\ell+m)!} P_\ell^m(x)$$

45. Use the relation between P_ℓ^m and P_ℓ^{-m} to write the orthogonality integral for the Legendre functions as

$$\int_{-1}^{1} P_\ell^m(x) P_{\ell'}^m(x)\, dx = (-1)^m \frac{(\ell-m)!}{(\ell+m)!} \int_{-1}^{1} P_\ell^m(x) P_{\ell'}^{-m}(x)\, dx$$

Then, use Rodrigues' formula for the Legendre functions to obtain

$$\int_{-1}^{1} P_\ell^m(x) P_{\ell'}^m(x) \, dx = \begin{cases} 0 & \text{if } \ell \neq \ell' \\ \dfrac{2}{2\ell+1} \dfrac{(\ell+m)!}{(\ell-m)!} & \text{if } \ell = \ell' \end{cases}$$

Hence deduce the normalization for the spherical harmonics in (6.69):

$$Y_{\ell,m}(\theta,\phi) = \sqrt{\frac{2\ell+1}{4\pi} \frac{(\ell-m)!}{(\ell+m)!}} P_\ell^m(\cos\theta) e^{im\phi}$$

Show that this normalization when combined with the relation (6.68) obtained in Problem 44 implies (6.70):

$$Y_{\ell,-m}(\theta,\phi) = (-1)^m Y_{\ell,m}^*(\theta,\phi)$$

Chapter 7

Transform Methods and Green's Functions

One of the most elegant and versatile uses of the superposition principle is the representation of solutions of partial differential equations in terms of Green's functions. This approach shares with the eigenfunction method a reliance on a linear combination of particular solutions to construct a more flexible and general solution to a differential equation. However, where the eigenfunction method expresses the solution as a discrete sum, the solution obtained by using Green's functions is expressed as an integral involving the Green's function and the initial and boundary conditions. This offers several advantages over eigenfunction expansions. For example, an integral representation provides a direct way of identifying the general analytical structure of a solution that may be obscured by an infinite series representation. On a more practical level, the evaluation of a solution from an integral representation may prove simpler than summing an infinite series, particularly near rapidly-varying features of a function, where the convergence of an eigenfunction expansion is expected to be slow. In fact, bearing in mind the Gibbs' phenomenon discussed in Chapter 4, the integral representation places less stringent requirements on the functions that describe the initial conditions or the values a solution is required to take on a boundary than expansions based on eigenfunctions.

In this chapter we will examine special solutions of the diffusion, Poisson and wave equations known as 'fundamental solutions' to illustrate the basic ideas behind the construction and implementation of Green's functions. The importance and utility of the fundamental solution can be seen most easily by examining its role in solving Poisson's equation in electrostatics. Consider a collection of charges confined to some region of space without any boundaries. The electrostatic potential due to the charges is simply the sum of the Coulomb potentials due to each charge individually, and the resulting potential is a solution of Poisson's equation for that charge distribution. The fundamental solution of Poisson's equation is the Coulomb potential for an individual *point* charge. The effects of boundaries can be included by applying the method of images to take an appropriate linear combination of fundamental solutions corresponding to the physical and image charges to satisfy the boundary condition. This linear combination produces the Green's function in the presence of the boundary.

The fundamental solutions of other linear partial differential equations play analogous roles. The equation is first solved for a point source without any boundaries. The source may be a single point charge, as in the electrostatic problem, a localized concentration or source of heat for the diffusion equation, or a localized disturbance for the wave equation. The solution for a more general source is then constructed by adding together the fundamental solutions for each component of the source. The effect of boundaries is included by using a generalized method of images to take appropriate linear combinations of the fundamental solution to obtain the Green's function that satisfies the required boundary conditions.

We begin this chapter with a brief discussion of Fourier and Laplace transforms. These integral transforms are useful for solving differential equations for two main reasons. First, differential operators are replaced by multiplicative operations, enabling the transform of the solution to be obtained by simple algebraic operations. The solution of the differential equation is then obtained in the original variables by inverting the transform. Second, the Fourier transform of the elementary source term used in determining the fundamental solution is simply a constant. Combined with the composition of products of Fourier transforms (the convolution theorem), this provides an elegant representation of the solution for initial-value and boundary-value problems. Examples in the main text and in the problems will be used to illustrate several applications to Green's functions, and to outline some extensions of the approach taken here.

7.1 The Fourier Transform

The Fourier series representation discussed in Chapter 4 is confined to functions that are periodic over a given interval. The Fourier transform provides a way of extending the class of functions to which the Fourier representation may be applied by removing the need for constructing a periodic extension. Thus, while the Fourier series representation is suitable for functions defined over finite regions, the Fourier transform can be used for functions defined over infinite and semi-infinite regions.

Consider a function $f(x)$ defined on the open interval $(-\frac{1}{2}L, \frac{1}{2}L)$. If f is extended periodically, then the extended function may be represented as a Fourier series, as described in Section 4.3. Using the complex form of the Fourier series (Problem 4.9), the periodic extension of $f(x)$ can be represented as

$$f(x) = \frac{1}{L} \sum_{n=-\infty}^{\infty} f_n\, e^{-ik_n x} \qquad (7.1)$$

where $k_n = 2\pi n/L$. By using the orthogonality of the functions $e^{-ik_n x}$ over the interval $(-\frac{1}{2}L, \frac{1}{2}L)$, as shown in Problem 4.2,

$$\int_{-L/2}^{L/2} e^{-ik_m x} e^{ik_n x}\, dx = \begin{cases} L & \text{if } m = n \\ 0 & \text{if } m \neq n \end{cases} \qquad (7.2)$$

expressions for the Fourier coefficients, f_n, are obtained as in Problem 4.9:

$$f_n = \int_{-L/2}^{L/2} f(x)\, e^{ik_n x}\, dx \qquad (7.3)$$

To derive the Fourier transform, we first introduce the quantity Δk_n as the distance between successive points k_n:

$$\Delta k_n \equiv k_n - k_{n-1} = \frac{2\pi}{L} \qquad (7.4)$$

Using this notation, the complex Fourier series in Equation (7.1) can be written in a more suggestive form as

$$f(x) = \sum_{n=-\infty}^{\infty} \frac{\Delta k_n}{2\pi} f_n\, e^{-ik_n x} \qquad (7.5)$$

The effect of increasing L is twofold. First, Δk_n becomes smaller, i.e. the k_n become more closely spaced. Second, the domain of integration in the definition of f_n in (7.3) approaches the entire real line. If the integral in (7.3) exists in the limit that $L \to \infty$, then (7.5) has the formal appearance of a Riemann sum for the integral obtained by letting k_n become a continuous variable k and letting Δk_n become the differential element of integration dk. Thus,

$$f(x) = \lim_{L \to \infty} \left[\sum_{n=-\infty}^{\infty} \frac{\Delta k_n}{2\pi} f_n\, e^{-ik_n x} \right] = \frac{1}{2\pi} \int_{-\infty}^{\infty} \tilde{f}(k)\, e^{-ikx}\, dk \qquad (7.6)$$

When the limit exists, the integral at right is the **Fourier integral representation** of f and the function $\tilde{f}(k)$ is called the **Fourier transform** of $f(x)$. We will use a tilde to distinguish the Fourier transform of a function from the original function throughout this book. Since the modulus of the complex exponential in (7.3) is unity, a *sufficient* condition for the limit in Equation (7.10) to exist is readily identified to be the absolute integrability of f over the real line,

$$\int_{-\infty}^{\infty} |f(x)|\,dx < \infty \tag{7.7}$$

though this condition may often be too stringent (Whittaker and Watson, 1963).

The Fourier transform $\tilde{f}(k)$ in Equation (7.6) is obtained from (7.3) in the limit $L \to \infty$ as

$$\tilde{f}(k) = \int_{-\infty}^{\infty} f(x)\,e^{ikx}\,dx \tag{7.8}$$

The condition in (7.7) insures that this integral exists. By substituting (7.8) into (7.6), we obtain the **Fourier integral formula**:

$$f(x) = \frac{1}{2\pi} \int_{-\infty}^{\infty} \int_{-\infty}^{\infty} e^{-ik(x-s)} f(s)\,ds\,dk \tag{7.9}$$

This representation of the function f shows that the signs of the arguments of the exponentials and the way the factors of 2π appear in (7.6) and (7.8) are largely a matter of convention and taste. It is important only to choose a particular convention and to use it consistently in moving back and forth between the Fourier transform of a function and the corresponding inverse Fourier transform to maintain the relation (7.9).

The convergence of the limit in Equation (7.6) can be examined within *Mathematica* by constructing the Fourier series representation in (7.1) over the interval $(-\frac{1}{2}L, \frac{1}{2}L)$ for increasing values of L. We consider the function

$$f(x) = e^{-x^2} \tag{7.10}$$

which clearly satisfies the integrability condition in Equation (7.7). The assembly of the various terms in (7.6) proceeds in very much the same way as the calculations in Example 4.2. The Fourier coefficients in (7.3) (written as a[n,L] below) are determined and substituted into Equation (7.6), which is represented as a partial sum:

```
a[n_,L_]:=NIntegrate[Exp[-t^2](Cos[(2 n Pi t)/L]+
    I Sin[(2 n Pi t)/L]),{t,-L/2,L/2}]

f[x_,n_,L_]:=Sum[(1/L)a[k,L](Cos[(2 k Pi x)/L]-
    I Sin[(2 k Pi x)/L]),{k,-n,n}]
```

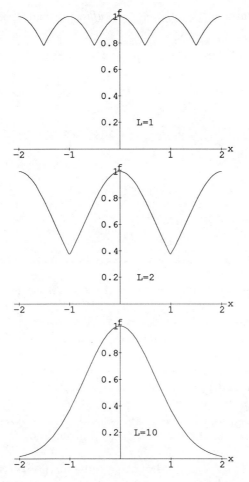

Figure 7.1 The Fourier series in (7.1) for the function $f(x)=\exp(-x^2)$ for $L=1$ (top), $L=2$ (center) and $L=10$ (bottom).

We have retained both the cosine and sine contributions in these expressions despite $f(x)$ being an even function. The quantity n in f[x,n,L] must be large enough to insure convergence to the desired threshold. We have used n=20 for the calculations performed here.

The Fourier representations for $L = 1$, $L = 2$ and $L = 10$ are shown in Figure 7.1. As L increases, successively larger domains of the function are included in the periodic unit of the Fourier representation and the limit in (7.6) is clearly seen to converge to the original function.

Fourier integral representations can also be constructed based on the sine and cosine functions. Thus, the **Fourier cosine transform** corresponding to (7.6) is (Problem 5)

$$f(x) = \frac{2}{\pi} \int_0^\infty \tilde{f}(k) \cos kx \, dk \qquad (7.11)$$

with the Fourier coefficient given by an expression similar to that in Equation (7.8)

$$\tilde{f}(k) = \int_0^\infty f(x) \cos kx \, dx \qquad (7.12)$$

The corresponding relations for the **Fourier sine transform** are

$$f(x) = \frac{2}{\pi} \int_0^\infty \tilde{f}(k) \sin kx \, dk \qquad (7.13)$$

and

$$\tilde{f}(k) = \int_0^\infty f(x) \sin kx \, dx \qquad (7.14)$$

Notice that the domain of integration of the Fourier sine and cosine transforms is the semi-infinite interval $[0, \infty)$, while that of the Fourier transform in Equations (7.6) and (7.8) is $(-\infty, \infty)$. The roles of these three types of Fourier transform in solving differential equations will be examined later in this chapter.

The generalization of the Fourier transform to scalar functions of vector variables and to vector functions of vector variables is straightforward. Suppose we have an n-component function $\mathbf{F}$ of a d-dimensional variable $\mathbf{x}$. Signifying the components of $\mathbf{F}$ by $F_1, \ldots, F_n$, the Fourier integral representation of the ith component is given by

$$\begin{aligned} F_i(x) &= \frac{1}{(2\pi)^d} \int_{-\infty}^\infty \cdots \int_{-\infty}^\infty \widetilde{F}_i(\mathbf{k}) \, e^{-ik_1 x_1} \cdots e^{-ik_d x_d} \, dk_1 \cdots dk_d \\ &= \frac{1}{(2\pi)^d} \int_{-\infty}^\infty \widetilde{F}_i(\mathbf{k}) \, e^{-i\mathbf{k}\cdot\mathbf{x}} \, d\mathbf{k} \end{aligned} \qquad (7.15)$$

where we have used the vector notation

$$\begin{aligned} \mathbf{x} &= (x_1, \ldots, x_d), \quad \mathbf{k} = (k_1, \ldots, k_d), \\ d\mathbf{x} &= dx_1 \cdots dx_d \end{aligned} \qquad (7.16)$$

The Fourier transform $\widetilde{F}_i$ is given by an expression analogous to (7.8):

$$\begin{aligned} \widetilde{F}_i(\mathbf{k}) &= \int_{-\infty}^\infty \cdots \int_{-\infty}^\infty F_i(\mathbf{x}) \, e^{ik_1 x_1} \cdots e^{ik_d x_d} \, dx_1 \cdots dx_d \\ &= \int_{-\infty}^\infty \widetilde{F}_i(\mathbf{x}) \, e^{i\mathbf{k}\cdot\mathbf{x}} \, d\mathbf{x} \end{aligned} \qquad (7.17)$$

The Fourier transform may also be used to decompose a function of time into frequency components:

$$\tilde{f}(\omega) = \int_{-\infty}^{\infty} f(t) e^{-i\omega t} \, dt \qquad (7.18)$$

where the sign of the exponential is again a matter of convention. This Fourier transform may be inverted as in (7.8), and the wavevector and frequency transforms may be combined for functions of both space and time:

$$\tilde{f}(\mathbf{k}, \omega) = \int_{-\infty}^{\infty} \left[\int_{-\infty}^{\infty} f(\mathbf{x}, t) e^{i\mathbf{k}\cdot\mathbf{x}} \, d\mathbf{x} \right] e^{-i\omega t} \, dt \qquad (7.19)$$

Example 7.1. Fourier transforms can be computed with *Mathematica* both symbolically and numerically. We consider as an example the function

$$f(x) = \begin{cases} 1 & \text{if } |x| \leq 1 \\ 0 & \text{if } |x| > 1 \end{cases} \qquad (7.20)$$

The Fourier transform of (7.20) is computed symbolically with the following entry:

```
Integrate[Cos[k x],{x,-1,1}]

   Sin[-k]     Sin[k]
-(-------) + ------
     k          k
```

The inverse transform is carried out with the commands

```
F[k_]:=Integrate[Cos[k x],{x,-1,1}]

f[t_,n_]:=NIntegrate[F[k]Cos[k t]/(2Pi),{k,-n Pi,n Pi}]
```

The second of these commands is seen to generate the equivalent in *Mathematica* of the Fourier integral theorem in Equation (7.9) as the range of integation is increased to span the entire real line. The plots of the inverse transform integrated over an increasing region of k is shown in Figure 7.2. The gross features of the function in (7.14) are obtained from the Fourier components with small values of k, which correspond to long wavelength variations of the function. However, reproducing the discontinuities at $x = \pm 1$ requires the higher Fourier components to be included in the range of integration. This is similar to the behavior found for Fourier series representations of functions with discontinuities.

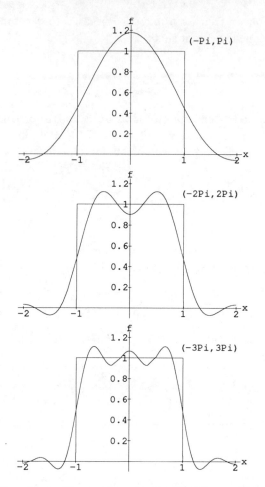

Figure 7.2 The effect of increasing the range of integration in the inverse Fourier transform for the function defined in (7.20). The ranges of integration are, from the top, $(-\pi, \pi)$, $(-2\pi, 2\pi)$ and $(-3\pi, 3\pi)$. The gross features of this function are obtained by including only the small values of k, corresponding to long wavelengths. However, reproducing the behavior near the discontinuities requires larger values of k to be included in the integration, just as was found for Fourier series.

A different type of behavior is seen for smooth functions. Consider again the function in (7.10). The Fourier transform can be determined by similar commands as those used above and the rate of convergence is displayed in Figure 7.3. The smooth form of (7.10) is recovered with a much smaller range of integration than that used above, as expected from the rapid decay of the Fourier transform with increasing k.

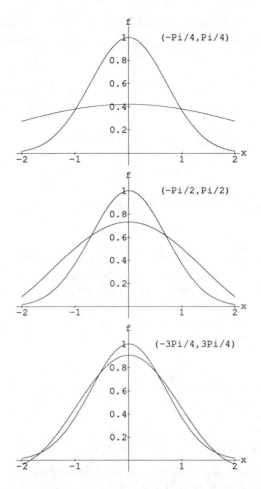

Figure 7.3 The effect of increasing the range of integration in the inverse Fourier transform for the function in (7.10). The ranges of integration are $(-\frac{1}{4}\pi, \frac{1}{4}\pi)$, $(-\frac{1}{2}\pi, \frac{1}{2}\pi)$ and $(-\frac{3}{4}\pi, \frac{3}{4}\pi)$, as indicated. The convergence of the Fourier representation to the function is seen to be much more rapid than that shown in Figure 7.2 for the function (7.20).

7.2 The Laplace Transform and the Bromwich Integral

The range of integration in the Fourier transform and the inverse Fourier transform is the entire real line. Even for the Fourier sine and cosine transforms the range can be extended over the entire real line, although the functions themselves are either even or odd with respect to the origin. Provided that the condition in (7.7) is satisfied, these Fourier transforms are well-defined.

Consider now an initial-value problem where conditions are specified at a particular time t_0 and a solution is required for later times $t > t_0$.

The Laplace transform is a modification of the Fourier transform that is used for solving such problems. One reason for the usefulness of the Laplace transform is that the initial conditions are incorporated directly into the transformed function. Thus, we can avoid having to first determine the general solution and then specializing to a particular problem. In this section we will describe the procedure used to determine both the Laplace transform and the inverse Laplace transform.

Given a function $f(t)$, consider the quantity

$$F(s) = \int_0^\infty f(t) e^{-st} \, dt \tag{7.22}$$

where s is a complex number with a positive real part:

$$s = u + iv, \quad u > 0 \tag{7.23}$$

The initial time t_0 has been taken to be zero without any loss of generality. Since the real part of s is positive, this integral converges only for $t > 0$. The function $F(s)$ is termed the **Laplace transform** of $f(t)$. In the following discussion, the Laplace transform of a function will be signified by the corresponding upper-case character that represents that function. In many elementary discussions of the Laplace transform, the transform variable s is taken to be a real variable. However, a complex value of s with a positive real part is necessary in order to be able to invert (7.22) to obtain an expression of $f(t)$ in terms of $F(s)$, as we now show.

We first write the Laplace transform in (7.22) as

$$F(s) = \int_0^\infty e^{-ivt} \left[e^{-ut} f(t) \right] dt \tag{7.24}$$

The right-hand side of this equation can now be interpreted as a Fourier transform of the function $e^{-ut} f(t)$. The orthogonality relation for the functions e^{ikx} (Problem 21) can then be used to obtain from (7.24) an integral representation for the quantity $e^{-ut} f(t)$:

$$e^{-ut} f(t) = \lim_{R \to \infty} \left[\frac{1}{2\pi} \int_{-R}^{R} F(s) e^{ivt} \, dv \right] \tag{7.25}$$

In applying this procedure, we have assumed that the condition in (7.7) is satisfied. For reasons that will become clear below, the infinite integral in (7.25) has been represented as the limit of a finite integral. Solving for $f(t)$, a few simple manipulations yield

$$f(t) = \lim_{R \to \infty} \left[\frac{1}{2\pi} \int_{-R}^{R} F(s) e^{ut} e^{ivt} \, dv \right]$$

$$= \lim_{R \to \infty} \left[\frac{1}{2\pi} \int_{-R}^{R} F(s) e^{st} \, dv \right] \tag{7.26}$$

The integration variable in the last integral in (7.26) is v, which is a real variable, while both factors in the integrand are functions of s, which is a complex quantity. To change the variable of integration from v to s, we first notice that for fixed u, we have the relation $ds = i\, dv$. Thus, changing the integration variable in (7.26) from v to $s = u + iv$, where u is a positive constant, and making the required changes in the limits of integration, produces for $f(t)$ the expression

$$f(t) = \lim_{R \to \infty} \left[\frac{1}{2\pi i} \int_{u-iR}^{u+iR} F(s)\, e^{st}\, ds \right] \tag{7.27}$$

This integral can be evaluated by using an appropriate contour in the complex plane, with the parameter u chosen to take into account the analytic structure of the integrand, such as the presence of any poles and branch cuts. The expression in Equation (7.27) is the **inverse Laplace transform** for the function F, and is often called the **Bromwich integral**.

Example 7.2. In this example, the use of the Laplace transform (7.22) and the inverse transform (7.27) will be demonstrated by solving the ordinary differential equation

$$\frac{d^2 y}{dt^2} + 2 \frac{dy}{dt} + 2y = 1 \tag{7.28}$$

with the general initial conditions

$$y(0) = y_0, \quad \left. \frac{dy}{dt} \right|_{t=0} = y_0' \tag{7.29}$$

To take the Laplace transform of (7.28) we multiply the equation by e^{st} and integrate over the range $0 \le t < \infty$. The term proportional to y produces a term proportional to the Laplace transform $Y(s)$. The term involving the first derivative of y is evaluated by performing an integration by parts:

$$\int_0^\infty \frac{dy}{dt} e^{-st}\, dt = y\, e^{-st} \Big|_0^\infty + s \int_0^\infty y\, e^{-st}\, dt$$

$$= -y_0 + s Y(s) \tag{7.30}$$

One of the initial conditions in (7.29) is seen to enter *explicitly* into the Laplace transform of this term. This results from the lower bound on the range of integration being the initial time $t = 0$. In addition, there is the contribution $sY(s)$, which results from the exponential form of the transformation function e^{-st}, by analogy with the Fourier transform. Notice also that the requirement (7.7) for the Laplace transform (7.22) to exist means that products of e^{-st} with y must vanish as $t \to \infty$.

272 Partial Differential Equations with Mathematica

The Laplace transform of the second derivative of y is evaluated by carrying out successive integrations by parts:

$$\begin{aligned}\int_0^\infty \frac{d^2y}{dt^2} e^{-st}\, dt &= \left.\frac{dy}{dt} e^{-st}\right|_0^\infty + s\int_0^\infty \frac{dy}{dt} e^{-st}\, dt \\ &= -y_0' + s\left[\left. y\, e^{-st}\right|_0^\infty + s\int_0^\infty y\, e^{-st}\, dt\right] \qquad (7.31) \\ &= -y_0' - sy_0 + s^2 Y(s)\end{aligned}$$

Both initial conditions are now seen to appear explicitly in the transformed expression. In fact, the Laplace transform of an n-order derivative of a function contains derivatives up to order $n-1$ evaluated at $t=0$ (Problem 7).

The Laplace transforms in (7.30) and (7.31) are now substituted into the original equation in (7.28) and the resulting expression is solved for $Y(s)$, the Laplace transform of the solution:

$$Y(s) = \frac{1 + (s^2 + 2s)y_0 + sy_0'}{s(s^2 + 2s + 2)} \qquad (7.32)$$

The solution, $y(t)$, to the initial-value problem in (7.28) and (7.29) can then be obtained in integral form by substituting (7.32) into the Bromwich integral (7.27):

$$y(t) = \lim_{R\to\infty}\left[\frac{1}{2\pi i}\int_{u-iR}^{u+iR} \frac{1 + (s^2 + 2s)y_0 + sy_0'}{s(s^2 + 2s + 2)} e^{st}\, ds\right] \qquad (7.33)$$

There are several points to consider in choosing the appropriate contour for the evaluation of the Bromwich integral in (7.33). First, the Laplace transform in (7.32) has simple poles at $s = -1 \pm i$. Second, the contour must be closed in the *left half-plane* to insure convergence of the integral, since $s > 0$. Finally, for the contour to enclose the poles of $Y(s)$, u must fall in the range $-1 < u < \infty$. The contour obtained by taking into account these requirements is shown in Figure 7.4. The integral can now be evaluated by a straightforward application of the Cauchy residue theorem. After some simple algebra, we obtain

$$y(t) = \tfrac{1}{2} - \tfrac{1}{2} e^{-t}(\cos t - \sin t) + y_0\, e^{-t}(\cos t + \sin t) + y_0'\, e^{-t}\sin t \qquad (7.34)$$

as the solution of (7.28) with the initial conditions (7.29).

The packages Laplace.m, InverseLaplace.m, and Integral-Tables.m can be used to perform Laplace and inverse Laplace transforms both for particular functions and symbolically. For example, the Laplace transform of e^{at} and the corresponding inverse transform are determined by the following commands:

```
Laplace[Exp[a t],t,s]
```

```
       1
-(-----)
     a - s
InverseLaplace[%,s,t]
```

```
 a t
E
```

These packages can also be used to perform symbolic Laplace transforms as in Equations (7.29) and (7.30) and applied to solving differential equations such as (7.28) with general initial conditions (7.29). The procedure is the same as in Equations (7.30)–(7.34). The Laplace transform Y[s] of y[t] is first defined as a generic function and then Laplace is applied to the differential equation (7.28):

```
Laplace[y[x_],x_,s_]=Y[s]
```

```
Y[s]
```

```
Laplace[#,x,s]& /@ (y''[x]+2y'[x]+2y[x]==1)
```

```
            2                                      1
2 Y[s] + s  Y[s] + 2 (s Y[s] - y[0]) - s y[0] - y'[0] == -
                                                         s
```

The Laplace-transformed equation is then solved for Y[s] as in Equation (7.32), and InverseLaplace is used to obtain the solution of the differential equation:

```
Y[s] /. Solve[%,Y[s]][[1]]
```

```
                    2
   -1 - 2 s y[0] - s  y[0] - s y'[0]
-(-----------------------------------)
                   2     3
              2 s + 2 s  + s
```

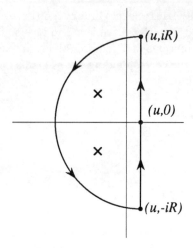

Figure 7.4 The complex contour used for performing the Bromwich integral in Equation (7.33). The poles are indicated by crosses.

```
InverseLaplace[%,s,x] // Expand

1     Cos[x]     Sin[x]     (Cos[x] + Sin[x]) y[0]     Sin[x] y'[0]
- -   ------  -  ------  +  ----------------------  +  ------------
2       x         x                    x                     x
       2 E       2 E                  E                     E
```

which agrees with the solution obtained in (7.34). Additional examples are provided in the problems at the end of the chapter.

∎

7.3 Fundamental Solution of the Diffusion Equation

As an illustration of the combined use of Fourier and Laplace transforms, as well as an introduction to Green's functions, we will solve the one-dimensional diffusion equation:

$$\frac{\partial u}{\partial t} - \frac{\partial^2 u}{\partial x^2} = 0 \qquad (7.35)$$

with the initial condition

$$u(x,0) = f(x) \qquad (7.36)$$

using the methods described in Sections 7.1 and 7.2. By performing a Fourier transform in space (because the spatial domain of the solution is unbounded) and a Laplace transform in time (because a solution is required for an initial-value problem), the transformed diffusion equation is obtained as

$$s\widetilde{U}(k,s) - \tilde{f}(k) + k^2 \widetilde{U}(k,s) = (s + k^2)\widetilde{U}(k,s) - \tilde{f}(k) = 0 \qquad (7.37)$$

where $\tilde{f}(k)$ is the Fourier transform of $f(x)$. As expected, the transformed equation includes the initial condition (7.36) explicitly. Solving for the transformed function $\widetilde{U}(k,s)$,

$$\widetilde{U}(k,s) = \frac{\tilde{f}(k)}{s + k^2} \qquad (7.38)$$

shows that $\widetilde{U}(k,s)$ has a simple pole at $s = -k^2$. The solution to (7.35) and (7.36) is now obtained by applying the inverse Fourier and Laplace operations to (7.38):

$$u(x,t) = \frac{1}{2\pi} \int_{-\infty}^{\infty} \left\{ \lim_{R \to \infty} \left[\frac{1}{2\pi i} \int_{u-iR}^{u+iR} \frac{\tilde{f}(k)}{s + k^2} e^{st} \, ds \right] \right\} e^{-ikx} \, dk \qquad (7.39)$$

The Laplace transform may be inverted directly, as in Example 7.2, to obtain the Fourier transform of $u(x,t)$:

$$\tilde{u}(k,t) = e^{-k^2 t} \tilde{f}(k) \qquad (7.40)$$

The solution of the diffusion equation in the original variables is now obtained by performing a Fourier transform with respect to k on $\tilde{u}(k,t)$:

$$u(x,t) = \frac{1}{2\pi} \int_{-\infty}^{\infty} \tilde{u}(k,t) e^{-ikx} \, dk$$

$$= \frac{1}{2\pi} \int_{-\infty}^{\infty} \tilde{f}(k) e^{-(ikx + k^2 t)} \, dk \qquad (7.41)$$

To proceed further to obtain an explicit form for this Fourier transform we require expressions for the Fourier components $\tilde{f}(k)$. We first consider the special case where the initial condition in (7.36) is given by a delta function (Problem 16), $f(x) = \delta(x)$, that is, where the initial concentration is completely localized at the origin. The reason behind this choice is that the Fourier components of this function are simply given by

$$\tilde{f}(k) = \int_{-\infty}^{\infty} \delta(x) e^{ikx} \, dx = 1 \qquad (7.42)$$

In other words, each Fourier component contributes to the delta function with the same weight. The Fourier transform of (7.40) can now be written in a form that is more amenable to a direct evaluation:

$$u(x,t) = \frac{1}{2\pi} \int_{-\infty}^{\infty} e^{-(k^2 t + ikx)} \, dk \qquad (7.43)$$

The integration over k can be carried out first by completing the square in the argument of the exponential,

$$k^2 t + ikx = \left(k\sqrt{t} + \frac{ix}{2\sqrt{t}}\right)^2 + \frac{x^2}{4t} \qquad (7.44)$$

and then changing the integration variable from k to $\kappa = k\sqrt{t} + ix/2\sqrt{t}$. With these changes, the integral in (7.43) becomes

$$\begin{aligned} u(x,t) &= \frac{e^{-x^2/4t}}{2\pi} \int_{-\infty}^{\infty} \exp\left[-\left(k\sqrt{t} + \frac{ix}{2\sqrt{t}}\right)^2\right] dk \\ &= \frac{e^{-x^2/4t}}{2\pi\sqrt{t}} \int_{-\infty}^{\infty} e^{-\kappa^2} \, d\kappa \end{aligned} \qquad (7.45)$$

Performing the Gaussian integration (Problem 24) yields finally

$$u(x,t) = \frac{1}{\sqrt{4\pi t}} \exp\left(-\frac{x^2}{4t}\right) \qquad (7.46)$$

For reasons that will become clear shortly, the expression (7.46) is known as the **fundamental solution** of the diffusion equation.

The time development of the solution (7.46) is plotted in Figure 7.5. The initially sharply peaked form is seen to gradually smoothen out. As we saw in Section 3.5, the action of diffusion is seen to be a homogenization of the distribution of temperature, in the case of heat conduction, or of a concentration profile, in the case of particle diffusion. The origin of this behavior can be traced back to Equation (7.40) for the time-dependence of the Fourier components. The larger the value of k, i.e. the smaller the wavelength of the component, the faster the decay. Only the $k=0$ component is independent of time, and so provides the only nonvanishing contribution in the limit $t \to \infty$. The $k=0$ Fourier component is a conserved quantity and is seen to correspond to the integral of u over all space. The invariance of this quantity stems from the fact that the diffusion equation is a conservation equation (Problem 3.20).

We now return to the evaluation of the Fourier transform in (7.41) for the case of a general initial function $f(x)$. The Fourier transform determining $u(x,t)$,

$$u(x,t) = \frac{1}{2\pi} \int_{-\infty}^{\infty} \tilde{f}(k) \, e^{-k^2 t} e^{-ikx} \, dk \qquad (7.47)$$

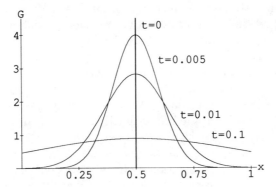

Figure 7.5 The evolution of the fundamental solution (7.45) of the diffusion equation from an initial distribution localized at $x = \frac{1}{2}$. The sharply-peaked profile of the initial condition is gradually smoothed out under the action of the diffusion equation, while the area under each curve is maintained constant as a function of time.

can be represented in a more transparent form by relating the Fourier transform of a product of functions $\tilde{f}(k)$ and $\tilde{g}(k)$ to a product of Fourier-transformed functions $f(x)$ and $g(x)$ (Problem 2):

$$\frac{1}{2\pi} \int_{-\infty}^{\infty} \tilde{f}(k) \tilde{g}(k) e^{-ikx} \, dk = \int_{-\infty}^{\infty} f(x') g(x - x') \, dx' \qquad (7.48)$$

The integral on the right-hand side of this equation is called the **convolution** of the functions $f(x)$ and $g(x)$ and this property of Fourier transforms is known as the **convolution theorem**. Thus, by identifying $\tilde{f}(k)$ with the Fourier components of the initial function in (7.36) and

$$\tilde{g}(k) = e^{-k^2 t} \qquad (7.49)$$

for the Fourier components of the fundamental solution, so that $f(x')$ corresponds to the initial condition in (7.36) and

$$g(x - x') = \frac{1}{\sqrt{4\pi t}} \exp\left[-\frac{(x-x')^2}{4t}\right] \qquad (7.50)$$

the convolution theorem can be used to write the Fourier transform of the solution in (7.47) as

$$u(x,t) = \frac{1}{\sqrt{4\pi t}} \int_{-\infty}^{\infty} f(x') \exp\left[-\frac{(x-x')^2}{4t}\right] dx' \qquad (7.51)$$

Thus, *any* solution of the initial-value problem for the one-dimensional diffusion equation with no boundaries can be written as a convolution of the initial function with the fundamental solution (7.46). Notice that the representation of the solution in integral form is more flexible than the corresponding series representation in Equations (4.64) and (4.65) in that initial functions that have slowly-convergent Fourier series can be handled with equal ease as comparatively smooth functions (see Example 7.3 below). This has particular importance for describing the early stages of the time development of the solution, though once the rapidly-varying Fourier components have decayed and the initial function has been smoothed out, the Fourier representation may in fact be easier to evaluate.

We now introduce some standard notation that is used for Green's functions and fundamental solutions. Green's functions are signified by G and fundamental solutions as G_0, with the subscript '0' indicating that the function is a fundamental solution, as well as being a particular type of Green's function. Thus, if in the solution of the diffusion equation the initial time is taken at a value t', the fundamental solution is written as

$$G_0(x, t; x', t') = \frac{1}{\sqrt{4\pi(t-t')}} \exp\left[-\frac{1}{4}\frac{(x-x')^2}{(t-t')}\right] \qquad (7.52)$$

In this notation, the fundamental solution in Equation (7.46) is written as $G_0(x, t; x', 0)$. The argument list of the fundamental solution is meant to emphasize the roles of the coordinates x' and t' as 'source' points and the points x and t as 'observation' points. In particular, the value of G_0 at the point x at time t is determined by the evolution of the distribution $\delta(x')$ at time t'. Thus, we can interpret (7.51) as the superposition of the evolutions of initial distributions of delta functions at points x' with amplitudes $f(x')$. Due to the equation being linear, the value of u at the point x at a later time t is obtained by summing the evolution of each of these initial distributions separately. This has several similarities with the corresponding solution of the Poisson equation, which is the subject of the next section.

Example 7.3. The solution (7.51) of the initial-value problem of the diffusion equation can be constructed by carrying out the corresponding operations in *Mathematica*. Consider the initial condition

$$f(x) = \begin{cases} 1, & \text{if } |x| \leq 1, \\ 0, & \text{if } |x| > 1 \end{cases} \qquad (7.53)$$

The solution in (7.51) is obtained by constructing the initial function in (7.53) and the fundamental solution (7.46) and then performing the convolution of these two functions using NIntegrate:

```
f[x_]:=If[Abs[x]<=1,1,0]
```

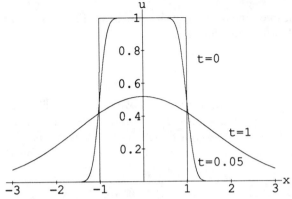

Figure 7.6 The evolution of the solution of the diffusion equation from the initial function given in (7.53). As is seen for the behavior of the fundamental solution in Figure 7.5, the sharply defined features of this initial function are smoothed out quickly as the profile of the solution spreads out while maintaining the area under the curve constant.

```
G[x_,t_]:=(1/Sqrt[4 Pi t])Exp[-(x^2)/(4 t)]

u[x_,t_]:=NIntegrate[f[y] G[x-y,t],{y,-1,1}]
```

The early stages of the time development of this solution is shown in Figure 7.6. The sharpest features in the profile of f (near the discontinuities) are smoothed out first as the profile spreads out as it decays to zero. This is to be expected from the behavior of the fundamental solution in Figure 7.5 and the time development of the individual Fourier components in (7.39).

■

7.4 Fundamental Solution of Poisson's Equation

The derivation of the fundamental solution of the diffusion equation in the preceding section illustrates the general methodology that can be used to obtain the fundamental solution of other equations. In this section, we will use this approach to obtain a fundamental solution in a somewhat more familiar setting, namely Poisson's equation. This equation is used in electrostatics to obtain the potential generated by an immobile distribution of charges. As in Section 7.3, the region occupied by the charges will be taken to have no boundaries.

In three spatial dimensions, Poisson's equation for the potential $V(\mathbf{x})$ due to a static charge distribution $\rho(\mathbf{x})$ is

$$\nabla^2 V(\mathbf{x}) = -4\pi \rho(\mathbf{x}) \tag{7.54}$$

This equation may be solved by using Equation (7.15) to perform a three-dimensional Fourier transform:

$$\int e^{i\mathbf{k}\cdot\mathbf{x}} \nabla^2 V(\mathbf{x})\, d\mathbf{x} = -4\pi \int e^{i\mathbf{k}\cdot\mathbf{x}} \rho(\mathbf{x})\, d\mathbf{x} \qquad (7.55)$$

The integrals in this equation are taken over all space. The right-hand side of this equation is given in terms of the Fourier transform of the charge distribution as $-4\pi\tilde{\rho}(\mathbf{k})$. To evaluate the left-hand side, we use the expression for the Laplacian in rectangular coordinates to write the Fourier transform of V as

$$\int e^{i\mathbf{k}\cdot\mathbf{x}} \nabla^2 V(\mathbf{x})\, d\mathbf{x} = \int \left(\frac{\partial^2 V}{\partial x^2} + \frac{\partial^2 V}{\partial y^2} + \frac{\partial^2 V}{\partial z^2} \right) e^{i\mathbf{k}\cdot\mathbf{x}}\, d\mathbf{x} \qquad (7.56)$$

Each of the integrals on the right-hand side can be evaluated by performing successive integrations by parts. For example, the integration over x in the first term yields

$$\int_{-\infty}^{\infty} \frac{\partial^2 V}{\partial x^2} e^{ikx}\, dx = \left. \frac{\partial V}{\partial x} e^{ikx} \right|_{-\infty}^{\infty} - \left. ikV e^{ikx} \right|_{-\infty}^{\infty} - k^2 \int_{-\infty}^{\infty} V e^{ikx}\, dx \quad (7.57)$$

The last integral on the right-hand side of this equation is recognized as the Fourier transform of the potential, $\tilde{V}(\mathbf{k})$. We now assume that the charge distribution is localized in space (cf. (7.7)), so that the potential and its derivatives decay to zero at infinity. Thus, the first two terms on the right-hand side of (7.57) vanish and (7.56) becomes

$$\int e^{i\mathbf{k}\cdot\mathbf{x}} \nabla^2 V(\mathbf{x})\, d\mathbf{x} = -(k_1^2 + k_2^2 + k_3^2) \int V e^{i\mathbf{k}\cdot\mathbf{x}}\, d\mathbf{x} \qquad (7.58)$$

The integral on the right-hand side of this equation is recognized as the Fourier transform of the potential. Thus, by writing $k^2 = k_1^2 + k_2^2 + k_3^2$, the Fourier transform of Poisson's equation becomes simply

$$-k^2 \tilde{V}(\mathbf{k}) = -4\pi \tilde{\rho}(\mathbf{k}) \qquad (7.59)$$

Solving for the Fourier transform of the potential, we obtain

$$\tilde{V}(\mathbf{k}) = \frac{4\pi}{k^2} \tilde{\rho}(\mathbf{k}) \qquad (7.60)$$

To obtain the solution $V(\mathbf{x})$ by inverting the Fourier transform (7.56), explicit expressions for the Fourier components $\tilde{\rho}(\mathbf{k})$ are required. By analogy with our solution of the diffusion equation in the preceding section, we can derive the corresponding fundamental solution of the Poisson equation by taking for $\rho(\mathbf{x})$ the special charge distribution $\rho(\mathbf{x}) = \delta(\mathbf{x})$. The physical significance of this choice is somewhat more apparent than for the diffusion equation: the charge distribution $\rho(\mathbf{x}) = \delta(\mathbf{x})$ corresponds to a unit point charge located at the origin. The Fourier components of this charge distribution are again given by (7.42), which again leads to a simple expression for the Fourier components of the corresponding solution:

$$\tilde{V}(\mathbf{k}) = \frac{4\pi}{k^2} \tag{7.61}$$

The inverse Fourier transform $\tilde{V}(\mathbf{k})$ can now be evaluated. The integration is performed most conveniently in spherical polar coordinates (k, θ, ϕ).

$$V(\mathbf{x}) = \frac{4\pi}{(2\pi)^3} \int \frac{e^{-i\mathbf{k}\cdot\mathbf{x}}}{k^2} d\mathbf{k}$$

$$= \frac{4\pi}{(2\pi)^3} \int_0^{2\pi} \left\{ \int_0^\pi \left[\int_0^\infty \frac{e^{-ikx\cos\theta}}{k^2} dk \right] d\theta \right\} d\phi \tag{7.62}$$

The magnitude of $\mathbf{x}$ has been denoted by x. The integrations over the polar and azimuthal angles are easily carried out:

$$V(\mathbf{x}) = \frac{2}{\pi} \int_0^\infty \frac{\sin kx}{kx} dk \tag{7.63}$$

The evaluation of the integral in (7.63) is discussed below and in Problem 33, with the result

$$\int_0^\infty \frac{\sin t}{t} dt = \frac{\pi}{2} \tag{7.64}$$

Substituting (7.64) into (7.63), we obtain the fundamental solution for Poisson's equation:

$$V(\mathbf{x}) = \frac{1}{|\mathbf{x}|} \tag{7.65}$$

This solution is simply the Coulomb potential at the point $\mathbf{x}$ produced by a unit point charge located at the origin. If the charge is placed at the point $\mathbf{x}'$, then the corresponding solution is written as

$$G_0(\mathbf{x}, \mathbf{x}') = \frac{1}{|\mathbf{x} - \mathbf{x}'|} \tag{7.66}$$

where by analogy with the notation used in (7.49), we have signified this solution by $G_0(\mathbf{x}, \mathbf{x}')$. This solution represents the potential at $\mathbf{x}$ due to a point source at $\mathbf{x}'$. However, as the form of (7.66) readily reveals, these identifications can be interchanged (Problem 36).

The integral in (7.64) can be evaluated by using *Mathematica* in two ways. The integral can be evaluated symbolically by using Integrate:

```
Integrate[Sin[x]/x,{x,0,Infinity}]
```

```
Pi
--
2
```

Alternatively, the numerical evaluation of (7.66) proceeds by the following command sequence. Notice that the upper limit of integration is treated as a variable.

```
Integral[x_]:=NIntegrate[Sin[t]/t,{t,0,x},
    MaxRecursion->20,WorkingPrecision->20,
    AccuracyGoal->7]
```

```
Do[Print[Integral[i^4]],{i,15}]
```

0.946083
1.631302
1.561306
1.570967
1.572371
1.570867
1.570513
1.5706
1.570764
1.570892
1.57077
1.570792
1.570821
1.570775
1.570793

To four decimal places the numerical value of this integral agrees with the exact result: $\pi/2 = 1.570796$. Listing the value of the integral with an increasing upper bound shows the slow convergence, which is due to the oscillatory nature of the integrand.

We now return to the solution of Poisson's equation for the actual charge distribution $\rho(x)$. The solution $V(x)$ is obtained by inverting the

Fourier transform in (7.61),

$$V(\mathbf{x}) = \frac{1}{(2\pi)^3} \int \frac{4\pi}{k^2} \tilde{\rho}(\mathbf{k}) e^{-i\mathbf{k}\cdot\mathbf{x}} d\mathbf{k} \qquad (7.67)$$

We now proceed as in Section 7.3 and apply the convolution theorem in (7.48) with the identifications

$$\tilde{f}(\mathbf{k}) = \tilde{\rho}(\mathbf{k}), \qquad \tilde{g}(\mathbf{k}) = \frac{4\pi}{k^2} \qquad (7.68)$$

Then, with the Fourier transform of $\tilde{g}(\mathbf{k})$ given by the fundamental solution in Equation (7.66), we obtain the solution of Poisson's equation (7.54) as

$$V(\mathbf{x}) = \int \frac{\rho(\mathbf{x}')}{|\mathbf{x}-\mathbf{x}'|} d\mathbf{x}' \qquad (7.69)$$

The potential V is thus obtained as a convolution between the static charge distribution ρ and the fundamental solution of the Poisson equation, in direct analogy with the form of the solution in (7.51) for the diffusion equation. In more physical terms, the potential at a point $\mathbf{x}$ due to a distribution of charge is the superposition of the potential due to the individual charges, which can be viewed as being located at each point $\mathbf{x}'$ with a strength determined by the value $\rho(\mathbf{x}')$ at that point.

7.5 Fundamental Solution of the Wave Equation

The initial-value problem for the wave equation can be obtained in closed form, as we have seen in Chapter 3. Thus, in this section we will concentrate on solving the initial-value problem for the *inhomogeneous* wave equation:

$$\frac{\partial^2 u}{\partial t^2} - \frac{\partial^2 u}{\partial x^2} = q(x,t) \qquad (7.70)$$

The term $q(x,t)$ represents the creation of a disturbance with intensity q at the position x at time t. We seek the solution to this equation with the usual Cauchy boundary conditions at $t=0$:

$$u(x,0) = f(x), \qquad u_t(x,0) = g(x) \qquad (7.71)$$

The solution to (7.70) and (7.71) will be obtained by performing a Fourier transform in space and a Laplace transform in time, by analogy with the approach taken for the diffusion equation in Section 7.3. By carrying out this procedure, substituting the Fourier components of initial conditions into the transformed equation, and solving for the Fourier–Laplace transform of $u(x,t)$, we obtain:

$$\widetilde{U}(k,s) = \frac{\widetilde{Q}(k,s)}{s^2 + k^2} + \frac{\widetilde{G}(k) + s\widetilde{F}(k)}{s^2 - k^2} \tag{7.72}$$

The second term on the right-hand side of this equation corresponds to the solution of the homogeneous wave equation, as shown in Problem 38. This solution can always be added to any solution of the inhomogeneous equation (7.70), so we set $f(x) = 0$ and $g(x) = 0$ for the problem at hand. The resulting solution to (7.70) is obtained by inverting the Fourier and Laplace transforms of the remaining terms in (7.72):

$$u(x,t) = \frac{1}{2\pi} \int_{-\infty}^{\infty} \left[\lim_{R\to\infty} \frac{1}{2\pi i} \int_{u-iR}^{u+iR} \frac{\widetilde{Q}(k,s)}{s^2 + k^2} e^{st}\, ds \right] e^{-ikx}\, dk \tag{7.73}$$

To obtain an explicit solution, we proceed as in Sections 7.3 and 7.4 and consider the special case of a disturbance of unit strength created at the origin, $x=0$, at the time $t=0$, i.e. $q(x,t) = \delta(x)\delta(t)$. The Fourier–Laplace components of this disturbance are readily verified to be $\widetilde{Q}(k,s) = 1$. Again using the notation G_0 to signify the fundamental solution, the expression to be evaluated is then given by

$$G_0(x,t) = \frac{1}{2\pi} \int_{-\infty}^{\infty} \left[\lim_{R\to\infty} \frac{1}{2\pi i} \int_{u-iR}^{u+iR} \frac{e^{st}}{s^2 + k^2}\, ds \right] e^{-ikx}\, dk \tag{7.74}$$

The evaluation of the Bromwich integral is straightforward and we obtain, after some simple rearrangements,

$$G_0(x,t) = \frac{1}{4\pi} \int_{-\infty}^{\infty} \frac{\sin k(x+t)}{k}\, dk - \frac{1}{4\pi} \int_{-\infty}^{\infty} \frac{\sin k(x-t)}{k}\, dk \tag{7.75}$$

By using the result of Problem 33,

$$\int_{-\infty}^{\infty} \frac{\sin(\pm \alpha s)}{s}\, ds = \pm \pi \tag{7.76}$$

for any quantity α that is independent of s, we see that for $t>0$, a non-zero solution is obtained only for values of x and t that satisfy the inequalities $x-t<0$ and $x+t>0$:

$$G_0(x,t) = \begin{cases} \frac{1}{2}, & \text{if } x - t < 0 \text{ and } x + t > 0 \\ 0, & \text{otherwise} \end{cases} \tag{7.77}$$

The solution (7.77) can be written in a more attractive form by using the Heaviside, or 'step' function, $H(x)$, defined by

$$H(x) = \begin{cases} 1 & \text{if } x > 0 \\ 0 & \text{if } x \leq 0 \end{cases} \tag{7.78}$$

Then (7.77) can be written as

$$G_0(x,t) = \tfrac{1}{2}H(t-x)H(t+x) \tag{7.79}$$

This solution is seen to correspond to a spreading wavefront moving with unit velocity to the left and right while leaving a wake that does not decay with time (see Figure 3.3(b)). This is in sharp contrast to the behavior of the corresponding solutions in two and three spatial dimensions (Problem 39). In two dimensions, the wake decays as the wave passes, while in three dimensions, the wavefront is sharp and leaves no wake. This phenomenon of sharp signals in three dimensions is known as **Huygens' principle**.

If the source $q(x,t)$ creates a disturbance not at the origin at time $t=0$ but at the point x' at time $t=t'$, then the Fourier–Laplace components of $q(x,t)$ are given by

$$\widetilde{Q}(k,s) = e^{ikx'-st'} \tag{7.80}$$

Thus, the integral that must be evaluated to obtain the corresponding solution is

$$G_0(x,t;x',t') = \lim_{R\to\infty}\left\{\frac{1}{2\pi}\int_{-\infty}^{\infty}\left[\frac{1}{2\pi i}\int_{u-iR}^{u+iR}\frac{e^{-ik(x-x')+s(t-t')}}{s^2+k^2}\,ds\right]dk\right\} \tag{7.81}$$

We have again used the notation introduced in Section 7.3 whereby the arguments of both the source point and the observation point are included in the arguments of the fundamental solution. By comparing this integral with that in Equation (7.74), we see that an expression for the solution is obtained by replacing x by $x-x'$ and t by $t-t'$ in (7.79):

$$\begin{aligned}G_0(x,t;x',t') &= \tfrac{1}{2}H[t-t'-(x-x')]H[t-t'+(x-x')] \\ &= G_0(x-x',t-t')\end{aligned} \tag{7.82}$$

Direct substitution (Problem 40) shows that (7.82) is indeed a solution of

$$\left(\frac{\partial^2}{\partial t^2}-\frac{\partial^2}{\partial x^2}\right)G_0(x,t;x',t') = \delta(x-x')\delta(t-t') \tag{7.83}$$

with the initial conditions

$$G_0(x-x',0) = 0, \qquad \left[\frac{\partial}{\partial t}G_0(x-x',t-t')\right]\bigg|_{t=t'} = 0 \tag{7.84}$$

Notice in particular that Equation (7.82) implies that the fundamental solution vanishes if $t \leq t'$, i.e. a disturbance can be transmitted through a medium only *after* it is initiated.

Having determined the fundamental solution of the wave equation, we can return to the solution of (7.70) and (7.71). We can obtain this solution directly by applying the convolution theorem for Fourier and Laplace transforms (Problems 2 and 11):

$$u(x,t) = \int_0^t \left[\int_{-\infty}^{\infty} G_0(x - x', t - t') q(x', t') \, dx' \right] dt'$$

$$+ \int_{-\infty}^{\infty} \left[g(x') G_0(x - x', t) - f(x') \frac{\partial}{\partial t'} G_0(x - x', t) \right] dx' \quad (7.85)$$

By substituting the fundamental solution (7.82) into (7.85) and evaluating the integral we obtain the solution of the initial-value problem of the inhomogeneous wave equation:

$$u(x,t) = \tfrac{1}{2} \int_0^t \left[\int_{x-(t-t')}^{x+(t-t')} q(x', t') \, dx' \right] dt'$$

$$+ \tfrac{1}{2}[f(x+t) + f(x-t)] + \tfrac{1}{2} \int_{x-t}^{x+t} g(s) \, ds \quad (7.86)$$

The second and third terms on the right-hand side of this equation are readily identified as d'Alembert's solution to the homogeneous wave equation in Equation (3.64). The first term is the particular solution of the inhomogeneous equation. It is instructive to observe that the solution (7.86) can also be obtained by direct integration of the inhomogeneous equation in characteristic coordinates (Chester, 1971).

7.6 Green's Functions in the Presence of Boundaries: The Method of Images

The fundamental solution of an equation was shown in the preceding section to be a particular solution of that equation for a point source. Solutions for more general source terms, including initial conditions, can then be represented as a linear combination of fundamental solutions in the form of an integral. In this section, we will examine how boundary conditions modify the construction and implementation of Green's functions. We will find that the Green's function in the presence of boundaries can also be obtained as a linear combination of fundamental solutions. In fact, the approach will be seen to be identical in philosophy to that used in the method of images in electrostatics, both from the mathematical and physical points of view.

Green's functions in the presence of boundaries can be obtained in several ways. We will adopt in this section a heuristic point of view and construct the appropriate linear combination of fundamental solutions by simply applying the reasoning used in the method of images in electrostatics. However, the same solution can always be obtained by solving the differential equations directly, and we will discuss several examples in the text and in the problems at the end of the chapter to demonstrate this. Depending on the number and types of boundaries, one approach or the other may be simpler. For example, if there is only a single boundary, the appropriate linear combination of fundamental solutions can usually be identified by inspection. However, if there are two boundaries, then there will be an infinite set of images, and it may prove simpler to determine the Green's function by solving the equation.

The basis of the method of images in electrostatics is contained in two properties of solutions to Poisson's equation. Suppose u is a solution of Poisson's equation for a particular charge distribution $\rho(\mathbf{x})$:

$$\nabla^2 u = -4\pi\rho \tag{7.87}$$

Then if ϕ is a solution of the associated homogeneous equation, i.e. Laplace's equation,

$$\nabla^2 \phi = 0 \tag{7.88}$$

the function $u+\phi$ is also a solution of Poisson's equation:

$$\nabla^2 (u + \phi) = -4\pi\rho \tag{7.89}$$

The second property is that the fundamental solution $G_0(\mathbf{x}, \mathbf{x}')$ is itself a solution of Laplace's equation if $\mathbf{x} \neq \mathbf{x}'$:

$$\nabla^2 G_0(\mathbf{x}, \mathbf{x}') = 0, \qquad \mathbf{x} \neq \mathbf{x}' \tag{7.90}$$

In particular, by combining (7.89) and (7.90) we see that any number of fundamental solutions can be added to a solution to Poisson's equation provided that the source charges of these additional solutions do not lie in the region of interest. The linear combinations are selected to satisfy the boundary conditions for the problem at hand. This is the philosophy behind the method of images and the functions so constructed are known as **Green's functions**, after the mathematician George Green, who was instrumental in developing this approach to solving inhomogeneous boundary value problems. The fundamental solution in Equation (7.66) is a particular example of a Green's function. It is easy to verify that the solutions of the diffusion and wave equations also have the corresponding properties in (7.89) and (7.90). Hence, the method of images can be applied to these equations as well.

To review the method of images in a familiar setting, we will begin by determining the Green's function for Poisson's equation in the presence of a planar boundary. Suppose that on the x-y plane we impose the Dirichlet boundary condition that the solution of Poisson's equation must vanish. Recalling that the physical interpretation of $G_0(\mathbf{x}, \mathbf{x}')$ is the potential at $\mathbf{x} = (x, y, z)$ due to a unit point charge placed at $\mathbf{x}' = (x', y', z')$, the Green's function G for this problem can be obtained by taking a superposition of fundamental solutions G_0 due to the physical charge at $\mathbf{x}'$ and that due to an image charge. In this case the image charge is a *negative* point charge, located at $\mathbf{x}'' = (x', y', -z')$, i.e. on the other side of the planar boundary. The Green's function obtained from the fundamental of the physical and image charges is

$$G(\mathbf{x}, \mathbf{x}') = G_0(\mathbf{x}, \mathbf{x}') - G_0(\mathbf{x}, \mathbf{x}'')$$
$$= \frac{1}{\sqrt{(x-x')^2 + (y-y')^2 + (z-z')^2}} \qquad (7.91)$$
$$- \frac{1}{\sqrt{(x-x')^2 + (y-y')^2 + (z+z')^2}}$$

This function clearly vanishes whenever $z = 0$, regardless of the values of x and y. Thus, the electrostatic potential $V(\mathbf{x})$ generated by a distribution of charge $\rho(\mathbf{x})$ with the requirement that the potential must vanish in the x-y plane is given by the sum of the Coulomb potential G_0 due to each charge and to each image charge:

$$V(\mathbf{x}) = \int G_0(\mathbf{x}; \mathbf{x}') \left[\rho(\mathbf{x}') + \rho(x', y', -z') \right] d\mathbf{x}'$$
$$= \int \left[G_0(\mathbf{x}; \mathbf{x}') - G_0(\mathbf{x}; x', y', -z') \right] \rho(\mathbf{x}') d\mathbf{x}' \qquad (7.92)$$
$$= \int G(\mathbf{x}; \mathbf{x}') \rho(\mathbf{x}') d\mathbf{x}'$$

The potential in the presence of the boundary on which the potential must vanish is therefore given by an expression that has the same appearance as that in Equation (7.69), but with the modified Green's function G replacing the fundamental solution G_0.

An analogous procedure may be applied to the case where the corresponding Neumann boundary conditions are imposed on the x-y plane: $\partial V / \partial z = 0$. Then, the appropriate linear combination of fundamental solutions is seen to be

$$G(\mathbf{x}, \mathbf{x}') = G_0(\mathbf{x}, \mathbf{x}') + G_0(\mathbf{x}, \mathbf{x}'')$$
$$= \frac{1}{\sqrt{(x-x')^2 + (y-y')^2 + (z-z')^2}} \qquad (7.93)$$
$$+ \frac{1}{\sqrt{(x-x')^2 + (y-y')^2 + (z+z')^2}}$$

The solution of Poisson's equation in this case can be obtained by taking steps analogous to those in (7.92).

The method of images can also be applied to solving the diffusion equation in the presence of boundaries on which Dirichlet or Neumann conditions are specified. Consider the one-dimensional diffusion equation (7.35) in the presence of a boundary at $x = 0$. We are interested in the solution for the semi-infinite half-plane $x > 0$. Since the functional form of the Green's function depends upon the nature of the boundary, we will consider two types of boundary condition, namely, absorbing and reflecting. For the case of an absorbing boundary, we imagine a particle undergoing diffusive motion. When the particle strikes the boundary it is removed from the system, i.e. the particle has been *absorbed* by the boundary. Thus, the density of particles at this boundary must vanish, so the Green's function must satisfy the Dirichlet boundary condition

$$G(0, t; x', t') = 0 \qquad (7.94)$$

Pursuing the analogy with the method of images, this Green's function can be constructed by adding a term to the fundamental solution to satisfy (7.94). The appropriate quantity is readily seen to be an image 'sink' term (or, alternatively, a source of particles with a 'negative' density) located symmetrically with respect to the origin in the region $x < 0$:

$$G(x, t; x', t') = G_0(x, t; x', t') - G_0(x, t; -x', t')$$

$$= \frac{1}{\sqrt{4\pi(t-t')}} \left[e^{-(x-x')^2/4(t-t')} - e^{-(x+x')^2/4(t-t')} \right] \qquad (7.95)$$

The evolution of this solution is plotted in Figure 7.7. Notice that the solution vanishes at $x = 0$ for all times, and that the solution becomes skewed with increasing time, as the influence of the boundary becomes more important or, in other words, when the negative contribution of the image source extends into the physical region ($x > 0$). The solution of the diffusion equation with an initial distribution $u(x, 0) = f(x)$ in the presence of an absorbing boundary at the origin can now be obtained by associating with each element of $f(x)$ a negative image element, and constructing the convolution of both sets of sources with the fundamental solution:

$$u(x, t) = \int_{-\infty}^{\infty} [f(x') - f(-x')] G_0(x, t; x', 0) \, dx'$$

$$= \int_{-\infty}^{\infty} f(x') G_0(x, t; x', 0) \, dx' - \int_{-\infty}^{\infty} f(x') G_0(x, t; -x', 0) \, dx'$$

$$= \int_{-\infty}^{\infty} f(x') G(x, t; x', 0) \, dx' \qquad (7.96)$$

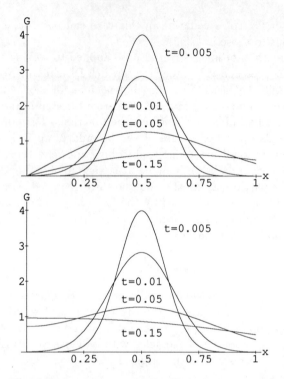

Figure 7.7 The time development of the Green's functions for the diffusion equation in the presence of an absorbing boundary at the origin (top) and with a reflecting boundary at the origin (bottom). For the case of an absorbing boundary, the Green's function must vanish at the origin, while for the case of a reflecting boundary, the spatial derivative of the Green's function must vanish at the origin.

Thus, the solution of the diffusion equation in the presence of boundaries has the same form as (7.48), but with the modified Green's function G replacing the fundamental solution G_0.

Now suppose there is a reflecting boundary at $x=0$. An incident diffusing particle striking the boundary is immediately reflected into the region $x > 0$, as though it was incident from the other side of the boundary. The presence of a reflecting boundary can be taken into account by adding to the fundamental solution an image 'source' term on the other side of the boundary again located symmetrically with respect to the origin. Accordingly, the Green's function in the presence of the reflecting boundary becomes

$$G(x,t;x',t') = G_0(x,t;x',t') + G_0(x,t;-x',t')$$

$$= \frac{1}{\sqrt{4\pi(t-t')}}\left[e^{-(x-x')^2/4(t-t')} + e^{-(x+x')^2/4(t-t')}\right] \quad (7.97)$$

The reflecting boundary can be cast as a boundary condition on the Green's function, by pursuing the interpretation based upon diffusing particles. Since every particle that is incident at the boundary is reflected by the boundary, the net current $\partial G/\partial x$ at the boundary is zero. The boundary condition for G is therefore

$$\left.\frac{\partial G}{\partial x}\right|_{x=0} = 0 \qquad (7.98)$$

The absorbing boundary thus corresponds to a Neumann problem. The evolution of this solution is shown in Figure 7.7. The construction of the solution to the diffusion equation in the presence of a reflecting boundary at the origin for an initial distribution $u(x,0) = F(x)$ proceeds in direct analogy with the steps taken in (7.96).

As a final illustration of the Green's function method in the presence of boundaries, we consider the diffusion equation in the *finite* region $0 \le x \le 1$, with absorbing boundaries at $x=0$ and at $x=1$. This problem can also be solved by the eigenfunction method discussed in Chapter 4, and the solution by the method of Green's functions provides a useful comparison of the two methods (Problem 48).

Consider the evolution of a point source initially localized at the point x' in the interval $(0,1)$. Without the presence of any boundaries, the time development is given by the fundamental solution, $G_0(x,t;x',0)$. The absorbing boundary at $x = 0$ means that an image source given by $-G_0(x,t;-x',0)$ must be added to the original fundamental solution to maintain the boundary condition. Similarly, the reflecting boundary at $x=1$ means that the image source $G_0(x,t;2-x',0)$ must be added as well. Because of the presence of the *two* boundaries, the image sources must themselves have images in order for the boundary conditions at the endpoints to be maintained for all times. Furthermore, these images must also have images, which in turn must also have images, and so on. The complete Green's function in the presence of both boundaries must therefore comprise an infinite sequence of fundamental solutions localized at the appropriate points and with the appropriate signs:

$$G(x,t;x',t') = \sum_{n=-\infty}^{\infty} G_0(x,t;2n+x',t') - \sum_{n=-\infty}^{\infty} G_0(x,t;2n-x',t') \qquad (7.99)$$

The time development of this solution is plotted in Figure 7.8. Also shown are the corresponding solutions obtained with reflecting boundary conditions at $x = 0$ and $x = 1$ and with an absorbing boundary at $x = 0$ and a reflecting boundary at $x = 1$. The profiles of all three solutions are similar for early times but develop distinctive differences as the presence of the boundaries begins to be felt.

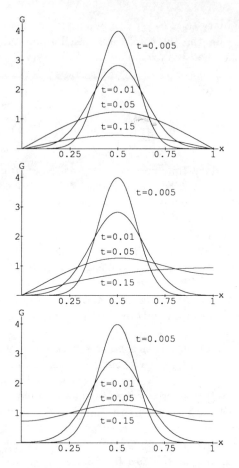

Figure 7.8 The early stages of the time development of Green's functions of the diffusion equation defined over the finite interval $0 \leq x \leq 1$ with absorbing boundaries at $x = 0$ and $x = 1$ (top), an absorbing boundary at $x = 0$ and a reflecting boundary at $x = 1$ (middle) and reflecting boundaries at $x = 0$ and $x = 1$ (bottom). The solutions have been calculated at the times $t = 0.005$, $t = 0.01$, $t = 0.05$ and $t = 0.15$.

In effect, what we done in the constructions for the solutions in Figure 7.8 is the same as for the periodic extensions used in Section 4.4 for different types of boundary conditions. For example, absorbing boundary conditions at $x = 0$ and $x = 1$ dictate that the periodic extension must be an *odd* function of x, and that is precisely what we have in Equation (7.99): a periodic odd extension of the fundamental solution. Similarly, reflecting boundary conditions at $x = 0$ and $x = 1$ would necessitate using an *even* periodic extension of the fundamental solution. These statements can be verified by determining the Green's functions directly from the diffusion equation. This is taken up in Problems 46 and 47.

Further Reading

Carslaw (1950) contains a readable mathematical treatment of Fourier series and integrals and Carslaw and Jaeger (1963) provide detailed discussions of Laplace transforms as applied to solving differential equations. Churchill (1972) provides a detailed discussion of the formulation and application of integral transforms. Green's functions for Poisson's equation are covered by Panofsky and Phillips (1962) and by Jackson (1975), including extensive discussions of the method of images. A comprehensive treatment of many aspects and applications of Green's functions may be found in Morse and Feshbach (1953).

References

Carslaw H. S. (1950). *An Introduction to the Theory of Fourier's Series and Integrals* 3rd edn. New York: Dover.
Carslaw H. S. and Jaeger J. C. (1963). *Operational Methods in Applied Mathematics.* New York: Dover.
Chester C. R. (1971). *Techniques in Partial Differential Equations.* New York: McGraw-Hill.
Churchill R. V. (1972). *Operational Mathematics.* New York: McGraw-Hill.
Courant R. (1962). *Partial Differential Equations.* New York: Wiley.
Jackson J. D. (1975). *Classical Electrodynamics* 2nd edn. New York: Wiley.
Lighthill M. J. (1958). *An Introduction to Fourier Analysis and Generalised Functions.* London: Cambridge University Press.
Morse P. M. and Feshbach H. (1953). *Methods of Theoretical Physics* Vol. 1. New York: McGraw-Hill.
Panofsky W. K. H. and Phillips M. (1962). *Classical Electricity and Magnetism* 2nd edn. Reading MA: Addison-Wesley.
Reif F. (1965). *Fundamentals of Statistical and Thermal Physics.* New York: McGraw-Hill.
Whittaker E. T. and Watson G. N. (1963). *A Course of Modern Analysis* 4th edn. Cambridge: Cambridge University Press.

Problems

1. Find the Fourier transforms of each of the following functions. The quantity a is a positive constant.

(a) $f(x) = \begin{cases} 1 & \text{if } |x| \leq 1 \\ 0 & \text{if } |x| > 1 \end{cases}$

(b) $f(x) = \begin{cases} 1 - |x| & \text{if } |x| \leq 1 \\ 0 & \text{if } |x| > 1 \end{cases}$

(c) $f(x) = \begin{cases} 1 - x^2 & \text{if } |x| \leq 1 \\ 0 & \text{if } |x| > 1 \end{cases}$

(d) $f(x) = \exp(-ax^2)$

(e) $f(x) = \exp(-a|x|)$

(f) $f(x) = x^2 \exp(-ax^2)$

(g) $f(x) = x \exp(-a|x|)$

Evaluate these integrals with *Mathematica* to verify your results.

2. The **convolution theorem** for Fourier transforms states

$$\frac{1}{2\pi} \int_{-\infty}^{\infty} \tilde{f}(k)\tilde{g}(k) e^{-ikx} \, dk = \int_{-\infty}^{\infty} f(x') g(x - x') \, dx'$$

Prove this theorem by writing the product $\tilde{f}(k)\tilde{g}(k)$ as

$$\tilde{f}(k)\tilde{g}(k) = \left[\int_{-\infty}^{\infty} f(x) e^{ikx} \, dx\right]\left[\int_{-\infty}^{\infty} f(x) e^{ikx} \, dx\right]$$

By combining these integrals and changing the integration variables, show that this expression can be written as

$$\tilde{f}(k)\tilde{g}(k) = \int_{-\infty}^{\infty} e^{ikx} \left[\int_{-\infty}^{\infty} f(x') g(x - x') \, dx'\right] dx$$

Hence, use Equation (7.8) to deduce the convolution theorem.

3. Repeat the steps taken in Problem 2 for the case that $g(k)$ is the complex conjugate of $f(k)$, i.e. $g(k) = f^*(k)$, to obtain

$$\frac{1}{2\pi} \int_{-\infty}^{\infty} |\tilde{f}(k)|^2 e^{iku} \, dk = \int_{-\infty}^{\infty} e^{-iku} \left[\int_{-\infty}^{\infty} f(x) f^*(x + u) \, dx\right] dk$$

By setting $u=0$, obtain Parseval's theorem for Fourier transforms:

$$\frac{1}{2\pi}\int_{-\infty}^{\infty}|\tilde{f}(k)|^2\,dk = \int_{-\infty}^{\infty}|f(x)|^2\,du$$

4. Plot the Fourier transforms of each of the functions in Problem 1 using the Plot command of *Mathematica*. Then investigate the convergence of Parseval's theorem for each of these functions by constructing the quantity

$$\mathcal{F}(K) = \frac{1}{2\pi\mathcal{N}}\int_{-K}^{K}|\tilde{f}(k)|^2\,dk$$

where $\mathcal{N}$ is given by

$$\mathcal{N} = \int_{-\infty}^{\infty}|f(x)|^2\,dx$$

Plot $\mathcal{F}(K)$ over an expanding range of K to see directly the convergence of this quantity toward unity. Comment on the *rate* of convergence of $\mathcal{F}(K)$ for each case.

5. The motivation behind the uses of the various Fourier transforms introduced in Section 7.1 can be traced back to the method of separation of variables discussed in Section 4.1. Consider as an example the eigenfunction equations obtained for the diffusion equation in Equation (4.59):

$$\frac{dT}{dt} + k^2 T = 0, \qquad \frac{d^2 X}{dx^2} + k^2 X = 0$$

where the solution $u(x,t)$ has been written as $u(x,t) = X(x)T(t)$. Suppose that there are no boundaries in the x direction. Then the appropriate general solution of the equation for X is

$$X_k(x) = A_k \cos kx + B_k \sin kx$$

or, in complex exponential form, as

$$X_k(x) = C_k\, e^{ikx}$$

With these functions as a basis and with the normalization derived in Section 7.1, we obtain the Fourier integral representation of a function $f(x)$ given in Equations (7.6) and (7.8).

Now suppose that a solution is required on the *semi-infinite* interval $[0,\infty)$ and that the boundary condition at $x=0$ is

$$u(0,t) = 0$$

i.e. the absorbing boundary conditions described in Section 7.6. Show that this boundary condition and the normalization of the functions $\exp(ikx)$ imply that the appropriate representation of a function is in terms of a **Fourier sine transform**:

$$\tilde{f}(k) = \int_0^\infty f(x) \sin kx \, dx, \qquad f(x) = \int_0^\infty \tilde{f}(k) \sin kx \, dx$$

Apply the same procedure for the case of the reflecting boundary conditions

$$u_x(0,t) = 0$$

to obtain the **Fourier cosine transform**:

$$\tilde{f}(k) = \int_0^\infty f(x) \cos kx \, dx, \qquad f(x) = \frac{2}{\pi} \int_0^\infty \tilde{f}(k) \cos kx \, dk$$

Obtain the integral theorem in Equation (7.9) for the sine and cosine transforms. Notice that these transforms are appropriate for *any* equation for which the equation determining the spatial eigenfunctions is given by $X'' + k^2 X = 0$.

6. Find the Laplace transforms of the following functions. The quantity a is a real constant

(a) $f(t) = 1$

(b) $f(t) = t^n, \qquad n = 0, 1, 2, \ldots$

(c) $f(t) = e^{-at}$

(d) $f(t) = \cos at$

(e) $f(t) = \sin at$

(f) $f(t) = \cosh at$

(g) $f(t) = \sinh at$

Use the *Mathematica* package **Laplace** to verify your results.

7. Determine the Laplace transforms of the following expressions. The quantity a is a real constant.

(a) $t^n f(t)$

(b) $\dfrac{1}{t} f(t)$

(c) $e^{at} f(t)$

(d) $\dfrac{d^n f}{dt^n}$

8. Find the inverse Laplace transforms of the following expressions by evaluating the Bromwich integral over the appropriate contour. The quantity a is a real constant.

(a) $F(s) = s$

(b) $F(s) = \dfrac{1}{s}$

(c) $F(s) = \dfrac{1}{s^n}, \quad n = 0, 1, 2, \ldots$

(d) $F(s) = \dfrac{1}{s} e^{-as}$

(e) $F(s) = \dfrac{1}{(s+a)^n}$

Use the *Mathematica* package InverseLaplace to verify your results.

9. Solve the following differential equations using the method of Laplace transforms given in Section 7.2 for general initial conditions. The primes indicate derivatives of y with respect to x and $f(x)$ is any function for which the Laplace transform exists.

(a) $y'' + 2y' + 2y = e^{-x}$

(b) $y'' + y' + y = \cos x$

(c) $y'' - y' - y = \sinh 2x$

(d) $y' - y = e^{-x}$

(e) $y'' + 4y = f(x)$

Apply the *Mathematica* procedure given in Section 7.2 to verify your solutions.

10. Use the *Mathematica* procedure given in Section 7.2 to obtain the solutions to the following differential equations for general initial conditions. The primes again indicate derivatives of y with respect to x, the quantities a, b and c are constants and $f(x)$ is any function for which the Laplace transform exists.

(a) $y''' + y'' + y' + y = \cos x$

(b) $y'' + 4y' + 4y = x \cos x$

(c) $y'''' + 4y = e^{-x} + x e^{-x}$

(d) $y'' + 4y = e^{-ax} + e^{-bx}$

(e) $ay'' + by' + cy = f(x)$

11. The convolution theorem for Laplace transforms is

$$\lim_{R \to \infty} \left[\frac{1}{2\pi i} \int_{u-iR}^{u+iR} F(s)G(s) e^{st} \, ds \right] = \int_0^t f(t')g(t-t') \, dt'$$

To prove this theorem, substitute the integral expression for the Laplace transform of f into the Bromwich integral, perform the integral over s, and observe that both f and g are defined only for *positive* values of their arguments.

12. Verify the following Laplace transform by integrating the series expansions of the Bessel function $J_0(t)$ term by term and resumming.

$$\int_0^\infty J_0(at) e^{-st} \, dt = \frac{1}{\sqrt{s^2 + a^2}}$$

where a is a positive constant. Use this result to deduce

$$\int_0^\infty I_0(at) e^{-st} \, dt = \frac{1}{\sqrt{s^2 - a^2}}$$

where $I_0(t)$ is a modified Bessel function of the first kind.

13. Evaluate the Bromwich integrals corresponding to the Laplace transforms obtained in Problem 12:

$$\lim_{R \to \infty} \left[\frac{1}{2\pi i} \int_{u-iR}^{u+iR} \frac{e^{st}}{\sqrt{s^2 + a^2}} \, ds \right] = J_0(at)$$

Assume that the integrand can be expanded into a Taylor series and evaluate the integral term by term. Similarly, show that

$$\lim_{R \to \infty} \left[\frac{1}{2\pi i} \int_{u-iR}^{u+iR} \frac{e^{st}}{\sqrt{s^2 - a^2}} \, ds \right] = I_0(at)$$

What are the restrictions on s in applying this procedure?

14. Use the results of Problems 11 and 12 to show that

$$\int_0^t J_0(s) J_0(t-s) \, ds = \sin t$$

15. Use the Laplace transform and the Bromwich integral to obtain the following integral representation of the Heaviside function:

$$H(x) = e^{ux} \lim_{R \to \infty} \left[\frac{1}{2\pi i} \int_{-R}^{R} \frac{e^{ivt}}{v - iu} \, dv \right]$$

By choosing appropriate contours in the complex v plane show that

$$H(x) = \begin{cases} 1 & \text{if } x > 0 \\ 0 & \text{if } x < 0 \end{cases}$$

16. Some of the properties of the delta function can be studied with the use of the convolution theorem for Fourier transforms (Problem 2). Thus, beginning with

$$\frac{1}{2\pi} \int_{-\infty}^{\infty} \tilde{f}(k)\tilde{g}(k) e^{-ikx} \, dk = \int_{-\infty}^{\infty} f(x')g(x-x') \, dx'$$

choose the Fourier components of $g(x)$ such that $\tilde{g}(k) = 1$ and introduce the notation $g(x) = \delta(x)$, i.e.

$$\delta(x) = \frac{1}{2\pi} \int_{-\infty}^{\infty} e^{-ikx} \, dk$$

Then, by substituting into the convolution relation obtain

$$f(x) = \int_{-\infty}^{\infty} f(x')\delta(x-x') \, dx'$$

Show that

$$\int_{-\infty}^{\infty} f(x')\delta^{(n)}(x-x') \, dx = f^{(n)}(x)$$

where

$$\delta^{(n)}(x) = \frac{d^n \delta(x)}{dx^n}, \qquad f^{(n)}(x) = \frac{d^n f(x)}{dx^n}$$

Deduce that

$$\int_{-\infty}^{\infty} f(x)\delta^{(n)}(x) \, dx = (-1)^n f^{(n)}(0)$$

where n is a nonnegative integer.

The delta function is an example of a **generalized function**, also sometimes called a **distribution**. Generalized functions are neither differentiable nor integrable in the conventional sense of defining these operations as limits of discretized quantities. This problem shows how derivatives and integrals of the delta function are to be interpreted. Comprehensive treatments of generalized functions may be found in the books by Courant (1962) and Lighthill (1958).

17. Show that the integral representation of the Heaviside function can be written as

$$H(x) = \frac{1}{2\pi} \int_{-\infty}^{\infty} \frac{e^{(u+iv)x}}{u+iv} \, dv$$

Then use the results of Problems 15 and 16 to deduce that the derivative of the Heaviside function can be represented in symbolic form as the delta function:

$$\frac{dH(x)}{dx} = \delta(x)$$

18. The fundamental solution of the diffusion equation can be used to provide another approach to defining the delta function. Consider the series of functions, $\phi_\epsilon(x)$, given by

$$\phi_\epsilon(x) = \frac{1}{\sqrt{4\pi\epsilon}} \exp\left(-\frac{x^2}{4\epsilon}\right)$$

where ϵ is a nonnegative quantity. To investigate the properties of this function in the limit as ϵ approaches zero, consider the following integral:

$$\int_{-\infty}^{\infty} f(x)\phi_\epsilon(x)\,dx = \frac{1}{\sqrt{4\pi\epsilon}} \int_{-\infty}^{\infty} f(x) \exp\left(-\frac{x^2}{4\epsilon}\right) dx$$

As ϵ decreases, $\phi_\epsilon(x)$ varies appreciably over successively smaller intervals about the origin (Figure 7.5). Thus, this integral can be evaluated by performing a Taylor series expansion of $f(x)$ about the origin. By so expanding $f(x)$ and substituting into the integral, obtain

$$\int_{-\infty}^{\infty} f(x)\phi_\epsilon(x)\,dx = \frac{1}{\sqrt{\pi}} \sum_{n=0}^{\infty} (4\epsilon)^{n/2} f^{(n)}(0) \int_{-\infty}^{\infty} s^n \exp(s^2)\,ds$$

where

$$f^{(n)}(0) = \left.\frac{d^n f}{dx^n}\right|_{x=0}$$

Evaluate the integrals on the right-hand side of this equation and deduce that

$$\lim_{\epsilon \to 0}\left[\int_{-\infty}^{\infty} f(x)\phi_\epsilon(x)\,dx\right] = f(0)$$

From this result and an appropriate number of integrations by parts, deduce that

$$\lim_{\epsilon \to 0}\left[\int_{-\infty}^{\infty} f(x)\phi_\epsilon^{(n)}(x)\,dx\right] = f^{(n)}(0)$$

Thus, the function $\phi_\epsilon(x)$ is said to approach the function $\delta(x)$ as ϵ approaches zero in the sense that these and other integral relations for these functions are the same.

19. Use *Mathematica* to evaluate numerically the quantity

$$\lim_{\epsilon \to 0}\left[\int_{-\infty}^{\infty} f(x)\phi_\epsilon(x)\,dx\right]$$

defined in Problem 18 for the functions $f(x)=x$ and $f(x)=\cos x$ as a function of ϵ. In particular, use Plot to display the values of these integrals as $\epsilon \to 0$. Apply the same procedure to evaluate

$$\lim_{\epsilon \to 0}\left[\int_{-\infty}^{\infty} f(x)\phi_{\epsilon}^{(n)}(x)\,dx\right]$$

for $n=1, 2$ and 3 for $f(x)=x^3$, $f(x)=\tanh x$ and $f(x)=\cos x$.

20. By substituting the delta function on the right-hand side of (7.8) show that the Fourier components of this function are given by

$$\tilde{\delta}(k) = 1$$

By substituting this expression into (7.6) obtain the following integral representation of the delta function

$$\delta(x) = \frac{1}{2\pi}\int_{-\infty}^{\infty} e^{ikx}\,dk$$

Since this integral does not converge, this definition as it stands should be regarded purely in 'formal' terms. We can examine this integral representation by considering the function

$$\Delta_L(x) \equiv \frac{1}{2\pi}\int_{-L}^{L} e^{ikx}\,dk$$

Calculate this integral and use Problem 33 to show that the integrated area under $\Delta_L(x)$ is unity, for all positive values of L. Then show that

$$\lim_{L\to\infty}\left[\int_{-\infty}^{\infty} f(x)\Delta_L(x)\,dx\right] = f(0)$$

In the sense that this and other properties of the delta function are reproduced by $\Delta_L(x)$ in the limit $L\to\infty$ we can make the identification

$$\lim_{L\to\infty}\left[\frac{1}{2\pi}\int_{-L}^{L} e^{ikx}\,dk\right] = \frac{1}{2\pi}\int_{-\infty}^{\infty} e^{ikx}\,dk = \delta(x)$$

21. Apply the result of Problem 20 to obtain the orthogonality relation for the functions e^{ikx}:

$$\int_{-\infty}^{\infty} e^{-ik'x} e^{ikx}\,dx = 2\pi\delta(k-k')$$

Compare this result with that obtained in Problem 4.2. Then obtain the completeness relation

$$\frac{1}{2\pi}\int_{-\infty}^{\infty} e^{-ikx'} e^{ikx}\,dk = \delta(x-x')$$

22. The behavior of the delta function under general coordinate transformations is important in many applications. Thus, consider the transformation from rectangular coordinates, (x,y,z), to another set of orthogonal curvilinear coordinates $u_i(x,y,z)$, for $i=1,2$ and 3. From Problem 18, we can write

$$\lim_{\epsilon \to 0} \left\{ \frac{1}{\sqrt{4\pi\epsilon}} \exp\left[-\frac{1}{4\epsilon}(x^2 + y^2 + z^2)\right] \right\} = \delta(x)\delta(y)\delta(z)$$

Observe that in the limit $\epsilon \to 0$, we need consider only the *infinitesimal* length

$$(dr)^2 = (dx)^2 + (dy)^2 + (dz)^2$$

From the discussion of Section 1.4, we can write this quantity in the (u_1, u_2, u_3) coordinate system in terms of the scale factors h_1, h_2 and h_3 as

$$(dr)^2 = h_1^2(du_1)^2 + h_2^2(du_2)^2 + h_3^2(du_3)^2$$

Thus, by substituting this expression for $(dr)^2$ into $\exp(-r^2/4\epsilon)$, deduce that

$$\delta(x)\delta(y)\delta(z) = \frac{1}{h_1 h_2 h_3} \delta(u_1)\delta(u_2)\delta(u_3)$$

$$= \frac{1}{J} \delta(u_1)\delta(u_2)\delta(u_3)$$

where $J = h_1 h_2 h_3$ is the Jacobian of the coordinate transformation.

23. The result of Problem 22 can be obtained for more general coordinate transformations by not appealing to any particular realization of the delta function but working instead from its general properties. Thus, consider the transformation from rectangular coordinates (x,y,z) to another coordinate system (ξ_1, ξ_2, ξ_3). We have from Problem 22 that if this new coordinate system is orthogonal, then

$$\delta(x)\delta(y)\delta(z)\,dx\,dy\,dz = \left[\frac{1}{J}\delta(u_1)\delta(u_2)\delta(u_3)\right] J\,du_1\,du_2\,du_3$$

$$= \delta(u_1)\delta(u_2)\delta(u_3)\,du_1\,du_2\,du_3$$

In fact, once we have identified $\delta(x)\,dx$ as the pertinent quantity from which to identify the transformation properties of the delta function, then we can write this equation for coordinates that are not necessarily orthogonal. Hence, deduce

$$\delta(x)\delta(y)\delta(z) = \frac{1}{J}\delta(\xi_1)\delta(\xi_2)\delta(\xi_3)$$

where the Jacobian J is given by determinant of the matrix with entries $\partial x_i/\partial \xi_j$, where $x_1 = x$, $x_2 = y$ and $x_3 = z$.

24. The Gaussian integral in Equation (7.45),

$$I = \int_{-\infty}^{\infty} \exp(x^2)\,dx$$

can be evaluated as follows. First construct the quantity

$$I^2 = \left[\int_{-\infty}^{\infty} \exp(x^2)\,dx\right]\left[\int_{-\infty}^{\infty} \exp(y^2)\,dy\right] = \int_{-\infty}^{\infty}\int_{-\infty}^{\infty} \exp(x^2 + y^2)\,dx\,dy$$

Then change the variables of integration to polar coordinates, $x = r\cos\phi$ and $y = r\sin\phi$ and, making the corresponding changes in the ranges of integration, obtain

$$I = \int_0^{2\pi}\left[\int_0^{\infty} \exp(r^2)r\,dr\right]d\phi$$

Evaluate this integral and so deduce that

$$\int_{-\infty}^{\infty} \exp(x^2)\,dx = \sqrt{\pi}$$

25. Evaluate the integral

$$\int_{-\infty}^{\infty} G_0(x, x'; t, t')\,dx$$

where $G_0(x, x'; t, t')$ is the fundamental solution of the diffusion equation with the initial condition $G_0(x, x'; t', t') = \delta(x - x')$. Give a physical interpretation of your result.

26. Consider the initial-value problem for the inhomogeneous diffusion equation:

$$\frac{\partial u}{\partial t} - \frac{\partial^2 u}{\partial x^2} = q(x, t)$$

with

$$u(x, 0) = f(x)$$

Show that the solution of this problem can be expressed in terms of the fundamental solution as

$$u(x, t) = \int_0^t \left[\int_{-\infty}^{\infty} q(x', t')G_0(x, t; x', t')\,dx'\right]dt' + \int_{-\infty}^{\infty} f(x)G_0(x, t;, x', 0)\,dx'$$

$$= \int_0^t \int_{-\infty}^{\infty} \frac{q(x', t')}{\sqrt{4\pi(t - t')}} \exp\left[-\frac{1}{4}\frac{(x - x')^2}{(t - t')}\right]dx'$$

$$+ \frac{1}{\sqrt{4\pi t}}\int_{-\infty}^{\infty} f(x')\exp\left[-\frac{(x - x')^2}{4t}\right]dx'$$

27. Obtain the fundamental solutions of the diffusion equation in $d = 1, d = 2$ and $d = 3$ spatial dimensions and show that they can be written as

$$G_0(x, t; x', t') = \frac{H(t - t')}{[4\pi(t - t')]^{d/2}}\exp\left[-\frac{1}{4}\frac{(\mathbf{x} - \mathbf{x}')^2}{(t - t')}\right]$$

where $H(x)$ is the Heaviside function defined in Equation (7.78).

28. The solution in (7.51) for the initial-value problem of the diffusion equation in Equations (7.35) and (7.36) can be written as

$$u(x,t) = \int_{-\infty}^{\infty} f(x-x')G_0(x',t')\,dx'$$

where $G_0(x,t)$ is the fundamental solution in (7.46):

$$G_0(x,t) = \frac{1}{\sqrt{4\pi t}} \exp\left(-\frac{x^2}{4t}\right)$$

By expanding this solution in a Taylor series in t and using the fact that it is a solution of the diffusion equation, invoke the results of Problem 18 to obtain

$$u(x,y) = \sum_{n=0}^{\infty} \frac{d^{2n} f(x)}{dx^n} \frac{t^n}{n!}$$

which is the solution obtained in (3.79) in an expansion around the initial condition. Notice that the series representation requires F to be infinitely differentiable, while the solution in (7.51) requires only that the convolution of F with the fundamental solution exists.

29. Consider the solution of the diffusion equation

$$\frac{\partial u}{\partial t} = \frac{\partial^2 u}{\partial x^2}$$

with the initial condition

$$u(x,0) = \begin{cases} 1 & \text{if } |x| \le 1 \\ 0 & \text{if } |x| > 1 \end{cases}$$

Express the solution of this initial-value problem in terms of the fundamental solution and use *Mathematica* to evaluate the integral numerically for $t > 0$ over a range of x and to display the results using Plot.

30. The fundamental solution of the diffusion equation in the form (7.46) can be obtained without invoking the Laplace transform. By performing a Fourier transform with respect to the spatial variable only, obtain a first-order equation for the Fourier components of the solution. Solve this equation for each Fourier component using the Fourier components of the initial condition (7.36) to obtain the expression in Equation (7.40).

31. Consider the following modification of the diffusion equation:

$$\frac{\partial u}{\partial t} = \frac{\partial}{\partial x}\left(\frac{\partial}{\partial x} + x\right)u$$

This is the Fokker–Planck equation used in statistical physics to describe the evolution of probability distribution functions (see Problem 6.1). A transformation can be made to bring this equation into a form more amenable to direct solution by the following steps (Reif, 1965). First observe that by neglecting the term u_{xx} the equation becomes a linear first-order equation that can be solved by the method of characteristics:

$$\frac{\partial u}{\partial t} - x\frac{\partial u}{\partial x} = u$$

Use the solution of this equation to motivate the change of variables

$$\xi = x\,e^t, \qquad v = u\,e^{-t}$$

Thus, by writing $u(x,t) = v(\xi,t)e^t$ and substituting into the Fokker–Planck equation, show that the equation satisfied by v is

$$\frac{\partial v}{\partial t} = e^{2t}\frac{\partial^2 v}{\partial \xi^2}$$

Make an appropriate change of the variable t to $\tau(t)$ to transform this equation into the diffusion equation:

$$\frac{\partial v}{\partial \tau} = \frac{\partial^2 v}{\partial \xi^2}$$

Thus, solutions of the Fokker–Planck equation with appropriate initial conditions can be obtained directly from an associated initial-value problem of the diffusion equation.

The fundamental solution of the Fokker–Planck equation is defined by the equation

$$\left[\frac{\partial}{\partial t} - \frac{\partial}{\partial x}\left(\frac{\partial}{\partial x} + x\right)\right] G_0(x, x'; t, t') = \delta(x - x')\delta(t - t')$$

By performing the variable transformations used above, show that this equation becomes

$$\left[\frac{\partial}{\partial \tau} - \frac{\partial^2}{\partial \xi^2}\right] G_0(\xi, \xi'; \tau, \tau') = \delta(\xi - \xi')\delta(\tau - \tau')$$

which is the defining equation for the fundamental solution for the diffusion equation. Hence, show that the fundamental solution of the Fokker–Planck equation is given by

$$G_0(x, x'; t, t') = \frac{1}{\sqrt{2\pi\{1 - \exp[-2(t-t')]\}}} \exp\left\{-\frac{1}{2}\frac{[x - x'\,e^{-(t-t')}]^2}{1 - e^{-2(t-t')}}\right\}$$

Show that

$$\lim_{t\to\infty} G_0(x, x'; t, t') = \frac{1}{\sqrt{2\pi}} \exp(-\tfrac{1}{2}x^2)$$

Then, if the initial conditions of the Fokker–Planck equation are given by

$$u(x,0) = f(x)$$

deduce that the behavior of the solution $u(x,t)$ as $t \to \infty$ is

$$\lim_{t \to \infty} u(x,t) = \frac{1}{\sqrt{2\pi}} \exp(-\tfrac{1}{2}x^2) \int_{-\infty}^{\infty} f(x')\,dx'$$

i.e. the probability distribution function u *always* approaches a Gaussian distribution.

32. To see the similarities and differences of solutions of the diffusion equation and the Fokker–Planck equation, use the appropriate fundamental solution to display the time development of the solutions from the following initial functions:

(a) $f(x) = \begin{cases} 1 & \text{if } |x| \leq 1 \\ 0 & \text{if } |x| > 1 \end{cases}$

(b) $f(x) = \begin{cases} \tfrac{1}{2} & \text{if } |x| \leq 1 \\ 0 & \text{if } |x| > 1 \end{cases}$

For each equation, notice that the solutions that are obtained from each of these initial functions evolve toward the same solution. Use *Mathematica* to construct these solutions with `NIntegrate` and to display and plot their evolution.

33. One way of evaluating the integral in (7.64),

$$\int_0^\infty \frac{\sin x}{x}\,dx$$

is by integrating the quantity $e^{-xy} \sin x$ over the square region $0 \leq x \leq L$ and $0 \leq y \leq L$. Specifically, this integral is evaluated by performing the integrations over x and y in two ways:

$$\int_0^L \left[\int_0^L e^{-xy} \sin x\,dx \right] dy = \int_0^L \left[\int_0^L e^{-xy}\,dy \right] \sin x\,dx$$

By evaluating these integrals and taking the limit $L \to \infty$, show that

$$\int_0^\infty \frac{\sin x}{x}\,dx = \tfrac{1}{2}\pi$$

and deduce that

$$\int_{-\infty}^\infty \frac{\sin x}{x}\,dx = \pi$$

34. Consider the solution of the Dirichlet problem for Poisson's equation

$$\nabla^2 V = -4\pi \rho(\mathbf{x})$$

in a region $\mathcal{V}$ bounded by a surface $\mathcal{S}$ on which the solution is required to take the values of a function f:

$$V(\mathbf{x}) = f(\mathbf{x})$$

for $\mathbf{x}$ lying on $\mathcal{S}$. The Green's function $G(\mathbf{x}, \mathbf{x}')$ for this problem satisfies

$$\nabla^2 G(\mathbf{x}, \mathbf{x}') = -4\pi \delta(\mathbf{x} - \mathbf{x}')$$

with the boundary condition

$$G(\mathbf{x}, \mathbf{x}') = 0$$

for $\mathbf{x}$ lying on $\mathcal{S}$. By multiplying the equation for V by G, multiplying the equation for G by V, and using Green's theorem and the boundary conditions for V and G, obtain the solution to the Dirichlet problem for V in terms of G as

$$V(\mathbf{x}) = \int_\mathcal{V} \rho(\mathbf{x}') G(\mathbf{x}, \mathbf{x}') \, d\mathbf{x}' + \frac{1}{4\pi} \int_\mathcal{S} f(\mathbf{x}') \nabla G(\mathbf{x}, \mathbf{x}') \cdot d\mathcal{S}$$

where $d\mathcal{S}$ is the unit of area of $\mathcal{S}$. How is this solution related to that in Equation (7.66)?

35. The equation for the fundamental solution of Poisson's equation in one spatial dimension is

$$\frac{d^2 G_0}{dx^2} = -2\delta(x - x')$$

Proceed as in Section 7.4 to obtain the fundamental solution as

$$G_0(x, x') = -|x - x'|$$

The inverse Fourier transform can be evaluated by first performing an integration by parts and using Problem 33. Substitute this expression into the differential equation and use Problem 17 to verify that G_0 is indeed a solution.

36. The fundamental solution (7.66) of Poisson's equation clearly has the property $G_0(\mathbf{x}, \mathbf{x}') = G_0(\mathbf{x}', \mathbf{x})$. This is an example of a **reciprocity relation**. To demonstrate the validity of this relation in more general circumstances, consider the Green's function within a volume $\mathcal{V}$ bounded by a surface $\mathcal{S}$. The boundary condition on $\mathcal{S}$ is

$$G(\mathbf{x}, \mathbf{x}') = 0$$

for $\mathbf{x}$ lying on $\mathcal{S}$. Then, beginning with the equations for $G(\mathbf{x}, \mathbf{x}')$ and $G(\mathbf{x}, \mathbf{x}'')$,

$$\nabla^2 G(\mathbf{x}, \mathbf{x}') = -4\pi \delta(\mathbf{x} - \mathbf{x}'), \qquad \nabla^2 G(\mathbf{x}, \mathbf{x}'') = -4\pi \delta(\mathbf{x} - \mathbf{x}'')$$

multiply the first equation by $G(\mathbf{x},\mathbf{x}'')$, the second by $G(\mathbf{x},\mathbf{x}')$ and perform steps analogous to those in Problem 34 to obtain

$$G(\mathbf{x}',\mathbf{x}'') = G(\mathbf{x}'',\mathbf{x}')$$

37. Determine the fundamental solution of Poisson's equation in spherical polar coordinates. Begin by transforming the equation defining $G_0(\mathbf{x},\mathbf{x}')$,

$$\nabla^2 G(\mathbf{x},\mathbf{x}') = -4\pi\delta(\mathbf{x}-\mathbf{x}')$$

into spherical polar coordinates using the result of Problem 23. Then, by expanding the fundamental solution as

$$G_0(r,\theta,\phi; r',\theta',\phi') = \sum_{\ell=0}^{\infty} \sum_{m=-\ell}^{\ell} A_{\ell,m}(r; r',\theta',\phi') Y_{\ell,m}(\theta,\phi)$$

substitute into the transformed equation to obtain the following differential equation for the coefficients $A_{\ell,m}$:

$$\frac{1}{r}\frac{\partial^2}{\partial r^2}(rA_{\ell,m}) - \frac{\ell(\ell+1)}{r^2}A_{\ell,m} = -\frac{4\pi}{r^2}\delta(r-r')Y_{\ell,m}^*(\theta',\phi')$$

Use this equation to deduce that the $A_{\ell,m}$ can be written as

$$A_{\ell,m}(r; r',\theta',\phi') = a_\ell(r,r') Y_{\ell,m}^*(\theta',\phi')$$

and show that the equation satisfied by a_ℓ is

$$\frac{1}{r}\frac{d^2}{dr^2}(ra_\ell) - \frac{\ell(\ell+1)}{r^2}a_\ell = -\frac{4\pi}{r^2}\delta(r-r')$$

For $r \neq r'$, show that requiring the solution to be finite everywhere, together with reciprocity (Problem 36), requires $a_\ell(r,r')$ to be of the form

$$a_\ell(r,r') = B_\ell \frac{r_<^\ell}{r_>^{\ell+1}}$$

where B_ℓ is a constant, possibly depending on ℓ. Notice that this function has a cusp at $r=r'$, so the first derivative has a jump discontinuity there and the second derivative is a delta function at that point. Thus, obtain the constant B_ℓ by integrating this equation from $r=r'-\epsilon$ (where $r_< = r$ and $r_> = r'$) to $r=r'+\epsilon$ (where $r_< = r'$ and $r_> = r$) to obtain $B_\ell = 4\pi$. Hence, deduce that

$$\frac{1}{|\mathbf{x}-\mathbf{x}'|} = 4\pi \sum_{\ell=0}^{\infty} \sum_{m=-\ell}^{\ell} \frac{r_<^\ell}{r_>^{\ell+1}} Y_{\ell,m}(\theta,\phi) Y_{\ell,m}^*(\theta',\phi')$$

which generalizes the result obtained in Equation (6.113). A complete discussion of Green's functions of Poisson's equation in different coordinate systems may be found in the book by Jackson (1975).

38. Derive the d'Alembertian form (3.67) of the general solution of the wave equation

$$\frac{\partial^2 u}{\partial x^2} - \frac{1}{c^2}\frac{\partial^2 u}{\partial t^2} = 0$$

with the initial conditions

$$u(x,0) = f(x), \qquad u_t(x,0) = g(x)$$

by Fourier transforming in space and Laplace transforming in time. In particular, show that the Fourier–Laplace transformation of the wave equation yields

$$\tilde{U}(k,s) = \frac{\tilde{g}(k) + s\tilde{f}(k)}{c^2 k^2 + s^2}$$

where $\tilde{f}(k)$ and $\tilde{g}(k)$ are the Fourier components of f and g, respectively, and that the inverse transformations, in conjunction with the convolution theorem, lead directly to the general solution.

39. The fundamental solutions of the wave equation in two and three spatial dimensions are solutions of

$$\left[\frac{\partial^2}{\partial t^2} - \nabla^2\right] G_0(\mathbf{x} - \mathbf{x}', t - t') = \delta(\mathbf{x} - \mathbf{x}')\delta(t - t')$$

with the boundary conditions

$$G_0(\mathbf{x} - \mathbf{x}', 0) = 0, \qquad \left[\frac{\partial}{\partial t} G_0(\mathbf{x} - \mathbf{x}', t - t')\right]\bigg|_{t-t'} = 0$$

where $\mathbf{x} = (x,y)$ and $\mathbf{x}' = (x',y')$ in two dimensions and $\mathbf{x} = (x,y,z)$, $\mathbf{x}' = (x',y',z')$ in three dimensions. Show that the fundamental solution in two spatial dimensions is

$$G_0(\mathbf{x} - \mathbf{x}', t - t') = \frac{1}{2\pi} \frac{H(t - t')}{\sqrt{(t - t')^2 - r^2}}$$

where $r^2 = (x-x')^2 + (y-y')^2$ and $H(x)$ is the Heaviside function, and in three spatial dimensions is

$$G_0(\mathbf{x} - \mathbf{x}', t - t') = \frac{\delta(t - t' - r)}{4\pi r}$$

where $r^2 = (x-x')^2 + (y-y')^2 + (z-z')^2$ and $\delta(x)$ is the Dirac delta function.

40. If the class of admissible solutions of the wave equation is expanded to include generalized functions, then the fundamental solution (7.79) can be shown to be a solution of the inhomogeneous wave equation in (7.70) with

$q(x,t) = \delta(x)\delta(t)$. To do so, first show from the definition of the Heaviside function that the fundamental solution satisfies the boundary conditions

$$G_0(x,0) = 0, \quad \left[\frac{\partial}{\partial t}G_0(x,t)\right]\bigg|_{t=0} = 0$$

Then, by performing the differentiations and using the results of Problem 17, show that $G_0(x,t)$ satisfies the equation

$$\left[\frac{\partial^2}{\partial t^2} - \frac{\partial^2}{\partial x^2}\right]G_0(x,t) = 2\delta(t-x)\delta(t+x)$$

Use the result of Problem 23 to show that

$$\delta(t-x)\delta(t+x) = \tfrac{1}{2}\delta(x)\delta(t)$$

and hence deduce that $G_0(x,t)$ is indeed a solution of the inhomogeneous wave equation.

41. Determine the fundamental solution that is given by the following equation:

$$\frac{\partial^2 G_0}{\partial t^2} - \frac{\partial^2 G_0}{\partial x^2} + G_0 = \delta(x-x')\delta(t-t')$$

with the initial conditions

$$G_0(x-x',0) = 0, \quad \left[\frac{\partial}{\partial t}G_0(x-x',t-t')\right]\bigg|_{t=t'} = 0$$

This is the fundamental solution of the **Klein–Gordon equation**. Obtain the fundamental solution in integral form as

$$G_0(x-x',t-t') = \tfrac{1}{2}\lim_{R\to\infty}\left[\frac{1}{2\pi i}\int_{u-iR}^{u+iR}\frac{\exp[-(x-x')\sqrt{s^2+1}]}{\sqrt{s^2+1}}e^{s(t-t')}ds\right]$$

Then, from Problem 13 deduce that

$$\lim_{R\to\infty}\left[\frac{1}{2\pi i}\int_{u-iR}^{u+iR}\frac{\exp(-b\sqrt{s^2+a^2})}{\sqrt{s^2+a^2}}e^{st}ds\right] = \begin{cases} 0 & \text{if } 0 < t < b \\ J_0[a\sqrt{t^2-b^2}] & \text{if } b < t \end{cases}$$

and so obtain the fundamental solution as

$$G_0(x-x',t-t') = \tfrac{1}{2}J_0\left[\sqrt{(t-t')^2-(x-x')^2}\right]H[t-t'-(x-x')]H[t-t'+(x-x')]$$

where H is the Heaviside function in Equation (7.78).

42. Consider the telegraph equation in the form

$$\frac{\partial^2 u}{\partial x^2} - \frac{\partial^2 u}{\partial t^2} - 2\frac{\partial u}{\partial t} = 0$$

By making the transformation $u = v e^{-t}$, show that the equation for v is

$$\frac{\partial^2 v}{\partial x^2} - \frac{\partial^2 v}{\partial^2 t} + v = 0$$

The fundamental solution of this equation is obtained by solving the following equation:

$$\frac{\partial^2 G_0}{\partial t^2} - \frac{\partial^2 G_0}{\partial x^2} - G_0 = \delta(x - x')\delta(t - t')$$

From the solution obtained in Problem 41, deduce that the fundamental solution of the transformed telegraph equation is given by

$$G_0(x-x', t-t') = \tfrac{1}{2} I_0\left[\sqrt{(t-t')^2 - (x-x')^2}\right] H[t-t' - (x-x')] H[t-t' + (x-x')]$$

Thus, if the initial conditions of the telegraph equation are

$$u(x, 0) = f(x), \qquad u_t(x, 0) = g(x)$$

show that the solution can be written in terms of the fundamental solution as

$$u(x,t) = \left\{ \frac{1}{2} \int_{x-t}^{x+t} g(x') I_0\left[\sqrt{t^2 - (x-x')^2}\right] dx' \right.$$

$$\left. + \frac{1}{2} \frac{\partial}{\partial t}\left[\int_{x-t}^{x+t} f(x') I_0\left[\sqrt{t^2 - (x-x')^2}\right] dx'\right] \right\} e^{-t}$$

By carrying out the derivative with respect to t in the second term, show that this solution can be written as

$$u(x,t) = \left\{ \tfrac{1}{2}[f(x+t) + f(x-t)] + \frac{1}{2} \int_{x-t}^{x+t} g(x') I_0\left[\sqrt{t^2 - (x-x')^2}\right] dx' \right.$$

$$\left. + \frac{t}{2} \int_{x-t}^{x+t} f(x')[t^2 - (x-x')^2]^{-1/2} I_1\left[\sqrt{t^2 - (x-x')^2}\right] dx' \right\} e^{-t}$$

which has some similarities with d'Alembert's solution of the wave equation in Equation (3.67).

43. Use *Mathematica* to plot the time development of the solution to the telegraph equation with the initial conditions

$$f(x) = \begin{cases} 1 & \text{if } |x| \leq 1 \\ 0 & \text{if } |x| > 1 \end{cases}, \qquad g(x) = 0$$

In performing the integral over the modified Bessel function, it proves useful to observe that $\lim_{x \to 0} I_1(x) = \tfrac{1}{2} x$. Comment on the similarities and differences between this solution and the corresponding solution of the wave equation.

44. An application of the Fourier sine and cosine transforms is to derive the form of the Green's functions in the presence of boundaries directly, rather than to argue heuristically as was done in Section 7.6. Thus, consider the initial-value problem of the diffusion equation,

$$\frac{\partial u}{\partial t} = \frac{\partial^2 u}{\partial x^2}$$

on the semi-infinite interval $(0, \infty)$. The boundary conditions are of the absorbing type:
$$u(0, t) = 0$$

The Green's function for this equation is the solution for the initial conditions

$$u(x, 0) = \delta(x - x')$$

Perform a Fourier sine transform of this equation, solve the first-order equation in t and invert the transform using the inversion formula in Problem 35 to obtain

$$G(x, t; x', 0) = \frac{1}{\sqrt{4\pi t}} \left\{ \exp[-(x - x')^2/4t] - \exp[-(x + x')^2/4t] \right\}$$

which is precisely the expression given in Equation (7.95).

45. Repeat Problem 44 with reflecting boundary conditions

$$u_x(0, t) = 0$$

to obtain the Green's function given in Equation (7.97):

$$G(x, t; x', 0) = \frac{1}{\sqrt{4\pi t}} \left\{ \exp[-(x - x')^2/4t] + \exp[-(x + x')^2/4t] \right\}$$

46. Laplace transforms can be used to obtain the solutions of partial differential equations defined over *bounded* intervals and expressed either as series expansions, as in Chapter 4, or in terms of Green's functions, as in Section 7.6. We will consider the former case in this problem and the latter in Problem 47. As an example, consider the diffusion equation defined over the interval $[0, L]$,

$$\frac{\partial u}{\partial t} = \frac{\partial^2 u}{\partial x^2}, \qquad 0 \leq x \leq L$$

with absorbing boundary conditions

$$u(0, t) = 0, \quad u(L, t) = 0$$

and with the initial condition
$$u(x,0) = f(x)$$

Perform a Laplace transform of the diffusion equation with respect to time to obtain the following differential equation for the Laplace transform $U(x,s)$ of $u(x,t)$:
$$\frac{\partial^2 U}{\partial x^2} - sU = -f(x)$$

with the boundary conditions $U(0,s) = 0$ and $U(L,s) = 0$. Show that the solution of this inhomogeneous second-order differential equation is

$$U(x,s) = \frac{1}{\sqrt{s}\sinh(L\sqrt{s})} \left\{ \sinh\left[(L-x)\sqrt{s}\right] \int_0^x \sinh\left(x'\sqrt{s}\right) F(x') \, dx' \right.$$
$$\left. + \sinh\left(x\sqrt{s}\right) \int_x^L \sinh\left[(L-x')\sqrt{s}\right] F(x') \, dx' \right\}$$

The solution of the diffusion is therefore given by

$$u(x,t) = \lim_{R \to \infty} \left[\frac{1}{2\pi i} \int_{u-iR}^{u+iR} e^{st} U(x,s) ds \right]$$

The integrand has simple poles at $s = 0$ and at $s = -k_n^2$, where $k_n = n\pi/L$ for every *positive* integer n. Thus, the contour of integration must be chosen to avoid these poles and u must be chosen to be positive: $u > 0$. Show that residue at $s=0$ vanishes and that the residue at $s=-k_n^2$ is

$$\frac{2}{L} \exp(-k_n^2 t) \sin k_n x \int_0^L f(x') \sin k_n x' \, dx'$$

Hence, obtain the solution in the form

$$u(x,t) = \sum_{n=1}^{\infty} \left[\frac{2}{L} \int_0^L f(x') \sin k_n x' \, dx' \right] \exp(-k_n^2 t) \sin k_n x$$

which is precisely the solution in (4.64) and (4.65).

47. Another way to evaluate the Laplace transform of $U(x,s)$ in Problem 46 is to use the geometric series

$$\frac{1}{1-x} = \sum_{n=0}^{\infty} x^n$$

for $|x| < 1$ to write $\sinh(L\sqrt{s})$ as

$$\sinh(L\sqrt{s}) = 2e^{-L\sqrt{s}} \sum_{n=0}^{\infty} e^{-2nL\sqrt{s}}$$

Notice that the real part of s must be chosen to insure that $\exp(-2L\sqrt{s}) < 1$. Then by expressing the hyperbolic functions in terms of exponentials, using the result that (Problem 8.31)

$$\int_0^\infty \frac{1}{\sqrt{\pi t}} \exp(-a^2/4t) e^{-st} \, dt = \frac{1}{\sqrt{s}} e^{-a\sqrt{s}}$$

where a is a constant, and assuming that the orders of integration and summation may be interchanged at will, obtain the solution in the form

$$u(x,t) = \int_0^L G(x,t;x',0) F(x') \, dx'$$

where the Green's function is given in terms of the fundamental solution by

$$G(x,t;x',0) = \sum_{n=-\infty}^\infty \Big[G_0(x,t;2nL+x',0) - G_0(x,t;2nL-x',0)\Big]$$

which agrees with (7.99).

48. Use *Mathematica* to construct the Green's function in Equation (7.99) with the procedure outlined in Problem 4.18 for periodic extensions. With this exact solution for the Green's function in a bounded domain, investigate the convergence of the sum in (7.99) with increasing time. In particular, show that for early times, this sum converges with very few terms, but that more terms are required with increasing time.

Then investigate the convergence of the eigenfunction expansion of this Green's function using the result obtained in Problem 46 with $f(x) = \delta(x)$. Show that the eigenfunction expansion converges slowly at early times, but that the convergence improves with increasing time.

Provide an explanation for the different regimes of convergence of these two expansions based on the properties of the diffusion equation.

49. Consider the problem of determining the Green's function of Laplace's equation in the presence of a sphere of radius R on which the potential is required to vanish. We will be concerned with determining the potential *within* the sphere. For simplicity in illustrating the general methodology, we will place the point charge along the z axis, so the expansion we will obtain will be suitable only for problems with azimuthal symmetry.

Apply the principle stated in Equations (7.87)–(7.89) to write the Green's function as

$$G(\mathbf{x}, \mathbf{x}') = G_0(\mathbf{x}, \mathbf{x}') + \phi$$

where $\mathbf{x} = (x, y, z)$, $\mathbf{x}' = (0, 0, z')$ and ϕ is to be determined. Show that $G_0(\mathbf{x}, \mathbf{x}')$ can be written as

$$G_0(\mathbf{x}, \mathbf{x}') = \frac{1}{\sqrt{x^2 + y^2 + (z-z')^2}} = \frac{1}{\sqrt{r^2 + z'^2 - 2rz'\cos\theta}}$$

where $r^2 = x^2 + y^2 + z^2$ and θ is the angle between the vector from the origin to the point charge and the vector from the origin to the observation point.

Use Equation (6.53) to write the solution ϕ as an expansion in Legendre polynomials,

$$\phi(r\theta) = \sum_{\ell=0}^{\infty} A_\ell r^\ell P_\ell(\cos\theta)$$

The expansion coefficients A_ℓ can be obtained by requiring the Green's function to vanish for $r = R$, i.e.

$$\frac{1}{\sqrt{R^2 + z'^2 - 2Rz'\cos\theta}} + \sum_{\ell=0}^{\infty} A_\ell R^\ell P_\ell(\cos\theta) = 0$$

Show that Equation (6.115) implies that the A_ℓ are given by

$$A_\ell = -\frac{z'^\ell}{R^{2\ell+1}}$$

and so sum the series for ϕ again using (6.115) to obtain the Green's function in the form

$$G(\mathbf{x}, \mathbf{x}') = \frac{1}{\sqrt{r^2 + z'^2 - 2rz'\cos\theta}}$$
$$- \frac{R}{z'} \frac{1}{\sqrt{r^2 + (z'/R^2)^2 - 2r(z'/R^2)\cos\theta}}$$

Interpret the solution ϕ as being due to an image charge of magnitude R/z' located at the point $(0, 0, z'/R^2)$.

50. Use `ContourPlot` to display the solutions in Equations (7.91) and (7.93) in any plane normal to the x-y plane to verify that the boundary conditions are satisfied.

51. Consider the diffusion equation with no boundaries with the initial condition

$$u(x, 0) = \exp(-x^2)$$

Evaluate the integral in (7.51) to obtain the solution for this initial-value problem.

Compare this exact solution with that obtained in Example 3.5 using the power series method in two ways. First use *Mathematica* to expand both solutions as Taylor series in t and verify that the terms are the same. Then investigate the number of terms in the expansion that are required to reproduce the exact value over the range $(-2, 2)$ to with a prescribed threshold (e.g. 5%).

Chapter 8

Integral Representations

Fourier and Laplace transforms are two of the most common examples of integral transforms, whereby the solution of a differential equation is represented as an integral from which an explicit expression of that solution can be obtained. In this chapter, we adopt an approach analogous to that taken in Chapter 7 to examine the properties of solutions of ordinary differential equations that have the form

$$y(x) = \int K(x,t)v(t)\,dt \qquad (8.1)$$

The function K is called the **kernel** of the transformation and $v(t)$ is the transform of $y(x)$ with respect to the kernel. The unknown quantities in (8.1) are the limits of integration and the transformed function v, both of which depend upon the particular differential equation to be solved. The kernel will also be found to depend on the form of the differential equation. The Fourier and Laplace transforms for equations with constant coefficients are particular examples of (8.1), with the kernels being simple exponential functions. An important point to keep in mind is that the Bromwich integral discussed in Section 7.2 shows that the path along which the integration is taken in Equation (8.1) need not be confined to the real axis.

The approach taken in this chapter is complementary to that taken in Chapters 5 and 6, where solutions of differential equations were expressed in series form. The method of Frobenius is straightforward to apply, but the solutions obtained are tied to the point about which the expansion of the solution is made. If an expansion is needed about a different point, the entire procedure must be repeated. On the other hand, an integral representation provides a closed-form solution that can be used to obtain the expansion of the solution about any point in the domain over which the solution is defined.

The closed-form representation (8.1) also provides other advantages. All of the solutions of a differential equation are represented by a single integral expression; the distinguishing feature of different types of solutions is contained in the evaluation of the integral in (8.1). Because this often relies on the use of contour integrals in the complex plane, the analytic structure of eigenfunctions in relation to that of the differential equation is more clearly displayed than in the series form. This property is particularly germane to the situation where solutions are required in regions separated by singular points of the differential equation. In the method of Frobenius, the validity of a solution terminates abruptly at the first singular point encountered from the expansion point. Obtaining a solution in contiguous regions separated by singular points requires a change of variables. However, if an integral representation is at hand, these solutions can be obtained simply by a change of the contour of integration. Another consequence of the analytic structure of (8.1) emerges naturally from Cauchy's integral formula to reveal the close relationship between the integral representations of a particular set of eigenfunctions and the generating functions for those eigenfunctions. Taken together, these features show that integral representations provide an elegant and practical way of solving differential equations that also consolidates and clarifies many of the properties of eigenfunctions discussed in Chapters 5 and 6.

8.1 The Laplace Transform

The first method that will be described is due to Laplace. It is analogous to the Laplace transform discussed in Chapter 7 and is simplest to apply to differential equations in which the coefficients are at most of first order in the independent variable. For the case of second-order equations, the general form of such an equation is

$$(a_2 + b_2 x)\frac{d^2 y}{dx^2} + (a_1 + b_1 x)\frac{dy}{dx} + (a_0 + b_0 x)y = 0 \qquad (8.2)$$

where the quantities a_n and b_n are constants. This form includes the Hermite, Laguerre and associated Laguerre equations as particular cases. To apply the Laplace transform to (8.2), it is convenient to rewrite this equation as

$$F\left(\frac{d}{dx}\right)y + xG\left(\frac{d}{dx}\right)y = 0 \qquad (8.3)$$

with the functions F and G being given by

$$F(s) = a_2 s^2 + a_1 s + a_0, \qquad G(s) = b_2 s^2 + b_1 s + b_0 \qquad (8.4)$$

These functions completely characterize any equation of the form (8.2). We now observe that if y is written as a Laplace transform, i.e. if $K(x,t) = e^{xt}$ in (8.1), then the functions F and G of the derivative operator acting on y can be replaced by the *same* functions of the transformation variable t. To see this, we first write the solution y of (8.2) as the Laplace transform of a function $Y(t)$:

$$y(x) = \int e^{xt} Y(t)\, dt \qquad (8.5)$$

Since derivatives of y with respect to x can be expressed as

$$\frac{d^n y}{dx^n} = \int t^n e^{xt} Y(t)\, dt \qquad (8.6)$$

substitution of (8.5) into (8.3) produces an integral equation for the unknown function Y in terms of F and G:

$$\int e^{xt} F(t) Y(t)\, dt + \int x e^{xt} G(t) Y(t)\, dt = 0 \qquad (8.7)$$

We again emphasize that the limits of integration have not yet been specified and that the domain of integration can either lie entirely on the real axis or be a contour in the complex plane.

The two integrals in (8.7) can be combined by performing an integration by parts on the second term on the left-hand side:

$$\int x e^{xt} G(t) Y(t)\, dt$$
$$= \int \frac{d}{dt}\left[e^{xt} G(t) Y(t) \right] dt - \int \left\{ e^{xt} \frac{d}{dt}\left[G(t) Y(t) \right] \right\} dt \qquad (8.8)$$

Substituting this expression into Equation (8.7), the transformed original equation in (8.2) becomes

$$\int \frac{d}{dt}\left[e^{xt} G(t) Y(t) \right] dt - \int e^{xt} \left\{ \frac{d}{dt}\left[G(t) Y(t) \right] - F(t) Y(t) \right\} dt = 0 \qquad (8.9)$$

This equation is seen to be composed of two parts. The first term on the left-hand side is the integral of a total derivative with respect to the transformation variable, and so depends only upon the end-points of the integration. The integrand of the second term on the right-hand side contains terms involving Y and the first derivative of Y. Thus, one way of satisfying

320   Partial Differential Equations with Mathematica

Equation (8.9) is to require each of the two terms to vanish *separately*. This can be achieved by choosing the limits of integration to make the first term on the left-hand side vanish,

$$\int \frac{d}{dt}\left[e^{xt}G(t)Y(t)\right]dt = 0 \qquad (8.10)$$

and by choosing the transformed function Y to make the integrand in the second term vanish, i.e.

$$\frac{d}{dt}\left[G(t)Y(t)\right] - F(t)Y(t) = 0 \qquad (8.11)$$

This is a first-order differential equation for Y involving the *known* functions F and G.

For real integrals, the condition in (8.10) requires the quantity in square brackets to take the same value at both the lower and upper limits of integration. For contour integrals in the complex plane, this requires the integrand to be unchanged in completing a circuit along the contour. In this case, the contour must be deformed to account for the presence of any singular points of the differential equation (which make their appearance in the functions F and G). Equations (8.10) and (8.11) illustrate the basic advantage of using the Laplace method for obtaining solutions to equations of the form (8.2), namely, that the solution of a *second*-order equation is replaced by the solution of a *first*-order equation with a condition determining the range of integration.

Equation (8.11) can be readily solved for Y:

$$Y(t) = \frac{1}{G(t)}\exp\left[\int^t \frac{F(s)}{G(s)}\,ds\right] \qquad (8.12)$$

This determines Y to within a multiplicative constant, so the lower limit of integration has been omitted. Substituting this expression into the integral representation in (8.5), the solution to (8.2) is obtained in the form

$$y(x) = \int \frac{1}{G(t)}\exp\left[xt + \int^t \frac{F(s)}{G(s)}\,ds\right]dt \qquad (8.13)$$

Substituting (8.12) into (8.10), the range of integration is now also determined entirely in terms of known functions by the condition

$$\int \frac{d}{dt}\left\{e^{xt}\exp\left[\int^t \frac{F(s)}{G(s)}\,ds\right]\right\}dt = 0 \qquad (8.14)$$

If the domain of integration is some region of the real axis with a lower bound 'i' and an upper bound 'f,' then (8.14) is equivalent to

$$\left\{e^{xt}\exp\left[\int^t \frac{F(s)}{G(s)}\,ds\right]\right\}\bigg|_i^f = 0 \qquad (8.15)$$

Otherwise, (8.14) is to be interpreted as a contour integral in the complex plane.

Example 8.1. To illustrate the use of Laplace's method, we will obtain integral representations of solutions to Hermite's equation:

$$\frac{d^2y}{dx^2} - 2x\frac{dy}{dx} + 2ny = 0 \qquad (8.16)$$

In this example, we will take n to be an integer, though the method can be applied to noninteger values as well. Comparing (8.4) with (8.16), the functions F and and G are readily identified as

$$F(s) = s^2 + 2n, \qquad G(s) = -2s \qquad (8.17)$$

Inserting these functions into (8.12) yields the function Y:

$$Y(t) = \frac{1}{2t^{n+1}} e^{-\frac{1}{4}t^2} \qquad (8.18)$$

Thus, the solution to Hermite's equation can be expressed as

$$y_n(x) = \int \frac{e^{xt - \frac{1}{4}t^2}}{t^{n+1}} dt \qquad (8.19)$$

where the limits of integration are determined by the requirement

$$\int \frac{d}{dt}\left[\frac{1}{t^n} e^{xt - \frac{1}{4}t^2}\right] dt = 0 \qquad (8.20)$$

To identify appropriate limits of integration to satisfy (8.20), we observe that the quantity being differentiated has an isolated pole of order n at the origin. Furthermore, the integrand in (8.19) has an isolated pole of order $n+1$, also at the origin. Thus, suppose we choose a contour that encircles the origin. The quantity $t^{-n} \exp(xt - \frac{1}{4}t^2)$ assumes its initial value multiplied by a factor $\exp[2\pi i(n-1)]$ upon completing the contour. For integer n, this factor is unity. Thus, (8.20) is satisfied, and a solution of the form (8.19) is obtained. For noninteger values of n this conclusion is not valid, since there is branch cut along the real axis, and a different contour must be chosen. Notice that a closed contour around *any* point would also lead to (8.20) being satisfied, but the corresponding solution (8.19) would be the trivial solution, $y=0$. Thus, the integral representation of the nontrivial solution to Hermite's equation (8.16) can be written as

$$y_n(x) = \oint \frac{e^{xt - \frac{1}{4}t^2}}{t^{n+1}} dt \qquad (8.21)$$

where the contour is a closed path that encloses the origin.

This representation of solutions to Hermite's equation has several consequences. First, the residue theorem can be applied to Equation (8.21) to deduce directly that

$$y_n(x) = \left[\frac{2\pi i}{n!} \frac{\partial^n}{\partial t^n}\left(e^{xt-\frac{1}{4}t^2}\right)\right]\bigg|_{t=0} \tag{8.22}$$

This expression identifies the quantity enclosed in parentheses in (8.22) as a generating function for the y_n (which are proportional to the Hermite polynomials, H_n), as discussed in Section 6.5. In particular, by multiplying both sides of Equation (8.22) by t^n and summing over n, we obtain

$$\frac{1}{2\pi i} \sum_{n=0}^{\infty} t^n y_n(x) = e^{xt-\frac{1}{4}t^2} \tag{8.23}$$

Moreover, by using the standard normalization of the Hermite polynomials, whereby the coefficient of x^n is 2^n (Section 6.1), we obtain the integral representation:

$$H_n(x) = \frac{2^n n!}{2\pi i} \oint \frac{e^{xt-\frac{1}{4}t^2}}{t^{n+1}} dt \tag{8.24}$$

Equation (8.24) can be used to obtain another useful representation of Hermite polynomials. By transforming the variable of integration from t to $s = x - \frac{1}{2}t$, (8.24) becomes

$$H_n(x) = (-1)^n \frac{n!}{2\pi i} e^{x^2} \oint \frac{e^{-s^2}}{(s-x)^{n+1}} ds \tag{8.25}$$

Applying the residue theorem to (8.25), we obtain an expression for the Hermite polynomials as a Rodrigues formula:

$$H_n(x) = (-1)^n e^{x^2} \frac{d^n}{dx^n} e^{-x^2} \tag{8.26}$$

This example, and particularly the results obtained in Equations (8.23), (8.24) and (8.26), provides a vivid illustration of the close connection that exists between generating functions for eigenfunctions of a particular differential equation and integral representations of solutions of that differential equation. ∎

8.2 The Euler Transform

The Laplace transform is most easily applied to equations of the form (8.2), where the coefficients of y and its derivatives are all linear functions of the independent variable. In this case, a second-order differential equation of the form (8.2) is replaced by a first-order equation for the transformed function, with the limits of integration chosen in order to satisfy (8.14). For equations with coefficients of higher order in the independent variable the Laplace transform is not as useful, since correspondingly higher-order derivatives will appear in the equation determining the transformed function Y. In such cases, there is no apparent advantage in using the Laplace transform, so we must seek a more appropriate kernel.

In this section we consider the application of the integral transforms to equations where the coefficient of each derivative appearing in the equation is a polynomial of the same order as the derivative. For second-order equations, this general form is

$$(a_2 + b_2 x + c_2 x^2)\frac{d^2 y}{dx^2} + (a_1 + b_1 x)\frac{dy}{dx} + a_0 y = 0 \qquad (8.27)$$

The quantities a_n, b_n and c_n are constants. Equation (8.27) includes the Chebyshev, ordinary Legendre and the hypergeometric equations as particular cases. For equations of the form (8.27), the appropriate kernel is

$$K(x,t) = \frac{1}{(x-t)^{\gamma+1}} \qquad (8.28)$$

This is the **Euler kernel** and the corresponding transformation (8.1) is the **Euler transform**. The quantity γ is chosen to suit the particular equation, as will be shown below. Thus, the unknown quantities using the Euler kernel in (8.1) include, in addition to the limits of integration and the transformed function, the exponent γ.

Although the Euler kernel can be derived for equations of the form (8.27) (Ince, 1956), we simply point out that (8.28) is more appropriate than the Laplace transform for equations with only regular singular points. The Laplace kernel has an essential singularity at infinity, and is thus better suited to solving Hermite's equation and other equations of the form of Equation (8.2), which have an irregular singular point at infinity, but are regular elsewhere.

To illustrate the use of the Euler transform, we will obtain integral representations of solutions of the ordinary Legendre equation:

$$(1-x^2)\frac{d^2 y}{dx^2} - 2x\frac{dy}{dx} + n(n+1)y = 0 \qquad (8.29)$$

where n is an integer. The Euler transform of y is

$$y(x) = \int \frac{v(t)}{(x-t)^{\gamma+1}} dt \qquad (8.30)$$

and the required derivatives of y are

$$\frac{dy}{dx} = -(\gamma+1)\int \frac{v(t)}{(x-t)^{\gamma+2}}\,dt$$

$$\frac{d^2y}{dx^2} = (\gamma+1)(\gamma+2)\int \frac{v(t)}{(x-t)^{\gamma+3}}\,dt \tag{8.31}$$

Substitution of Equations (8.30) and (8.31) into (8.29) yields

$$\int \left[(1-x^2)(\gamma+1)(\gamma+2) + 2x(\gamma+1)(x-t)\right.$$

$$\left. + n(n+1)(x-t)^2\right]\frac{v(t)}{(x-t)^{\gamma+3}}\,dt = 0 \tag{8.32}$$

To proceed further, and in analogy with the step taken in passing from (8.7) to (8.8), we rewrite Equation (8.32) in such a way that x appears only in factors involving powers of $x-t$. To do so, we make use of the identities

$$x = (x-t) + t$$

$$x^2 = [(x-t) + t]^2 = (x-t)^2 + 2t(x-t) + t^2 \tag{8.33}$$

which we substitute into Equation (8.32). After some simple algebraic steps, we obtain an equation of the form

$$\int \frac{Av(t)}{(x-t)^{\gamma+1}}\,dt + \int \frac{B(t)v(t)}{(x-t)^{\gamma+2}}\,dt + \int \frac{C(t)v(t)}{(x-t)^{\gamma+3}}\,dt = 0 \tag{8.34}$$

where the quantities A, B and C are given by

$$A = n(n+1) - \gamma(\gamma+1)$$

$$B = -2t(\gamma+1)^2 \tag{8.35}$$

$$C = (1-t^2)(\gamma+1)(\gamma+2)$$

Notice, in particular, that A is independent of t. The three integrals on the left-hand side of (8.34) can be combined into a single expression by performing an integration by parts once on the second integral and twice on the third integral. This procedure produces two sets of boundary terms and a *second*-order ordinary differential equation for v, and so does not offer an obvious advantage over solving the original equation, which is also of second order. However, if we use our freedom in choosing γ to make A vanish, then the first integral in (8.34) also vanishes, only a single integration by parts is required to combine the third integral with the second integral, and a *first*-order differential equation is obtained for v with only a single boundary term.

For Legendre's equation in (8.29), we choose γ by the requirement

$$n(n+1) - \gamma(\gamma+1) = 0 \tag{8.36}$$

This is a quadratic equation for γ which has the two solutions $\gamma = n$ and $\gamma = -n - 1$. Since the form of the differential equation is unaltered by replacing n by $-n-1$ (see the discussion in Section 6.2), we will choose the root $\gamma = n$. With this value of γ substituted into (8.35), the equation in (8.34) becomes, after some rearrangement

$$\int \frac{n+2}{(x-t)^{n+3}}(1-t^2)v(t)\,dt - \int \frac{2t(n+1)v(t)}{(x-t)^{n+2}}\,dt = 0 \tag{8.37}$$

In obtaining this result we have cancelled a factor of $n+1$ from the integrand in each term. The first term on the left-hand side of (8.37) is now integrated by parts:

$$\int \frac{n+2}{(x-t)^{n+3}}(1-t^2)v(t)\,dt$$
$$= \int \frac{d}{dt}\left[\frac{(1-t^2)v(t)}{(x-t)^{n+2}}\right]dt - \int \left\{\frac{1}{(x-t)^{n+2}}\frac{d}{dt}\left[(1-t^2)v(t)\right]\right\}dt \tag{8.38}$$

Substituting this expression into (8.37), we obtain

$$\int \frac{d}{dt}\left[\frac{(1-t^2)v(t)}{(x-t)^{n+2}}\right]dt$$
$$- \int \frac{1}{(x-t)^{n+2}}\left\{\frac{d}{dt}\left[(1-t^2)v(t)\right] - 2t(n+1)v(t)\right\}dt = 0 \tag{8.39}$$

This equation is similar in form to that in (8.9) obtained using the Laplace kernel and we can proceed with correspondingly similar steps to those taken in (8.10)–(8.15). Equation (8.39) can be satisfied by requiring each of the two terms to vanish separately. Thus, the domain of integration in the integral transform must be chosen to insure that

$$\int \frac{d}{dt}\left[\frac{(1-t^2)v(t)}{(x-t)^{n+2}}\right]dt = 0 \tag{8.40}$$

while the transformed function v is chosen to be a solution of the differential equation

$$\frac{d}{dt}\left[(1-t^2)v(t)\right] - 2t(n+1)v(t) = 0 \tag{8.41}$$

so that the second term on the left-hand side of (8.39) also vanishes. This equation is of the form of the equation in (8.11) and so can solved using (8.12):

$$v(t) = (1-t^2)^n \qquad (8.42)$$

Thus, the integral representation of solutions to Legendre's equation with the Euler kernel in (8.28) is given by

$$y(x) = \int \frac{(1-t^2)^n}{(x-t)^{n+1}} \, dt \qquad (8.43)$$

and the domain of integration is determined by the requirement

$$\int \frac{d}{dt}\left[\frac{(1-t^2)^{n+1}}{(x-t)^{n+2}}\right] dt = 0 \qquad (8.44)$$

To choose an appropriate domain of integration in (8.43) and (8.44), we observe that the numerator in each expression vanishes at the points $t = \pm 1$ and that the denominator vanishes whenever $x = t$. In particular, if $x = 1$ (where the equation has regular singular points), the integrands in (8.43) and (8.44) have simple poles at $t = 1$. Thus, if the range of x is restricted to values such that $|x| > 1$, then the domain of integration can extended over the *closed* interval $[-1, 1]$, which insures that Equation (8.44) is satisfied and that there are no singularities in the integrands in (8.43) and (8.44). The solution of Legendre's equation that is obtained is written in standard form as

$$Q_n(x) = \frac{1}{2^{n+1}} \int_{-1}^{1} \frac{(1-t^2)^n}{(x-t)^{n+1}} \, dt \qquad (8.45)$$

The functions $Q_n(x)$ are the **Legendre functions of the second kind** of order n encountered previously in Problem 6.14. For $|x| > 1$, $Q_n(x)$ is linearly independent of the Legendre polynomial $P_n(x)$. Thus, since the Legendre polynomials are finite for any finite value of x, the two solutions $P_n(x)$ and $Q_n(x)$ form a fundamental set for the ordinary Legendre equation in the region $|x| > 1$ (for integer n).

Example 8.2. Expressions for the Legendre functions of the second kind can be obtained in *Mathematica* by evaluating the integral in (8.45) directly using Integrate. The $Q_n(x)$ are first defined in terms of the operations in (8.45):

```
Q[n_,x_]:=2^(-n-1)Integrate[(1-t^2)^n/(x-t)^(n+1),{t,-1,1}]
```

To obtain compact expressions for the $Q_n(x)$, we must explicitly declare certain algebraic relations involving the Log function. This is done by using the commands Protect and Unprotect for the required operations:

```
(Unprotect[Log,Plus,Subtract,Times];
(a_)Log[x_]+(a_)Log[y_]:=(a)Log[x y];
(a_)Log[x_]-(a_)Log[y_]:=(a)Log[x/y];
(a_)Log[x_]+(b_)Log[x_]:=(a+b)Log[x];
Protect[Log,Plus,Subtract,Times])

{Log, Plus, Subtract, Times}
```

Expressions for the Legendre functions can now be obtained for the required values of n. For example,

```
Q[0,x]

     1 + x
Log[------]
     -1 + x
-----------
      2

Apart[Q[1,x]]

                -1 - x
        x Log[------]
                1 - x
-1 + --------------
              2

Apart[Q[2,x]]

                     2
 -3 x      1   3 x       1 - x
 ---- + (- - ----) Log[------]
  2        4    4        -1 - x
```

Notice that the factors multiplying the logarithms in the $Q_n(x)$ are proportional to the $P_n(x)$. There is in fact a very close relationship between the Legendre polynomials and the Legendre functions of the second kind. This will be examined below and in the problems at the end of the chapter.

If $|x| < 1$, then the interval of integration in (8.45) cannot be used because of the singularity at $t = x$. We must therefore choose a contour in the complex t plane for which the value of (8.44) is the same at the beginning and at the end of the contour. Following our procedure for the Hermite polynomials in the preceding section, we choose any closed contour that surrounds the point $t = x$. This insures that (8.44) is satisfied and that a nontrivial expression is obtained from (8.43). The resulting integral representation of the Legendre polynomials is due to Schläfli (Copson, 1935; Whittaker and Watson, 1963):

$$P_n(x) = \frac{1}{2\pi i} \oint \frac{(1-t^2)^n}{2^n(x-t)^{n+1}} \, dt \qquad (8.46)$$

The factor 2^n in the integrand is in keeping with the standard condition that $P_n(1) = 1$. Using Cauchy's integral formula the Legendre polynomials can be expressed in terms of derivatives with respect to the numerator of the integrand:

$$P_n(x) = \frac{1}{2^n n!} \frac{d^n}{dx^n} \left[(x^2-1)^n\right] \qquad (8.47)$$

This is Rodrigues' formula for the Legendre polynomials (Problem 6.43). The relationship between the integral representation (8.46) and the generating function derived in Section 6.5 is treated in Problem 7.

We can obtain a simple integral relationship between the P_n and the Q_n by performing n integrations by parts in (8.45) and using (8.47) (Problem 10). The resulting expression is known as **Neumann's formula**:

$$Q_n(x) = \frac{1}{2} \int_{-1}^{1} \frac{P_n(t)}{x-t} \, dt \qquad (8.48)$$

Notice that $|x| > 1$ and $|t| \leq 1$, so the denominator of the integrand does not vanish. Among the several important consequences and uses of this relationship (Whittaker and Watson, 1963) is that many of the recurrence relationships that can be derived for the Legendre polynomials are also valid for the Legendre functions of the second kind.

8.3 Hypergeometric Functions

The hypergeometric function, $F(\alpha, \beta; \gamma; x)$, is a solution to the hypergeometric or Gauss equation which, in standard form, is

$$x(1-x)\frac{d^2y}{dx^2} + [\gamma - (\alpha+\beta+1)x]\frac{dy}{dx} - \alpha\beta y = 0 \qquad (8.49)$$

where α, β and γ are constants. This equation was derived in Problem 5.14 and the solution was obtained using the method of Frobenius in Problem 5.15. Since this equation is readily identified as being of the form (8.27), we can obtain solutions using the Euler transform. The procedure for obtaining an integral representation of solutions to (8.49) is the same as that used to solve the Legendre equation, so we will summarize only the important steps.

The solution y is again written as in Equation (8.30) which, together with the derivatives in (8.31), is substituted into (8.49). Requiring the coefficient of x^2 to vanish again yields a quadratic equation whose roots are found to be $\gamma = \alpha - 1$ and $\gamma = \beta - 1$. Since the hypergeometric equation is invariant under the interchange of α and β, either root can be chosen; we will take $\gamma = \alpha - 1$. With this choice, the transformed hypergeometric equation is

$$\int \frac{(\alpha+1)t(1-t)}{(x-t)^{\alpha+2}} v(t)\,dt + \int \frac{\alpha - \gamma + 1 - (\alpha - \beta + 1)t}{(x-t)^{\alpha+1}} v(t)\,dt = 0 \quad (8.50)$$

Following the steps in (8.38) and (8.39), we find that $v(t)$ solves the differential equation

$$\frac{d}{dt}\Big[t(1-t)v(t)\Big] - \Big[\alpha - \gamma + 1 - (\alpha - \beta + 1)t\Big]v(t) = 0 \quad (8.51)$$

with the requirement that

$$\int \frac{d}{dt}\left[\frac{t(1-t)v(t)}{(x-t)^{\alpha+1}}\right]dt = 0 \quad (8.52)$$

Solving (8.51) by using (8.12) yields

$$v(t) = t^{\alpha-\gamma}(1-t)^{\gamma-\beta-1} \quad (8.53)$$

Thus, to within a proportionality constant, solutions of the hypergeometric equation can be expressed as

$$y(x) = \int t^{\alpha-\gamma}(1-t)^{\gamma-\beta-1}(x-t)^{-\alpha}\,dt \quad (8.54)$$

provided that

$$\int \frac{d}{dt}\left[t^{\alpha-\gamma+1}(1-t)^{\gamma-\beta}(x-t)^{-\alpha-1}\right]dt = 0 \quad (8.55)$$

To determine an appropriate range of integration in (8.54), we first observe that the hypergeometric function has regular singular points at $x = \pm 1$. Thus, suppose that we confine the range of x to the values $|x| < 1$. Then, if $\gamma > \beta$, the quantity in square brackets vanishes at $t = 1$. We now examine the behavior of this quantity in the limit as $t \to \infty$:

$$\lim_{t\to\infty}\left[t^{\alpha-\gamma+1}(1-t)^{\gamma-\beta}(x-t)^{-\alpha-1}\right]$$

$$= t^{\alpha-\gamma+1}(-t)^{\gamma-\beta}(-t)^{-\alpha-1}$$

$$= (-1)^{\gamma-\beta-\alpha-1}t^{-\beta} \quad (8.56)$$

This limit vanishes if $\beta > 0$. Thus, if $|x| < 1$ and if $\gamma > \beta > 0$, the limits of integration can be chosen to be $t = 1$ and $t = \infty$, in which case (8.55) is satisfied, and the range of integration in (8.54) is the semi-infinite interval $[1, \infty)$:

$$y(x) = \int_1^\infty t^{\alpha-\gamma}(1-t)^{\gamma-\beta-1}(x-t)^{-\alpha}\, dt \tag{8.57}$$

This integral representation can be related to the series representation for the hypergeometric function obtained in Problem 5.15. We first expand the factor $(x-t)^\alpha$ and integrate each term in the series. This series is found to be the same as that in Problem 5.15 to within a multiplicative constant of proportionality. The details of this procedure are discussed in Problem 15. Thus, the integral representation of the standard definition of the hypergeometric function is

$$F(\alpha, \beta; \gamma; x) = \frac{\Gamma(\gamma)}{\Gamma(\beta)\Gamma(\gamma-\beta)} \int_1^\infty t^{\alpha-\gamma}(t-1)^{\gamma-\beta-1}(t-x)^{-\alpha}\, dt \tag{8.58}$$

An equivalent form of the representation (8.48) can be obtained by changing the variable of integration from t to t^{-1}:

$$F(\alpha, \beta; \gamma; x) = \frac{\Gamma(\gamma)}{\Gamma(\beta)\Gamma(\gamma-\beta)} \int_0^1 t^{\beta-1}(1-t)^{\gamma-\beta-1}(1-xt)^{-\alpha}\, dt \tag{8.59}$$

The validity of both this integral representation and that in (8.59) is contingent upon the inequality $\gamma > \beta > 0$, as discussed above.

Example 8.3. The integral representation in Equation (8.59) can be used to show directly the relationship between the hypergeometric function and several elementary functions by performing the integral in *Mathematica* and evaluating the resulting expression with particular values of α, β and γ. We define $F(\alpha, \beta; \gamma; x)$ by performing the definite integral in (8.59) using Integrate and the in-built function Gamma:

```
F[a_,b_,c_,x_]:=((Gamma[c]/(Gamma[b]Gamma[c-b])) 
    Integrate[t^(b-1) (1-t)^(c-b-1) (1- x t)^(-a),
    {t,0,1}])
```

The function F[a,b,c,x] can now be used to derive some well-known cases where hypergeometric function reduces to an elementary function. A comprehensive list of such cases has been compiled by Abramowitz and Stegun (1965):

```
F[1,1,2,x]

    Log[1 - x]
-(-----------)
        x

F[1/2,1/2,3/2,x^2]

ArcSin[x]
---------
    x

F[1/2,1/2,3/2,-x^2]

                  2
Log[x + Sqrt[1 + x ]]
---------------------
          x
```

If α is equal to a negative integer then, in the notation of Problem 5.15, F reduces to an nth-order polynomial:

$$F(-n,\beta;\gamma;x) = \sum_{k=0}^{n} \frac{(-n)_k(\beta)_k}{(\gamma)_k} \frac{x^k}{k!}$$

We can use the function F[a,b,c,x] to obtain particular expressions for these polynomials from the integral representation. For example,

`Expand[F[-8,1,2,x]]`

```
              2                       5                      8
          28 x         3         4   28 x        6       7  x
1 - 4 x + ----- - 14 x  + 14 x - ----- + 4 x  - x  + --
            3                      3                  9
```

The equivalence of this expression to the summation given above can be verified easily.

8.4 Bessel Functions

Integral representations of Bessel functions can be derived in several ways. Beginning with the series solution (6.86) an integral representation can be obtained by using the integral representation of the reciprocal of the

Gamma function (Copson, 1935). Alternatively, Bessel's equation can be transformed into a confluent hypergeometric equation (Problem 5.20) and the integral representation of the confluent hypergeometric function can be used to obtain directly the corresponding integral representation of the Bessel function. In the latter approach (Morse and Feshbach, 1953), the solution y of Bessel's equation,

$$\frac{d^2y}{dx^2} + \frac{1}{x}\frac{dy}{dx} + \left(1 - \frac{\nu^2}{x^2}\right)y = 0 \tag{8.60}$$

is written in terms of a reduced function F given by $y(x) = x^\nu e^{-ix} F(x)$. The quantity ν is not restricted to being an integer. The equation for F is

$$x\frac{d^2F}{dx^2} + (2\nu + 1 - 2ix)\frac{dF}{dx} - (2\nu + 1)iF = 0 \tag{8.61}$$

This equation is a confluent hypergeometric equation and the solution can be expressed in terms of the series representation obtained in Problem 5.20 or the corresponding integral representation (Morse and Feshbach, 1953).

An even more direct approach to obtaining an integral representation of solutions to Bessel's equation is to observe that Equation (8.61) is of the form of Equation (8.2) and so can be solved by using Laplace's method. For (8.61) the quantities F and G in (8.4) are

$$F(s) = (2\nu + 1)(s - i), \qquad G(s) = s^2 - 2is \tag{8.62}$$

Substituting these expressions into (8.13) and (8.14), solutions to Equation (8.60) are obtained as

$$y_\nu(x) = x^\nu e^{-ix} \int e^{xt}(t^2 - 2it)^{\nu - \frac{1}{2}}\, dt \tag{8.63}$$

with the requirement

$$\int \frac{d}{dt}\left[e^{xt}(t^2 - 2it)^{\nu + \frac{1}{2}}\right] dt = 0 \tag{8.64}$$

Since x takes on values only along the positive real axis (or, if x is complex, values only with a positive real part), the quantity in square brackets in (8.64) vanishes in the limit $t \to -\infty$ (or, more generally, in the limit $t \to -\infty + i\alpha$ for any finite real number α). In addition, if $\nu > -\frac{1}{2}$, then this quantity also vanishes at the points $t=0$ and $t=2i$. Pairs of these points can thus be taken as the limits of integration in (8.63) to obtain various solutions of Bessel's equation.

The first solution we examine is

$$y_\nu(x) = x^\nu \int_0^{2i} e^{x(t-i)}(t^2 - 2it)^{\nu - \frac{1}{2}}\, dt \tag{8.65}$$

For $\nu > \frac{1}{2}$, the value of the integral is independent of the path taken between the limits of integration. If $\nu - \frac{1}{2} < 0$, then a 'figure-eight' contour must be chosen that encircles the points $t = 0$ in the negative sense and $t = 2i$ in the positive sense in order to satisfy (8.64). The solution (8.65) remains finite at the origin and so must be proportional to J_ν. The constant of proportionality can be determined by expanding (8.65) in a power series about the origin and comparing with (6.86) (Problem 21). The resulting integral representation of J_ν can be written in several ways:

$$\begin{aligned}
J_\nu(x) &= \frac{\left(\frac{1}{2}x\right)^\nu}{i\Gamma\left(\nu + \frac{1}{2}\right)\Gamma\left(\frac{1}{2}\right)} \int_0^{2i} e^{x(t-i)}(t^2 - 2it)^{\nu-\frac{1}{2}}\, dt \\
&= \frac{\left(\frac{1}{2}x\right)^\nu}{i\Gamma\left(\nu + \frac{1}{2}\right)\Gamma\left(\frac{1}{2}\right)} \int_{-i}^{i} e^{xs}(1 + s^2)^{\nu-\frac{1}{2}}\, ds \\
&= \frac{\left(\frac{1}{2}x\right)^\nu}{\Gamma\left(\nu + \frac{1}{2}\right)\Gamma\left(\frac{1}{2}\right)} \int_{-1}^{1} e^{ixu}(1 - u^2)^{\nu-\frac{1}{2}}\, du \\
&= \frac{\left(\frac{1}{2}x\right)^\nu}{\Gamma\left(\nu + \frac{1}{2}\right)\Gamma\left(\frac{1}{2}\right)} \int_0^{\pi} e^{ix\cos\theta} \sin^{2\nu}\theta\, d\theta
\end{aligned} \qquad (8.66)$$

Example 8.4. The integral representations in (8.66) can be used to generate expressions for the spherical Bessel functions, which were discussed in Problems 6.35 and 6.36. The spherical Bessel function of order n, signified by $j_n(x)$, is related to the half-order Bessel function $J_{n+\frac{1}{2}}(x)$ by

$$j_n(x) = \sqrt{\frac{\pi}{2x}} J_{n+\frac{1}{2}}(x)$$

We can obtain expressions for the $j_n(x)$ by evaluating the integral representations in (8.66) using *Mathematica* just as in Example 8.3 for the hypergeometric function. We construct $J_n(x)$ from the third integral representation in (8.66) by using Integrate and the in-built function Gamma:

```
J[x_,n_]:=2(((x/2)^n/(Gamma[n+1/2]Gamma[1/2]))
    Integrate[Cos[x u](1-u^2)^(n-1/2),{u,0,1}])
```

The spherical Bessel functions are now constructed in accordance with the definition given above:

```
j[x_,n_]:=Sqrt[Pi/(2x)]J[x,n+1/2]
```

The spherical Bessel functions may now be determined by calling the function j[x,n]. The first few functions are

```
j[x,0]
```

```
Sin[x]
------
  x
```

```
Apart[j[x,1]]
```

```
   Cos[x]     Sin[x]
-(------) +  ------
     x          2
                x
```

```
Apart[j[x,2]]
```

```
 -3 Cos[x]    3 Sin[x]    Sin[x]
 --------- +  -------- -  ------
      2           3          x
     x           x
```

To explore further the consequences and uses of the integral representations in (8.66), we observe that the last of these expressions takes on a particularly simple form if $\nu=0$:

$$J_0(x) = \frac{1}{\pi}\int_0^\pi e^{ix\cos\theta}\,d\theta = \frac{1}{\pi}\int_0^\pi \cos(x\cos\theta)\,d\theta \qquad (8.67)$$

By evaluating the same representation for $\nu = 1, 2, 3$ and 4 by performing successive integrations by parts and using standard trigonometric identities, we obtain the following suggestive sequence of integral representations for the corresponding $J_\nu(x)$:

$$J_1(x) = \frac{1}{\pi}\int_0^\pi \sin(x\cos\theta)\cos\theta\,d\theta$$

$$J_2(x) = -\frac{1}{\pi}\int_0^\pi \cos(x\cos\theta)\cos 2\theta\,d\theta$$

$$J_3(x) = -\frac{1}{\pi}\int_0^\pi \sin(x\cos\theta)\cos 3\theta\,d\theta \qquad (8.68)$$

$$J_4(x) = \frac{1}{\pi}\int_0^\pi \cos(x\cos\theta)\cos 4\theta\,d\theta$$

These integrals and that in Equation (8.67) can be consolidated into a single expression as

$$J_n(x) = \frac{1}{\pi} \int_0^\pi \cos\left(x \cos\theta - \tfrac{1}{2}n\pi\right) \cos n\theta \, d\theta \qquad (8.69)$$

or in an even more concise form as

$$J_n(x) = \frac{1}{\pi} \int_0^\pi \cos(x \sin\theta - n\theta) \, d\theta \qquad (8.70)$$

The integral representations in Equations (8.69) and (8.70), which we obtained by explicit evaluation, can be established for any positive integer n either by generalizing the procedure used above (Copson, 1935), or more easily by induction with the aid of the recurrence relation

$$\frac{d}{dx}\left[x^{-n} J_n(x)\right] = -x^{-n} J_{n+1}(x) \qquad (8.71)$$

which can be derived directly from the series solution (6.86). Notice also that an immediate consequence of both (8.69) and (8.70) is the relationship $J_{-n}(x) = (-1)^n J_n(x)$, so both integral representations are valid for *all* integers n.

The representation in (8.70) can be used to derive a generating function for the integer-order Bessel functions. We first notice that changing the integration variable from θ to $-\theta$ does not alter the value of the integrand, but causes the quantity $\sin(n\theta - x \sin\theta)$ to change sign. Thus, the integrand in (8.70) can be written as a complex exponential if the lower limit of integration is extended to $\theta = -\pi$:

$$J_n(x) = \frac{1}{2\pi} \int_{-\pi}^\pi e^{ix \sin\theta - in\theta} \, d\theta \qquad (8.72)$$

An equivalent form of (8.72) that will be useful in making comparisons with integral representations of other solutions of Bessel's equation is obtained by transforming the real variable of integration θ into an imaginary variable through $\phi = i\theta$:

$$J_n(x) = \frac{1}{2\pi i} \int_{-i\pi}^{\pi i} e^{x \sinh\phi - n\phi} \, d\phi \qquad (8.73)$$

To derive the generating function for the Bessel functions from (8.72), the variable of integration in (8.72) is first changed to $u = e^{i\theta}$. The range of integration is then transformed to the unit circle in the complex plane (with the direction of integration taken in the positive sense), and the integral becomes

$$J_n(x) = \frac{1}{2\pi i} \oint \exp\left[\tfrac{1}{2} x(u - u^{-1})\right] u^{-n-1} \, du \qquad (8.74)$$

The exponential in the integrand of this expression is an analytic function of u whose only singular point is an essential singularity at the origin. We can apply Laurent's theorem to obtain an expansion of this function that is valid in any annular region that does not contain the origin. The expansion coefficients in the Laurent series are given precisely by (8.74), which then yields the generating function for the Bessel functions:

$$e^{\frac{1}{2}x(u-u^{-1})} = \sum_{n=-\infty}^{\infty} u^n J_n(x) \qquad (8.75)$$

Another form of the integral representation in Equation (8.74) that is widely used for many formal and practical applications is obtained by introducing the variable $t = 2u/x$. The representation then becomes

$$J_n(x) = \frac{1}{2\pi i}\left(\frac{x}{2}\right)^n \int \exp\left(t - \frac{x^2}{4t}\right) t^{-n-1}\, dt \qquad (8.76)$$

Examples of the uses of the integral representations in (8.74) and (8.76) are discussed in the problems at the end of the chapter. We now consider a few simple examples of series of Bessel functions that can be obtained from (8.75).

Example 8.5. A number of elementary functions can be expressed in terms of Bessel functions by performing a few simple manipulations on the generating function in (8.75). For example, by setting $u = 1$, we obtain the 'sum rule'

$$1 = J_0(x) + 2\sum_{n=1}^{\infty} J_{2n}(x) \qquad (8.77)$$

Making the substitution $u = e^{i\phi}$ produces an alternative form of the generating function:

$$e^{ix \sin \phi} = \sum_{n=-\infty}^{\infty} J_n(x) e^{in\phi} \qquad (8.78)$$

Setting $\phi = \frac{1}{2}\pi$ in this expression and then equating separately the real and imaginary parts yields expansions for the sine and cosine functions:

$$\cos x = J_0(x) + 2\sum_{n=1}^{\infty}(-1)^n J_{2n}(x)$$

$$\sin x = 2\sum_{n=1}^{\infty}(-1)^n J_{2n-1}(x) \qquad (8.79)$$

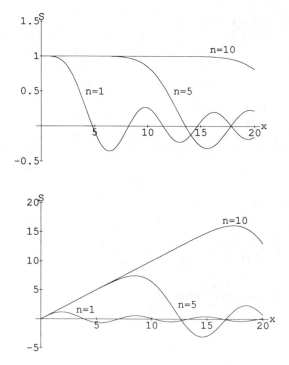

Figure 8.1 The convergence of the series in (8.77) (top) and (8.80) (bottom). For both series the partial summations are shown after 1, 5 and 10 terms. As with all series of Bessel functions, the two series shown here converge first for small values of the argument, with an increasing number of terms required as the argument increases.

Similarly, equating imaginary parts in (8.78), differentiating with respect to ϕ and then setting $\phi=0$, produces

$$x = 2\sum_{n=0}^{\infty}(2n+1)J_{2n+1}(x) \qquad (8.80)$$

In obtaining (8.77), (8.79) and (8.80) we have made use of the relationship $J_{-n}(x)=(-1)^n J_n(x)$ for integral-order Bessel functions, which can be established directly from the series representation in Equation (6.86).

The convergence of the expansions in Equations (8.77) and (8.80) is shown in Figure 8.1. Both series are seen to converge for small values of x by taking only the first few terms in the series, while an increasing number of terms is required to converge for larger values of x. This is to be expected from our earlier study of the series representation of Bessel functions, since the higher-order Bessel functions attain appreciable magnitude at increasingly larger distances from the origin. Similar results are obtained for the series in (8.78). ∎

8.5 Hankel Functions

We now return to the integral representation in Equation (8.63) and the condition in Equation (8.64) for the range of integration. In the preceding section Bessel functions of the first kind were obtained by taking the limits of integration to be $t=0$ and $t=2i$. However, in choosing appropriate limits of integration for the integral representation of the $J_n(x)$, we also found that the integrand in Equation (8.64) vanishes in the limit $t \to -\infty + i\alpha$ for any value of α. In this section, we will obtain integral representations of another type of solution by taking the upper and lower limits in turn to infinity in (8.65) and (8.66). The solutions can be written in various forms. By analogy with (8.86), we have the representations:

$$H_\nu^{(1)}(x) = \frac{2\left(\frac{1}{2}x\right)^\nu}{i\Gamma\left(\nu+\frac{1}{2}\right)\Gamma\left(\frac{1}{2}\right)} \int_{-\infty+i}^{2i} e^{x(t-i)}(t^2 - 2it)^{\nu-\frac{1}{2}}\, dt$$

$$= \frac{2\left(\frac{1}{2}x\right)^\nu}{i\Gamma\left(\nu+\frac{1}{2}\right)\Gamma\left(\frac{1}{2}\right)} \int_{-\infty}^{i} e^{xs}(1+s^2)^{\nu-\frac{1}{2}}\, ds \quad (8.81)$$

$$= \frac{2\left(\frac{1}{2}x\right)^\nu}{i\Gamma\left(\nu+\frac{1}{2}\right)\Gamma\left(\frac{1}{2}\right)} \int_{-\infty}^{\frac{1}{2}i\pi} e^{x \sinh u} \cosh^{2\nu} u\, du$$

and

$$H_\nu^{(2)}(x) = \frac{2\left(\frac{1}{2}x\right)^\nu}{i\Gamma\left(\nu+\frac{1}{2}\right)\Gamma\left(\frac{1}{2}\right)} \int_0^{-\infty+i} e^{x(t-i)}(t^2 - 2it)^{\nu-\frac{1}{2}}\, dt$$

$$= \frac{2\left(\frac{1}{2}x\right)^\nu}{i\Gamma\left(\nu+\frac{1}{2}\right)\Gamma\left(\frac{1}{2}\right)} \int_{-i}^{-\infty} e^{xs}(1+s^2)^{\nu-\frac{1}{2}}\, ds \quad (8.82)$$

$$= \frac{2\left(\frac{1}{2}x\right)^\nu}{i\Gamma\left(\nu+\frac{1}{2}\right)\Gamma\left(\frac{1}{2}\right)} \int_{-\frac{1}{2}i\pi}^{-\infty} e^{x \sinh u} \cosh^{2\nu} u\, du$$

The functions $H_\nu^{(1)}(x)$ and $H_\nu^{(2)}(x)$ are the **Hankel functions of the first and second kinds** of order ν, respectively, often called simply the **Hankel functions**, and sometimes are referred to as **Bessel functions of the third kind**. There are several motivations for the choices of the prefactors and the limits of integration that will become apparent below.

Although (8.81) and (8.82) are adequate for many purposes, a representation analogous to that obtained in (8.72) has several advantages, and has been adopted as a 'standard' integral representation for the Hankel functions. Such a representation can be derived by following the steps taken in (8.68)–(8.72) with only a few slight modifications to account for the presence of the hyperbolic functions in (8.81) and (8.82) in place of the trigonometric functions in (8.66). Using the integral representations in (8.81) as an example, we carry out the required number of integrations by parts to obtain the following representations for the first few integer-order Hankel functions:

$$H_0^{(1)}(x) = \frac{2}{i\pi} \int_{-\infty}^{\frac{1}{2}i\pi} e^{x \sinh u} \, du$$

$$H_1^{(1)}(x) = -\frac{2}{i\pi} \int_{-\infty}^{\frac{1}{2}i\pi} e^{x \sinh u} \sinh u \, du$$

$$H_2^{(1)}(x) = \frac{2}{i\pi} \int_{-\infty}^{\frac{1}{2}i\pi} e^{x \sinh u} \cosh 2u \, du$$

$$H_3^{(1)}(x) = -\frac{2}{i\pi} \int_{-\infty}^{\frac{1}{2}i\pi} e^{x \sinh u} \sinh 3u \, du$$

(8.83)

Analogous representations can be established for any positive integer n, and the result can be written as

$$H_n^{(1)}(x) = \frac{1}{i\pi} \int_{-\infty}^{\frac{1}{2}i\pi} e^{x \sinh u} \left[e^{-nu} + (-1)^n e^{nu} \right] du$$

$$= \frac{1}{i\pi} \int_{-\infty}^{\frac{1}{2}i\pi} e^{x \sinh u} e^{-nu} \, du + \frac{(-1)^n}{i\pi} \int_{-\infty}^{\frac{1}{2}i\pi} e^{x \sinh u} e^{nu} \, du \quad (8.84)$$

By performing a few simple transformations on the integration variable in the second integral on the right-hand side of (8.84), the two integrals can be combined into a single expression as

$$H_n^{(1)}(x) = \frac{1}{i\pi} \int_{-\infty}^{\infty + i\pi} e^{x \sinh u - nu} \, du \quad (8.85)$$

which is the desired result. The corresponding representation for the Hankel functions of the second kind can be obtained by following an analogous procedure:

$$H_n^{(2)}(x) = -\frac{1}{i\pi} \int_{-\infty}^{\infty - i\pi} e^{x \sinh u - nu} \, du \quad (8.86)$$

The representations in (8.85) and (8.86) are the most commonly used integral representations of the Hankel functions, although those in Equations (8.81) and (8.82) are equally valid. The representations in (8.85) and (8.86) can be shown to be valid even if n is not an integer (Problem 24).

The integral representations in (8.85) and (8.86) can be used to obtain a number of useful interrelationships among the various Bessel functions. First, an immediate consequence of these representations is the connection between the Hankel functions of positive and negative order:

$$H^{(1)}_{-\nu}(x) = e^{i\nu\pi} H^{(1)}_{\nu}(x), \qquad H^{(2)}_{-\nu}(x) = e^{-i\nu\pi} H^{(2)}_{\nu}(x) \qquad (8.87)$$

We have used the index ν to emphasize that these functions need not be of integer order. The relationship,

$$J_{\nu}(x) = \tfrac{1}{2}\left[H^{(1)}_{\nu}(x) + H^{(2)}_{\nu}(x)\right] \qquad (8.88)$$

which follows from the integral representations in Equations (8.66), (8.81) and (8.82), as discussed in Problem 28, can also be established from (8.85) and (8.86) (Copson, 1935). Either in this way, or by taking the steps described in Problem 18, a generalization of the integral representation in (8.74) is obtained:

$$J_{\nu}(x) = \frac{1}{2\pi i} \oint_{\mathcal{C}} \exp\left[\tfrac{1}{2}x(u - u^{-1})\right] u^{-n-1}\, du \qquad (8.89)$$

The contour $\mathcal{C}$ begins at $u = -\infty$, approaches the origin along the negative real axis, encircles the origin in the positive sense, and returns to $u = -\infty$ along the negative real axis. The evaluation of (8.89) is carried out in Problem 19, and leads to an integral representation for $J_{\nu}(x)$ valid even when ν is not an integer:

$$J_{\nu}(x) = \frac{1}{\pi}\int_0^{\pi} \cos(\nu\theta - x\sin\theta)\, d\theta - \frac{\sin\nu\pi}{\pi}\int_0^{\infty} e^{-\nu\theta - x\sinh\theta}\, d\theta \qquad (8.90)$$

When ν is an integer, the second term on the right-hand side of this equation vanishes, in which case this representation reduces to the simpler form in Equation (8.70).

We now introduce a second solution of Bessel's equation to complement the expressions in (8.81) and (8.82):

$$Y_{\nu}(x) = \frac{1}{2i}\left[H^{(1)}_{\nu}(x) - H^{(2)}_{\nu}(x)\right] \qquad (8.91)$$

This is the Bessel function of the second kind of order ν that we have already introduced in Section 6.4. This equation and the equation in (8.88), show that the Bessel function of the first kind, $J_{\nu}(x)$, and the Bessel function of the second kind, $Y_{\nu}(x)$, are the real and imaginary parts, respectively, of the Hankel functions:

$$H^{(1)}_{\nu}(x) = J_{\nu}(x) + iY_{\nu}(x), \qquad H^{(2)}_{\nu}(x) = J_{\nu}(x) - iY_{\nu}(x) \qquad (8.92)$$

To show that the function $Y_\nu(x)$ is the same as that introduced in Section 6.4 (Problem 28), we combine (8.88) and (8.91) with

$$J_{-\nu}(x) = \tfrac{1}{2}\left[e^{i\nu\pi}H_\nu^{(1)}(x) + e^{-i\nu\pi}H_\nu^{(2)}(x)\right] \tag{8.93}$$

to obtain

$$Y_\nu(x) = \frac{\cos\nu\pi J_\nu(x) - J_{-\nu}(x)}{\sin\nu\pi} \tag{8.94}$$

This is indeed the definition of the $Y_\nu(x)$ in (6.96). If ν is an integer n then, as shown in Equation (6.97), the expression in Equation (8.94) must be evaluated by taking the limit $\nu \to n$. What we have shown in this section and in Section 8.4 is that the motivation for this definition, which is generally accepted as the standard for the Bessel functions of the second kind, has its roots in the integral representations of solutions to Bessel's equation.

An integral representation for the $Y_\nu(x)$ analogous to the representation in (8.90) for $J_n(x)$ can be derived by substituting (8.90) into (8.94), as discussed in Problem 30. For the case that ν is equal to an integer n, we obtain the representation

$$Y_n(x) = \frac{1}{\pi}\int_0^\pi \sin(x\sin\theta - n\theta)\,d\theta - \int_0^\infty e^{-x\sinh\theta}\left[e^{n\theta} + (-1)^n e^{-n\theta}\right]d\theta \tag{8.95}$$

The second integral on the right-hand side of this equation is seen to contain the terms responsible for the singular behavior of these functions at $x=0$, as we first saw in Equation (6.98).

The integral representations derived in this section for the Hankel functions, and the methods used to evaluate them, show that these functions remain valid when the arguments take on complex values. In fact, the Hankel functions exhibit a very rich structure in the complex plane and considerable insight into their behavior can be gained by examining various graphical representations of these functions evaluated with complex arguments. An noteworthy example of this approach is the book by Jahnke and Emde (1945), where contour and surface plots of many special functions may be found.

Plot3D and ContourPlot can be used to evaluate the Hankel functions for complex arguments and the results displayed in one of several ways. Since the functions are complex-valued for complex arguments, we can plot either the real and imaginary parts (using Real and Im), the modulus and phase of the function (using Abs and Arg) or the modulus and phase *together* (using VectorField).

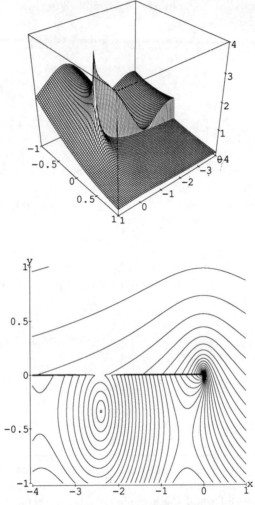

Figure 8.2. Surface plot (top) and contour plot (bottom) of the *absolute value* of the zeroth-order Hankel function of the first kind as a function of the complex argument $z = x + iy$. The surface plot shows clearly the singularity at the origin and both plots reveal the presence of a branch cut along the negative real axis.

Two examples of such plots are shown in Figure 8.2 for the Hankel function of the first kind of order zero. At the top of the figure a surface plot of the modulus is shown, and the corresponding contour plot is displayed at the bottom. Most evident in the structure of this function are the singularity at the origin, the branch cut along the negative real axis and the local minimum near $z = -2.4 - 0.7i$.

Further Reading

Extensive treatments of integral representations may be found in the books by Morse and Feshbach (1953), Whittaker and Watson (1963) and Lebedev (1972). Watson (1952) provides a comprehensive discussion of integral representations for Bessel functions. Introductions to integral representations are given by Ince (1956) and Copson (1935), with the book by Copson also providing all of the necessary background in complex analysis. Tables of integral representations are compiled by Abramowitz and Stegun (1965) and Erdelyi et al. (1953). Both tables and illustrations of special functions with complex arguments are provided by Jahnke and Emde (1945).

References

Abramowitz M. and Stegun I. A. (1965). *Handbook of Mathematical Functions.* New York: Dover.
Andrews L. C. (1985). *Special Functions for Engineers and Applied Mathematicians.* New York: Macmillan.
Carslaw H. S. and Jaeger J. C. (1963). *Operational Methods in Applied Mathematics.* New York: Dover.
Copson E. T. (1935). *An Introduction to the Theory of Functions of a Complex Variable.* Oxford: Oxford University Press.
Erdelyi A., Magnus W., Oberhettinger F. and Tricomi F. G. (1953). *Higher Transcendental Functions* (Bateman Manuscript Project) Vol. 2. New York: McGraw-Hill.
Ince E. L. (1956). *Ordinary Differential Equations.* New York: Dover.
Jahnke E. and Emde F. (1945). *Tables of Functions.* New York: Dover.
Lebedev N. N. (1972). *Special Functions and their Applications* (translated and edited by R. A. Silverman). New York: Dover.
Morse P. M. and Feshbach H. (1953). *Methods of Theoretical Physics* Vol. 1. New York: McGraw-Hill.
Watson G. N. (1952). *A Treatise on the Theory of Bessel Functions* 2nd edn. Cambridge: Cambridge University Press.
Whittaker E. T. and Watson G. N. (1963). *A Course of Modern Analysis* 4th edn. Cambridge: Cambridge University Press.

Problems

1. Show that Equations (8.23) and (8.24) imply that the generating function for the Hermite polynomials can be written as

$$\exp(2xt - t^2) = \sum_{n=0}^{\infty} \frac{t^n}{n!} H_n(x)$$

2. Consider the problem of determining integral representations of Laguerre's equation

$$x\frac{d^2 y}{dx^2} + (1-x)\frac{dy}{dx} + ny = 0$$

where n is an integer. Obtain the representation

$$y_n(x) = \int \frac{(t-1)^n}{t^{n+1}} e^{xt}\, dt$$

with the limits of integration determined by the condition

$$\int \frac{d}{dt}\left[\frac{(t-1)^{n+1}}{t^n} e^{xt}\right] dt = 0$$

By making the transformation of variables in these integrals from t to

$$s = \frac{t}{t-1}$$

show that the integral representation of y_n becomes

$$y_n(x) = \int \frac{1}{s^{n+1}} \frac{\exp[-xs/(1-s)]}{1-s}\, ds$$

and that the condition determining the domain of integration becomes

$$\int \frac{d}{ds}\left\{\frac{1}{s^n} \frac{\exp[-xs/(1-s)]}{1-s}\right\} ds$$

Hence, by choosing the contour of integration to be a closed path around the origin, and imposing the condition that $y_n(0) = 1$, obtain the generating function for Laguerre polynomials of order n, $L_n(x)$:

$$\frac{\exp[-xs/(1-t)]}{1-s} = \sum_{k=0}^{\infty} s^k L_k(x)$$

3. Beginning with the integral representation of the Laguerre polynomials obtained in Problem 2, change the integration variable from s to

$$u = \frac{x}{1-s}$$

to transform this integral representation to

$$L_n(x) = \frac{1}{2\pi i} e^x \oint \frac{u^n e^{-u}}{(u-x)^{n+1}} du$$

Deduce from this expression the Rodrigues formula for the Laguerre polynomials:

$$L_n(x) = \frac{e^x}{n!} \frac{d^n}{dx^n} (x^n e^{-x})$$

4. Use *Mathematica* to obtain expressions for the first few Laguerre polynomials from the generating function in Problem 2 and the Rodrigues formula in Problem 3. For example, the Laguerre polynomials corresponding to $n = 0, 1, 2$ and 3 are given by

$$L_0(x) = 1$$

$$L_1(x) = 1 - x$$

$$L_2(x) = \frac{1}{2}(2 - 4x + x^2)$$

$$L_3(x) = \frac{1}{6}(6 - 18x + 9x^2 - x^3)$$

5. Repeat the calculations in Problems 2 and 3 for the associated Laguerre equation:

$$x\frac{d^2y}{dx^2} + (m + 1 - x)\frac{dy}{dx} + ny = 0$$

In particular, obtain the generating function for the associated Laguerre polynomials,

$$\frac{\exp[-xt/(1-t)]}{(1-t)^m} = \sum_{n=0}^{\infty} t^n L_n^m(x)$$

and the Rodrigues formula,

$$L_n^m(x) = x^{-m} \frac{e^x}{n!} \frac{d^n}{dx^n} (x^{n+m} e^{-x})$$

6. Using the generating function in Problem 2, obtain the normalization integral for the Laguerre polynomials,

$$\int_0^\infty e^{-x} L_m(x) L_n(x)\, dx = \begin{cases} 1 & \text{if } m = n \\ 0 & \text{if } m \neq n \end{cases}$$

and the recursion relations

$$(2n+1-x)L_n(x) = nL_{n-1}(x) + (n+1)L_{n+1}(x)$$

$$xL'_n(x) = nL_n(x) - nL_{n-1}(x)$$

where the prime in the second equation indicates differentiation with respect to x.

7. Beginning with the integral representation (8.43), change the variable of integration to

$$u = 2\frac{x-t}{1-t^2}$$

and use the standardization $P_n(1)=1$ to obtain the integral representation of the Legendre polynomials

$$P_n(x) = \frac{n!}{2\pi i}\oint \frac{1}{u^{n+1}\sqrt{1+u^2-2ux}}\,du$$

where the integration contour is a closed path around the origin. Hence, obtain the generating function for the Legendre polynomials:

$$\frac{1}{\sqrt{1+u^2-2ux}} = \sum_{n=0}^{\infty} u^n P_n(x)$$

8. Use the integral representation (8.43), the condition (8.44) and the standardization $P_n(1)=1$ to deduce the Rodrigues formula for the Legendre polynomials:

$$P_n(x) = \frac{1}{2^n n!}\frac{d^n}{dx^n}(x^2-1)^n$$

9. Show that the integral representation of solutions of the differential equation

$$(1-x^2)\frac{d^2y}{dx^2} - (2\lambda+1)x\frac{dy}{dx} + n(n+2\lambda)y = 0$$

can be written as

$$y(x) = \int \frac{(1-t^2)^{\lambda+n-1/2}}{(x-t)^{2\lambda+n}}\,dt$$

with the domain of integration determined by the condition

$$\int \frac{d}{dt}\left[\frac{(1-t^2)^{\lambda+n+1/2}}{(x-t)^{2\lambda+n+1}}\right]dt = 0$$

10. By n repeated applications of integration by parts show that the integral representation of the Legendre functions of the second kind in Equation (8.45) can be written as

$$Q_n(x) = \frac{1}{2^{n+1}} \int_{-1}^{1} \frac{(1-t^2)^n}{(x-t)^{n+1}} \, dt = \frac{1}{2^{n+1} n!} \int_{-1}^{1} \frac{1}{x-t} \left\{ \frac{d^n}{dt^n} \left[(t^2-1)^n \right] \right\} dt$$

Then, by using Rodrigues' formula (8.47), obtain Neumann's formula:

$$Q_n(x) = \frac{1}{2} \int_{-1}^{1} \frac{P_n(t)}{x-t} \, dt$$

11. Show that Neumann's formula (8.48) together with the orthogonality of the Legendre polynomials in Problem 6.42, implies the relation

$$\frac{1}{x-t} = \sum_{n=0}^{\infty} (2n+1) Q_n(x) P_n(t)$$

where $x > 1$ and $t < 1$.

12. Beginning with the Rodrigues formula for the Legendre functions obtained in Problem 6.43

$$P_n^m(x) = \frac{1}{2^n n!} (1-x^2)^{m/2} \frac{d^{n+m}}{dx^{n+m}} \left[(x^2-1)^n \right]$$

together with the integral representation in Equation (8.46), use Cauchy's integral formula to deduce the following integral representation of the Legendre functions $P_n^m(x)$:

$$P_n^m(x) = \frac{(n+m)!}{2^n n!} (x^2-1)^{m/2} \frac{1}{2\pi i} \oint \frac{(t^2-1)^n}{(t-x)^{n+m+1}} \, dt$$

13. By the same reasoning as that used in Problem 6.13, the associated Legendre functions of the second kind $Q_n^m(x)$ can be expressed in terms of the Legendre functions $Q_n(x)$ by

$$Q_n^m(x) = (x^2-1)^{m/2} \frac{d^m}{dx^m} Q_n(x)$$

Combine this relation with the integral representation for the $Q_n(x)$ in Equation (8.45) to obtain the following integral representation for the $Q_n^m(x)$:

$$Q_n^m(x) = (-1)^m \frac{(n+m)!}{2^{n+1} n!} (x^2-1)^{m/2} \int_{-1}^{1} \frac{(1-t^2)^n}{(x-t)^{n+m+1}} \, dt$$

14. Use Neumann's formula to calculate the $Q_n(x)$ directly using the inbuilt function LegendreP in *Mathematica*. Then apply the relation derived in Problem 13 to obtain expressions for the associated Legendre functions $Q_n^m(x)$.

15. The integral representation in Equation (8.57) for the regular solution of the hypergeometric equation is

$$y(x) = \int_1^\infty t^{\alpha-\gamma}(1-t)^{\gamma-\beta-1}(x-t)^{-\alpha}\,dt$$

To compare this with the series representation of the hypergeometric function $F(\alpha,\beta;\gamma;x)$ obtained in Problem 5.15,

$$F(\alpha,\beta;\gamma;x) \equiv \sum_{k=0}^\infty \frac{(\alpha)_k(\beta)_k}{(\gamma)_k}\frac{x^k}{k!}$$

where

$$(p)_k = p(p+1)\cdots(p+k-1)$$

and $(p)_0 = 1$, expand the factor $(x-t)^{-\alpha}$ in powers of x to obtain

$$y(x) = \sum_{k=0}^\infty (\alpha)_k \frac{x^k}{k!} \int_1^\infty t^{-\gamma-k}(1-t)^{\gamma-\beta-1}\,dt$$

The integral in this expression can be evaluated using the result of Problem 6.22 by making the substitution $t = \cos^2\theta$, with the result

$$\int_1^\infty t^{-\gamma-k}(1-t)^{\gamma-\beta-1}\,dt = (-1)^{\gamma-\beta-1}\frac{\Gamma(\beta+k)\Gamma(\gamma-\beta)}{\Gamma(\gamma+k)}$$

Thus, deduce that the integral representation of $F(\alpha,\beta;\gamma;x)$ is

$$F(\alpha,\beta;\gamma;x) = \frac{\Gamma(\gamma)}{\Gamma(\beta)\Gamma(\gamma-\beta)}\int_1^\infty t^{\alpha-\gamma}(t-1)^{\gamma-\beta-1}(t-x)^{-\alpha}\,dt$$

which is Equation (8.58).

16. One application of the integral representations in Equations (8.58) and (8.59) is the evaluation of the quantity $F(\alpha,\beta;\gamma;1)$. With the restriction $\gamma > \beta > 0$, show that Equation (8.59) implies

$$F(\alpha,\beta;\gamma;1) = \frac{\Gamma(\gamma)}{\Gamma(\beta)\Gamma(\gamma-\beta)}\int_0^1 t^{\beta-1}(1-t)^{\gamma-\beta-\alpha-1}\,dt$$

Then, by using the result of Problem 6.22, obtain the result

$$F(\alpha,\beta;\gamma;1) = \frac{\Gamma(\gamma)\Gamma(\gamma-\alpha-\beta)}{\Gamma(\gamma-\alpha)\Gamma(\gamma-\beta)}$$

For this expression to be well-defined, neither γ nor the quantity $\gamma-\alpha-\beta$ can be negative integers or zero.

17. Deduce from *each* of the three integral representations in (8.81) and (8.82) that
$$J_\nu(x) = \tfrac{1}{2}\left[H_\nu^{(1)}(x) + H_\nu^{(2)}(x)\right]$$

18. Beginning with the integral representation of the Bessel function $J_n(x)$ in Equation (8.74), but with an unspecified integration contour,
$$J_n(x) = \frac{1}{2\pi i}\int \exp\left[\tfrac{1}{2}x\left(u - u^{-1}\right)\right] u^{-n-1}\,du$$
make the change of variables to $\xi = 2u/x$ to obtain
$$J_n(x) = \frac{1}{2\pi i}\left(\frac{x}{2}\right)^n \int \exp\left(\xi - \frac{x^2}{4\xi}\right) \xi^{-n-1}\,d\xi$$
By using *Mathematica*, or otherwise, show that this expression satisfies Bessel's equation for *any* value of n, not simply integer values, provided the integration contour is chosen such that
$$\int \frac{d}{d\xi}\left[\exp\left(\xi - \frac{x^2}{4\xi}\right)\xi^{-\nu-1}\right] d\xi = 0$$
where we have written ν in place of n.

If ν is an integer, then the integrand returns to its initial value along a closed path that encircles the origin, and we obtain the representation in Equation (8.74). To find an appropriate contour when ν is not an integer, we observe that the integrand vanishes in the limit $\xi \to -\infty$. Thus, we can obtain a solution of Bessel's equation by choosing the contour to begin at $\xi = -\infty$, approach the origin along the negative real axis, encircle the origin in the positive sense and return to $\xi = -\infty$ along the negative real axis.

19. Let C signify the contour in Problem 18. To evaluate the integral representation of $J_\nu(x)$ obtained in Problem 18, C is divided into three parts: C_1, the path along the negative real axis from $u = -\infty$ to $u = -1$, C_2, the circle $|u| = 1$, taken in the positive sense, and C_3, the path along the negative real axis from $u = -1$ to $u = -\infty$. By writing u as $u = t e^{i\theta}$ and choosing the appropriate value(s) for θ for each of the contours, show that

$$\frac{1}{2\pi i}\int_{C_1} \exp\left[\tfrac{1}{2}x(u - u^{-1})\right] u^{-n-1}\,du = \frac{e^{(n+1)\pi i}}{2\pi i}\int_0^\infty e^{-\nu\theta - x\sinh\theta}\,d\theta$$

$$\frac{1}{2\pi i}\int_{C_2} \exp\left[\tfrac{1}{2}x(u - u^{-1})\right] u^{-n-1}\,du = \frac{1}{2\pi}\int_{-\pi}^{\pi} e^{ix\sin\theta - in\theta}\,d\theta$$

$$\frac{1}{2\pi i}\int_{C_3} \exp\left[\tfrac{1}{2}x(u - u^{-1})\right] u^{-n-1}\,du = -\frac{e^{-(n+1)\pi i}}{2\pi i}\int_0^\infty e^{-\nu\theta - x\sinh\theta}\,d\theta$$

Thus, obtain the integral representation in (8.90):

$$J_\nu(x) = \frac{1}{\pi}\int_0^\pi \cos(\nu\theta - x\sin\theta)\,d\theta - \frac{\sin\nu\theta}{\pi}\int_0^\infty e^{-\nu\theta - x\sinh\theta}\,d\theta$$

which is valid for both integer and noninteger values of ν.

20. Obtain the power series expansion of the Bessel function $J_n(x)$ in Equation (6.86) directly from the integral representation

$$J_n(x) = \frac{1}{2\pi i}\left(\frac{x}{2}\right)\oint \exp\left(\xi - \frac{x^2}{4\xi}\right)\xi^{-n-1}\,d\xi$$

where the contour C may be taken to be the circle $|\xi| = 1$. First expand the exponential containing the factor proportional to ξ^2 to obtain

$$J_n(x) = \frac{1}{2\pi i}\sum_{k=0}^\infty \frac{(-1)^k}{k!}\left(\frac{x}{2}\right)^{n+2k}\oint \xi^{-n-k-1}e^\xi\,d\xi$$

Evaluate the integral in this expression to obtain, for $n \geq 0$, the expansion

$$J_n(x) = \sum_{k=0}^\infty \frac{(-1)^k}{k!(n+k)!}\left(\frac{x}{2}\right)^{n+2k}$$

which is precisely (6.86). What do you obtain for negative values of n?

21. To determine the constant of proportionality between the solution in Equation (8.65) and the series solution of Bessel's equation in Equation (6.86), it proves most convenient to work with the representation

$$y_\nu(x) = x^\nu \int_{-1}^1 e^{ixu}(1-u^2)^{\nu-1/2}\,du$$

in Equation (8.66). Expand the exponential in the integrand to obtain

$$y_\nu(x) = 2x^\nu \sum_{k=0}^\infty (-1)^k \frac{x^{2k}}{(2k)!}\int_0^1 u^{2k}(1-u^2)^{\nu-1/2}\,du$$

By making a suitable change of variables in the integrand, evaluate the resulting integral as in Problem 15 to obtain

$$y_\nu(x) = 2x^\nu \sum_{k=0}^\infty (-1)^k \frac{x^{2k}}{(2k)!}\,\frac{\Gamma\left(k+\frac{1}{2}\right)\Gamma\left(\nu+\frac{1}{2}\right)}{\Gamma(k+\nu+1)}$$

Hence, by comparing with the series in Equation (6.86), deduce that

$$J_\nu(x) = \frac{\left(\frac{1}{2}x\right)^\nu}{\Gamma\left(\nu+\frac{1}{2}\right)\Gamma\left(\frac{1}{2}\right)}\int_{-1}^1 e^{ixu}(1-u^2)^{\nu-1/2}\,du$$

22. By combining the relationship between the modified Bessel function of the first kind of order n, $I_n(x)$ and $J_n(x)$,

$$I_n(x) = i^{-n} J_n(ix)$$

and the appropriate integral representation in (8.66), deduce the integral representation

$$I_n(x) = \frac{(\tfrac{1}{2}x)^n}{\Gamma(n+\tfrac{1}{2})\Gamma(\tfrac{1}{2})} \int_{-1}^{1} e^{-xt}(1-t^2)^{n-1/2}\, dt$$

Beginning with the generating function in (8.75) for the $J_n(x)$, obtain the corresponding generating function for the $I_n(x)$:

$$\exp\!\left[\tfrac{1}{2}x\left(s+s^{-1}\right)\right] = \sum_{n=-\infty}^{\infty} s^n I_n(x)$$

23. The integral representations of the modified Bessel functions can be used to obtain expressions for the asymptotic forms of the Bessel functions and modified Bessel functions for large arguments. Consider first the integral representation for I_n derived in Problem 22. By changing the integration variable first to $s = t+1$ and then to $u = sx$ show that this integral representation is transformed to

$$I_n(x) = \frac{1}{\Gamma(n+\tfrac{1}{2})\Gamma(\tfrac{1}{2})} \frac{e^x}{\sqrt{2x}} \int_0^{2x} e^{-u} u^{n-1/2} \left(1 - \frac{u}{2x}\right)^{n-1/2} du$$

Use *Mathematica* to graph the integral in this expression as a function of x and compare the result with the value obtained by setting $x \to \infty$:

$$\lim_{x\to\infty} \int_0^{2x} e^{-u} u^{n-1/2} \left(1 - \frac{u}{2x}\right)^{n-1/2} du = \int_0^\infty e^{-u} u^{n-1/2} du = \Gamma(n+\tfrac{1}{2})$$

Hence obtain the asymptotic form of the modified Bessel function of the first kind for large argument:

$$I_n(x) \sim \frac{1}{\sqrt{2\pi x}} e^x, \qquad x \to \infty$$

24. Show by direct substitution that the integral representations

$$H_\nu^{(1)}(x) = \frac{1}{i\pi} \int_{-\infty}^{\infty + i\pi} e^{x \sinh u - \nu u}\, du$$

$$H_\nu^{(2)}(x) = -\frac{1}{i\pi} \int_{-\infty}^{\infty - i\pi} e^{x \sinh u - \nu u}\, du$$

are solutions of Bessel's equation even for the noninteger values of ν.

25. The modified Bessel function of the second kind of order ν, $K_\nu(x)$, is given in terms of the modified Bessel function of the first kind as

$$K_\nu(x) = \frac{\pi}{2} \frac{I_{-\nu}(x) - I_\nu(x)}{\sin \nu\pi}$$

From this definition and that in Equation (8.94) deduce the relation

$$K_\nu(x) = \tfrac{1}{2}i\pi e^{i\nu\pi/2} H_\nu^{(1)}(ix)$$

By using the appropriate integral representation in Equation (8.81), and taking careful note of Equations (8.63) and (8.64), obtain the following integral representation for $K_\nu(x)$:

$$K_\nu(x) = \frac{\sqrt{\pi}\left(\tfrac{1}{2}x\right)^\nu}{\Gamma(\nu+\tfrac{1}{2})} \int_1^\infty e^{-xt}(t^2-1)^{\nu-1/2}\, dt$$

26. By applying a suitable modification of the procedure used in Problem 23 for the integral representation of the Bessel function of the second kind derived in Problem 25, show that the asymptotic form of $K_n(x)$ for large argument is

$$K_n(x) \sim \left(\frac{\pi}{2x}\right)^{1/2} e^{-x}, \qquad x \to \infty$$

Use this result to deduce the corresponding asymptotic forms as $x \to \infty$ for the Bessel functions of the first and second kind,

$$J_n(x) \sim \left(\frac{2}{\pi x}\right)^{1/2} \cos\left(x - \tfrac{1}{2}n\pi - \tfrac{1}{4}\pi\right)$$

$$Y_n(x) \sim \left(\frac{2}{\pi x}\right)^{1/2} \sin\left(x - \tfrac{1}{2}n\pi - \tfrac{1}{4}\pi\right)$$

and for the Bessel functions of the third kind,

$$H_n^{(1)}(x) \sim \left(\frac{2}{\pi x}\right)^{1/2} \exp\left[i\left(x - \tfrac{1}{2}n\pi - \tfrac{1}{4}\pi\right)\right]$$

$$H_n^{(2)}(x) \sim \left(\frac{2}{\pi x}\right)^{1/2} \exp\left[-i\left(x - \tfrac{1}{2}n\pi - \tfrac{1}{4}\pi\right)\right]$$

27. Use *Mathematica* to display the Bessel functions and the respective asymptotic forms on the same graph for several values of n.

28. Show that (8.87) and (8.88) imply (8.93):

$$J_{-\nu}(x) = \tfrac{1}{2}\left[e^{i\nu\pi} H_\nu^{(1)}(x) + e^{-i\nu\pi} H_\nu^{(2)}(x)\right]$$

Then, use this equation and

$$J_\nu(x) = \tfrac{1}{2}\left[H_\nu^{(1)}(x) + H_\nu^{(2)}(x)\right]$$

to solve for the Hankel functions in terms of the Bessel functions. Substitute the resulting expressions into (8.91) to obtain (8.94):

$$Y_\nu(x) = \frac{\cos\nu\pi\, J_\nu(x) - J_{-\nu}(x)}{\sin\nu\pi}$$

29. Use the integral representations in Equations (8.84) and (8.85),

$$H_\nu^{(1)}(x) = \frac{1}{i\pi}\int_{-\infty}^{\infty+i\pi} e^{x\sinh u - \nu u}\, du$$

$$H_\nu^{(2)}(x) = -\frac{1}{i\pi}\int_{-\infty}^{\infty-i\pi} e^{x\sinh u - \nu u}\, du$$

to deduce the relationships between the positive-order and negative-order Hankel functions in (8.86):

$$H_{-\nu}^{(1)}(x) = e^{i\nu\pi} H_\nu^{(1)}(x), \qquad H_{-\nu}^{(2)}(x) = e^{-i\nu\pi} H_\nu^{(2)}(x)$$

Establish this result by changing ν to $-\nu$ in the integral representations and making an appropriate change of variables.

30. Substitute the integral representation for $J_\nu(x)$ into the Equation (8.94) to obtain the following representation for $Y_\nu(x)$:

$$Y_\nu(x) = \frac{1}{\pi}\frac{\cos\nu\pi}{\sin\nu\pi}\int_0^\pi \cos(\nu\theta - x\sin\theta)\, d\theta - \frac{\cos\nu\pi}{\pi}\int_0^\infty e^{-\nu\theta - x\sinh\theta}\, d\theta$$

$$-\frac{1}{\pi}\frac{1}{\sin\nu\pi}\int_0^\pi \cos(\nu\theta + x\sin\theta)\, d\theta - \frac{1}{\pi}\int_0^\infty e^{\nu\theta - x\sinh\theta}\, d\theta$$

Then, change the integration variable in the third term on the right-hand side from θ to $\pi - \theta$, and show that this representation becomes

$$Y_\nu(x) = \frac{1}{\pi}\int_0^\pi \sin(x\sin\theta - \nu\theta)\, d\theta - \frac{1}{\pi}\int_0^\infty e^{-x\sinh\theta}\left[e^{\nu\theta} + e^{-\nu\theta}\cos\nu\pi\right] d\theta$$

By taking the limit $\nu \to n$, obtain the representation in Equation (8.95) for $Y_n(x)$.

31. Apply the procedure used in Section 8.4 to obtain the integral representation in (8.76) to obtain the following integral representation for the modified Bessel functions $K_\nu(x)$ from that in Problem 25:

$$K_\nu(x) = \frac{1}{2}\left(\frac{x}{2}\right)^\nu \int_0^\infty \exp[-t - (x^2/4t)] t^{-\nu-1}\, dt$$

32. Integral representations can be used for obtaining Laplace transforms of functions involving Bessel functions. As an illustration of this, use the integral representation derived in Problem 31 and the functional form of $K_{1/2}(x)$,

$$K_{1/2}(x) = \sqrt{\frac{\pi}{2x}}\, e^{-x}$$

which can be deduced from Problems 6.32 and 6.35, to show that

$$\int_0^\infty \exp(a^2 x^2 - b^2/x^2)\, dx = \frac{\sqrt{\pi}}{2a}\, e^{-2ab}$$

where a and b are constants. Use this result to verify the following Laplace transform:

$$\int_0^\infty \frac{1}{\sqrt{\pi t}} \exp(-a^2/4t)\, e^{-st}\, dt = \frac{1}{\sqrt{s}}\, e^{-a\sqrt{s}}$$

Other examples of using integral representations to evaluate Laplace transforms can be found in the book by Carslaw and Jaeger (1963).

Chapter 9

Introduction to Nonlinear Partial Differential Equations

Almost all of the methods that have been described in the preceding chapters for solving differential equations rely on the superposition principle to express the solution of a particular equation with specified initial and boundary conditions as a linear combination of particular elementary solutions. The various methods presented, such as eigenfunction expansions, integral transforms and Green's functions differ only in the basis that is used in forming the solution. For nonlinear equations, however, linear superposition cannot be applied to generate new solutions, so general approaches to finding solutions are far less abundant than for linear equations. A change of variables can sometimes be found that transforms a nonlinear equation into a linear equation, or some other *ad hoc* technique may yield a solution for a particular equation, but finding the solutions of most nonlinear equations generally requires new techniques.

Nonlinear partial differential equations appear in many branches of physics, engineering and applied mathematics, including nonlinear optics, quantum field theory, fluid mechanics, elasticity theory and condensed-matter physics. Nonlinear wave equations in particular have provided several examples of solutions that are strikingly different from those obtained for linear wave propagation. The best-known of these are solitary waves, or solitons. Characteristic properties of solitons include a localized wave form that is retained upon interaction with other solitons, giving solitons a 'particle-like' quality. This and other intriguing properties of solutions to nonlinear partial differential equations will be explored in the following two chapters.

In this chapter the physical origins of several well-known nonlinear equations will be outlined briefly, and then some of the methods used for obtaining solutions will be described. We first discuss the Hopf–Cole transformation of the Burgers equation as an example where a nonlinear equation can be transformed into a linear equation. The solution of this linear equation can be obtained by using the methods described in previous chapters, and then transformed back to produce the solution of the original equation.

In a somewhat more general setting, Bäcklund transformations provide a framework relating solutions of particular pairs of equations, or even solutions of the same equation. In the latter case, the Bäcklund transformation will be used to show that new solutions of the nonlinear equation can be generated by purely algebraic operations involving known solutions. This generalizes the superposition principle for linear equations and leads to the notion of nonlinear superposition associated with a particular equation. As an application of this method, solutions with a successively increasing number of solitons will be obtained systematically by iterating the nonlinear superposition rule. These solutions show that solitons can be viewed as 'building blocks' for more general solutions for certain types of equation. Further evidence of these observations, and their generalizations, is the subject of Chapter 10.

9.1 Nonlinearity in Partial Differential Equations

In many physical problems the governing equations appear to be linear, but require constitutive equations to obtain a closed set of equations. Familiar examples include the diffusion equation, which is based upon applying Fourier's law to the heat conservation equation (Problem 1), and Maxwell's equations for a dielectric medium, where the dielectric constant relates the displacement to the electric field.

One important application that is of considerable current interest and illustrates how constitutive relations can lead to nonlinearities is the **nonlinear Schrödinger equation**,

$$i\frac{\partial u}{\partial t} + \frac{\partial^2 u}{\partial x^2} + |u|^2 u = 0 \qquad (9.1)$$

In nonlinear optics, u is the envelope function of the electric field, and the cubic term results from the refractive index of the medium being dependent upon the intensity of the field (the Kerr effect):

$$n = n_0 + n_2 |E|^2 \qquad (9.2)$$

Hasegawa and Tappert showed theoretically that an optical pulse in a dielectric fiber forms an envelope soliton solution of (9.1), i.e. the pulses of light either do not change shape or change shape periodically. Mollenauer later demonstrated this effect experimentally. An introductory discussion of these developments has been given by Hasegawa (1990).

To derive (9.1) for the propagation of electromagnetic waves in nonlinear media, we first observe that the wave equation obtained from Maxwell's equations for the electric field **E** with the nonlinearity in the refractive index (9.2) is given by

$$\nabla^2 \mathbf{E} - \frac{1}{c}\frac{\partial^2 \mathbf{D}}{\partial t^2} = \nabla^2 \mathbf{E} - \frac{\epsilon_0}{c}\frac{\partial^2 \mathbf{E}}{\partial t^2} - \frac{\epsilon_2}{c}\frac{\partial^2}{\partial t^2}(|E|^2 \mathbf{E}) = 0 \qquad (9.3)$$

where the dielectric constant is given by

$$\epsilon = \epsilon_0 + \epsilon_2 |E|^2 \qquad (9.4)$$

If the medium is linear, then $\epsilon_2 = 0$, and Equation (9.3) reduces to the usual wave equation for electromagnetic waves. Equation (9.4) identifies the quantity

$$\delta = \frac{\epsilon_2 |E|^2}{\epsilon_0} \qquad (9.5)$$

as a measure of the strength of the nonlinearity in (9.3). In the following discussion, we will take this quantity to be small, i.e. $\delta \ll 1$.

The electric field **E** is now represented as the product of an envelope function $\mathcal{E}$ and a carrier wave $e^{ikx-i\omega t}$. To derive one form of the nonlinear Schrödinger equation, we suppose that a steady state has been attained wherein the envelope function does not depend upon the time. The envelope function is taken to be slowly varying as a function of position in comparison with the carrier wave. Accordingly, we introduce the slowly varying space variables ξ and η through (Ablowitz and Segur, 1981)

$$\begin{aligned}\xi &= (\xi_x, \xi_y, \xi_z) = (x\delta, y\delta, z\delta) \\ \eta &= (\eta_x, \eta_y, \eta_z) = (x\delta^2, y\delta^2, z\delta^2)\end{aligned} \qquad (9.6)$$

Since δ is assumed to be small, the variation in ξ and η occurs over much longer regions of space than that in x, y or z. The electric field **E** is now written as

$$\mathbf{E}(\mathbf{x},t) = \hat{\mathbf{e}}\delta\mathcal{E}(\xi,\eta)\, e^{ikx-i\omega t} \qquad (9.7)$$

where $\mathbf{x} = (x, y, z)$, $\hat{\mathbf{e}}$ is the linear polarization vector and $k = (\omega/c)\sqrt{\epsilon_0}$. This expression is substituted into the wave equation in (9.3) and an expansion in powers of δ is carried out. For (9.7) to be a solution of (9.3) the coefficients of each power of δ must vanish separately. The first-order terms are found to cancel because of the relation between k and ω. Requiring the second-order terms to vanish leads to $\mathcal{E}$ being independent of ξ_x. At third order, we obtain the equation satisfied by the envelope function which, upon rescaling, produces the 'dimensionless' form of the nonlinear Schrödinger equation:

$$i\frac{\partial \psi}{\partial \tau} + \nabla_\perp^2 \psi + |\psi|^2 \psi = 0 \tag{9.8}$$

where ψ is the rescaled envelope function and τ is the rescaled variable η_1. The quantity $\nabla_\perp^2$ is defined by

$$\nabla_\perp^2 = \frac{\partial^2}{\partial \xi_2^2} + \frac{\partial^2}{\partial \xi_3^2} \tag{9.9}$$

The name 'nonlinear Schrödinger equation' comes from the interpretation of the nonlinear term $|\psi|^2 \psi$ as a 'potential' for an equation which is then similar in appearance to the time-dependent Schrödinger equation of quantum mechanics. The underlying equations leading to (9.8) are, of course, classical.

The **sine-Gordon Equation**,

$$\frac{\partial^2 u}{\partial x^2} - \frac{\partial^2 u}{\partial t^2} = \sin u \tag{9.10}$$

has been used in many different applications, including the propagation of ultra-short optical pulses in resonant laser media (Lamb, 1971), a unitary theory of elementary particles (Skyrme, 1958, 1961; Enz, 1963) and the propagation of magnetic flux in Josephson junctions (Josephson, 1965). Another form of the sine-Gordon equation frequently encountered is obtained by putting the wave operator on the left-hand side of (9.10) into the normal form for a hyperbolic equation (Section 3.1) by transforming to characteristic coordinates $\xi = \frac{1}{2}(x+t)$, $\eta = \frac{1}{2}(x-t)$:

$$\frac{\partial^2 \phi}{\partial \eta \partial \xi} = \sin \phi \tag{9.11}$$

Although (9.10) and (9.11) are often used interchangeably, care must be taken in transforming between the two equations if an initial-value problem is to be solved, since the line $t=0$ in (9.10) is transformed to the line $\xi=\eta$ in (9.11).

One way of deriving the sine-Gordon equation that is easy to visualize is to consider the arrangement shown in Figure 9.1, which consists of a chain of pendula connected by springs (Scott, 1969). The pendula are allowed to rotate only in the direction *perpendicular* to the chain, with the angular deviation from the vertical denoted by ϕ. The forces acting on each pendulum are the torque due to the relative angular displacements of neighboring pendula and the gravitational restoring torque. If interactions

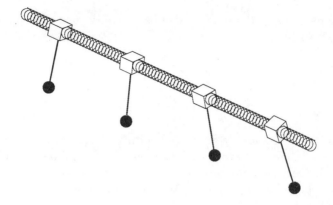

Figure 9.1 One-dimensional chain of pendula used for the derivation of the sine-Gordon equation. The displacement of the pendula is in a plane perpendicular to the chain, and is measured by the angular deviation from the vertical.

among the pendula are restricted to be only between nearest neighbors, then the equation of motion for the ith pendulum can be written in terms of ϕ_i as

$$m\frac{d^2\phi_i}{dt^2} = k[(\phi_{i-1} - \phi_i) + (\phi_{i+1} - \phi_i)] - t\sin\phi_i \qquad (9.12)$$

In this equation m is the moment of inertia of the pendulum, k is the torque constant of the spring and $t\sin\phi$ is the gravitational restoring torque. To convert the difference equation in (9.12) into a differential equation, we must take the limit of (9.12) as the distance Δx between nearest-neighbor pendula becomes smaller and approaches zero. In doing so, the quantities m, k and t must be replaced by the respective densities, where m becomes a moment of inertia *per unit length*, and so on. Thus, dividing both sides of (9.12) by $(\Delta x)^2$, we obtain, after a slight rearrangement, the equation

$$\frac{1}{\Delta x}\left[\frac{\phi_{i+1} - \phi_i}{\Delta x} - \frac{\phi_i - \phi_{i-1}}{\Delta x}\right] - \frac{m/\Delta x}{k\Delta x}\frac{d^2\phi_i}{dt^2} = \frac{t/\Delta x}{k\Delta x}\sin\phi_i \qquad (9.13)$$

With $\Delta x = 1$, (9.13) is the same as (9.12). The terms in brackets on the left-hand side of (9.13) are readily identified as the discrete forms of the left and right derivatives of ϕ with respect to x taken at nearest-neighbor points. The difference of these two quantities divided by Δx is therefore the discrete form of the second derivative of ϕ with respect to x. In the limit $\Delta x \to 0$, the quantity $m/\Delta x$ becomes the moment of inertia per unit length, μ, $1/k\Delta x$ becomes the inverse spring constant per unit length, κ^{-1}, $t/\Delta x$ becomes the torque per unit length, τ, and the index i becomes a continuous spatial variable along the chain, $\phi_i(t) \to \phi(x,t)$. In this limit, the difference equation in (9.13) becomes

$$\frac{\partial^2\phi}{\partial x^2} - \mu\kappa^{-1}\frac{\partial^2\phi}{\partial t^2} = \tau\kappa^{-1}\sin\phi \qquad (9.14)$$

A simple rescaling of the space and time variables yields the sine-Gordon equation in (9.10).

The **Korteweg–de Vries equation,**

$$\frac{\partial u}{\partial t} + 6u\frac{\partial u}{\partial x} + \frac{\partial^3 u}{\partial x^3} = 0 \qquad (9.15)$$

is perhaps the best-known equation exhibiting solitary wave solutions, and many of the techniques developed for studying such solutions of other equations originated with studies of this equation. The other standard way of writing this equation is with a negative nonlinear term:

$$\frac{\partial u}{\partial t} - 6u\frac{\partial u}{\partial x} + \frac{\partial^3 u}{\partial x^3} = 0 \qquad (9.16)$$

The solutions of (9.15) and (9.16) are readily seen to be negatives of one another, so there is no essential distinction between these two forms of the Korteweg–de Vries equation. The Korteweg–de Vries equation was originally derived to describe shallow water waves, but later was applied to anharmonic oscillations in lattices and to problems in plasma physics (Zabusky and Kruskal, 1965). We will obtain the Korteweg–de Vries equation from a heuristic argument, and leave an alternative derivation based upon a chain of nonlinear oscillators to the problems at the end of the chapter. A detailed derivation for fluid mechanical applications has been given by Drazin and Johnson (1989).

We saw in Chapter 2 that the equation

$$\frac{\partial u}{\partial t} + u\frac{\partial u}{\partial x} = 0 \qquad (9.17)$$

exhibits a solution that develops a discontinuity in the spatial dependence of u. If we regard the quantity u as the height of a water level, then the behavior shown in Figures 2.6 and 2.7 is analogous to the breaking of a wave.

The effect of dispersion on the evolution of u can be included by adding to (9.17) terms proportional to successively higher-order spatial derivatives of u. Beginning with the first derivative, a term proportional to u_x can be absorbed into an additive constant to u, leaving the form of (9.17) invariant. This does not lead to any qualitative difference in the solution, but we will return to this term below. Adding a term proportional to u_{xx} produces

$$\frac{\partial u}{\partial t} + u\frac{\partial u}{\partial x} = \frac{\partial^2 u}{\partial x^2} \qquad (9.18)$$

which is the **Burgers equation**. This equation describes the competition between the sharpening of the profile due to the nonlinear term, and the broadening of sharp features in u through dissipation by the linear diffusive term, u_{xx}. The solution of (9.18) will be derived in the next section.

The lowest-order derivative of u which leads to a nonlinearity in the dispersion relation for wave propagation that is non-dissipative is the third-order term u_{xxx}. The resulting equation

$$\frac{\partial u}{\partial t} + u\frac{\partial u}{\partial x} = \frac{\partial^3 u}{\partial x^3} \qquad (9.19)$$

is the Korteweg–de Vries equation. Equation (9.19) can be put into the form shown in Equation (9.15) by a rescaling of u, x and t. This equation describes the competition between the sharpening due to the nonlinear term and the dispersive effects of the third-order derivative.

To see the effect of the dispersion, but without dissipation, we add the first and third-order derivatives to the right hand side of (9.17):

$$\frac{\partial u}{\partial t} + u\frac{\partial u}{\partial x} = \frac{\partial u}{\partial x} + \frac{\partial^3 u}{\partial x^3} \qquad (9.20)$$

If we consider only small-amplitude disturbances, then this equation can be 'linearized' by retaining only the linear terms in (9.20). Seeking travelling wave solutions of the form $e^{i(kx-\omega t)}$ yields the dispersion relation for small-amplitude waves:

$$\omega = k - k^3 \qquad (9.21)$$

The phase velocity, $\omega/k = 1 - k^2$, is seen to decrease with decreasing wavelength $\lambda = 2\pi/k$. Thus, the short wavelength disturbances that are required to develop the discontinuous behavior of solutions travel with a slower phase velocity than the longer wavelength components and so fall behind. Under the appropriate circumstances, this produces a balance between the tendency toward sharpening by the term uu_x and the dispersion by the term u_{xxx}. This may be viewed as the underlying mechanism behind the formation of solitary waves.

9.2 The Burgers Equation—An Exact Solution

The Burgers equation,

$$\frac{\partial u}{\partial t} + u\frac{\partial u}{\partial x} = \frac{\partial^2 u}{\partial x^2} \qquad (9.22)$$

was used as a simple model of turbulence in an extensive study by Burgers (1974). With u regarded as the velocity field of a fluid, the essential ingredient of (9.22) to this study is the competition between the dissipative term, u_{xx}, the coefficient of which is the kinematic viscosity, and the nonlinear term uu_x.

Hopf (1950) and Cole (1951) showed that the Burgers equation could be transformed into the diffusion equation. Since the initial value problem for the diffusion equation can be solved by any of several methods discussed in the preceding chapters, this provides a way of solving the initial-value problem of (9.22) exactly. In this section, we will solve the initial value problem for the Burgers equation by applying the Hopf–Cole transformation to the fundamental solution of the diffusion equation derived in Section 7.3.

The initial-value problem for the Burgers equation is the solution satisfying (9.22) with the profile of u at time $t=0$ specified by a function u_0:

$$u(x,0) = u_0(x) \tag{9.23}$$

The Hopf–Cole transformation will be developed in two steps. We first introduce the function ϕ defined in terms of u by

$$u(x,t) = -\frac{\partial}{\partial x}\phi(x,t) \tag{9.24}$$

By writing the nonlinear term in (9.22) as $u u_x = \frac{1}{2}(u^2)_x$, the differential equation for ϕ is found to be

$$\frac{\partial}{\partial t}\left(-\frac{\partial \phi}{\partial x}\right) + \frac{1}{2}\frac{\partial}{\partial x}\left(\frac{\partial \phi}{\partial x}\right)^2 + \frac{\partial^3 \phi}{\partial x^3}$$

$$= -\frac{\partial}{\partial x}\left[\frac{\partial \phi}{\partial t} - \frac{1}{2}\left(\frac{\partial \phi}{\partial x}\right)^2 - \frac{\partial^2 \phi}{\partial x^2}\right] = 0 \tag{9.25}$$

Notice that the order of the differentiations with respect to x and t has been interchanged in obtaining the right-hand side of (9.25). Integrating (9.25) with respect to x, we obtain

$$\frac{\partial \phi}{\partial t} = \frac{1}{2}\left(\frac{\partial \phi}{\partial x}\right)^2 + \frac{\partial^2 \phi}{\partial x^2} + f(t) \tag{9.26}$$

where $f(t)$ is a function of integration. By introducing the quantity $\psi(x,t)$, defined by

$$\phi(x,t) = \psi(x,t) + \int_0^t f(s)\,ds \tag{9.27}$$

f can be eliminated from (9.26), and we obtain the Burgers equation in the form

$$\frac{\partial \psi}{\partial t} = \frac{1}{2}\left(\frac{\partial \psi}{\partial x}\right)^2 + \frac{\partial^2 \psi}{\partial x^2} \tag{9.28}$$

The next step of the transformation is carried out by introducing a function z defined in terms of ψ by

$$\psi(x,t) = 2\ln[z(x,t)] \tag{9.29}$$

Performing the required differentiations of (9.29), and substituting into (9.19), we obtain

$$\frac{\partial \psi}{\partial t} - \frac{1}{2}\left(\frac{\partial \psi}{\partial x}\right)^2 - \frac{\partial^2 \psi}{\partial x^2}$$

$$= \frac{2}{z}\frac{\partial z}{\partial t} - \frac{2}{z^2}\left(\frac{\partial z}{\partial x}\right)^2 - \frac{2}{z^2}\left[z\frac{\partial^2 z}{\partial x^2} - \left(\frac{\partial z}{\partial x}\right)^2\right]$$

$$= \frac{2}{z}\left[\frac{\partial z}{\partial t} - \frac{\partial^2 z}{\partial x^2}\right] = 0 \qquad (9.30)$$

When the transformation between z and ϕ in (9.29) is valid, i.e. when $z(x,t)$ is a positive function, Equation (9.30) states that z satisfies the linear diffusion equation:

$$\frac{\partial z}{\partial t} = \frac{\partial^2 z}{\partial x^2} \qquad (9.31)$$

The initial condition for (9.31),

$$z(x,0) = z_0(x) \qquad (9.32)$$

can be related to the initial condition ψ_0 of (9.28) and the initial condition u_0 of (9.23) by successive application of the transformations (9.29) and (9.24):

$$z_0(x) = \exp\left[\tfrac{1}{2}\psi_0\right] \qquad (9.33a)$$

$$= \exp\left[-\frac{1}{2}\int_{-\infty}^{x} u_0(x')\,dx'\right] \qquad (9.33b)$$

Several comments are in order concerning the transformation of (9.22) into (9.31). First, the obvious advantage of the procedure is that a nonlinear equation is transformed into a linear equation. The diffusion equation can be solved by any one of several standard methods, including the effects of boundaries. Second, this serves to illustrate the power of *ad hoc* methods, whereby a particular equation may be transformed to a much simpler equation by finding the 'right' transformation. In addition, we will see in Section 9.4 that the Hopf–Cole transformation may be regarded as a prototype for other more general transformations between nonlinear equations. Third, since the starting point of finding a solution of (9.22) is the diffusion equation (9.31), to which the inverse transformations in (9.24) and (9.29) are applied, and since $\psi_x = \phi_x$, the function of integration f appearing in (9.26) can be neglected entirely. Thus, if z is a solution of the diffusion equation (9.31) with initial condition (9.32), then the quantity

$$u(x,t) = -2\frac{\partial}{\partial x}\ln[z(x,t)] = -2\frac{z_x(x,t)}{z(x,t)} \qquad (9.34)$$

is a solution of the Burgers equation with boundary condition (9.33b).

We will obtain the solution to the diffusion equation for the case where there are no boundaries for the initial condition (9.25) using the fundamental solution of the diffusion equation derived in Section 7.3. The appropriate Green's function is the fundamental solution (7.46),

$$G_0(x,t) = \frac{e^{-x^2/4t}}{\sqrt{4\pi t}} \qquad (9.35)$$

The solution of (9.31) and (9.32) can now be expressed as a convolution between (9.35) and (9.32), as shown in (7.51):

$$z(x,t) = \frac{1}{\sqrt{4\pi t}} \int_{-\infty}^{\infty} z_0(x') e^{-(x-x')^2/4t} \, dx' \qquad (9.36)$$

It is now straightforward to obtain the solution of the Burgers equation in either of the forms (9.22) or (9.28). Applying the transformation (9.29) to (9.36) yields

$$\phi(x,t) = 2\ln\left[\frac{1}{\sqrt{4\pi t}} \int_{-\infty}^{\infty} z_0(x') e^{-(x-x')^2/4t} \, dx'\right] \qquad (9.37)$$

Combining (9.36) and (9.33a) produces the solution to the initial value problem for (9.28) in the form

$$\phi(x,t) = 2\ln\left\{\frac{1}{\sqrt{4\pi t}} \int_{-\infty}^{\infty} \exp\left[-\frac{(x-x')^2}{4t} + \tfrac{1}{2}\phi_0(x')\right] dx'\right\} \qquad (9.38)$$

The solution to (9.22) can now be obtained by applying the transformation (9.24) to (9.38) and using (9.33b):

$$u(x,t) = \frac{\int_{-\infty}^{\infty} [(x-x')/t] \exp\left\{-[(x-x')^2/4t] - \tfrac{1}{2}\int_{-\infty}^{x'} u_0(x'') \, dx''\right\} dx'}{\int_{-\infty}^{\infty} \exp\left\{-[(x-x')^2/4t] - \tfrac{1}{2}\int_{-\infty}^{x'} u_0(x'') \, dx''\right\} dx'} \qquad (9.39)$$

An alternative form to (9.39) can be obtained by performing an integration by parts in the numerator:

$$u(x,t) = \frac{\int_{-\infty}^{\infty} u_0(x') \exp\left\{-[(x-x')^2/4t] - \tfrac{1}{2}\int_{-\infty}^{x'} u_0(x'') \, dx''\right\} dx'}{\int_{-\infty}^{\infty} \exp\left\{-[(x-x')^2/4t] - \tfrac{1}{2}\int_{-\infty}^{x'} u_0(x'') \, dx''\right\} dx'} \qquad (9.40)$$

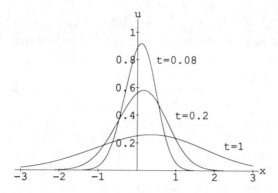

Figure 9.2 The early time development of the solution of the initial-value problem of the Burgers equation in (9.42). The initial condition is given by $u_0(x) = \delta(x)$. The increasingly asymmetric form of the profile is due to the nonlinear term in the equation. The profiles of this solution should be compared with those in Figure 2.7, where the absence of the diffusion term leads to the appearance of sharp structures in the solution.

Example 9.1. To illustrate the application of (9.39) or (9.40) to calculating solutions of the Burgers equation, as well as to show the effect of the nonlinear term, we consider the initial-value problem for which

$$u(x,0) = \delta(x) \tag{9.41}$$

where $\delta(x)$ is the Dirac delta function discussed in Problems 7.16–7.20. We will use the solution in the form shown in (9.40). To evaluate the integral in the numerator, we break up the range of integration into the two semi-infinite regions $(-\infty, -\epsilon)$ and $(-\epsilon, \infty)$. Then, performing the integration over the initial function and taking the limit $\epsilon \to 0$, we obtain, after some rearrangement, the solution to the Burgers equation with the initial condition (9.41):

$$u(x,t) = \frac{2}{[1/(\sqrt{e}-1)] + \frac{1}{2}\operatorname{erfc}(x/\sqrt{4t})} \left[\frac{1}{\sqrt{4\pi t}} e^{-x^2/4t} \right] \tag{9.42}$$

where $\operatorname{erfc}(x)$ is the complementary error function:

$$\operatorname{erfc}(x) = \frac{2}{\sqrt{\pi}} \int_x^\infty e^{-t^2}\,dt \tag{9.43}$$

The solution (9.42) is seen to be comprised of the product of the fundamental solution of the diffusion equation in (9.35) and a prefactor which represents the effect of the nonlinear term in the Burgers equation.

The early stages of the time development of the solution (9.42) are shown in Figure 9.2. For small values of the time t, the variation in the prefactor is correspondingly small, so (9.42) resembles the fundamental solution of the diffusion equation (Figure 7.5). With increasing time, however, the effect of the complementary error function becomes stronger and the profile of the solution becomes increasingly skewed, though the solution continues to spread out under the influence of the diffusion term.

The effect of the competition between the nonlinear term and the diffusion is seen by comparing Figure 9.2 with Figure 2.7. Figure 2.7 shows the time development of the equation in (2.53), which does not contain the second derivative on the right-hand side of (9.22). The nonlinearity uu_x in that equation causes the solution to 'break.' For the solution in (9.42) this breaking is suppressed by the second derivative, which acts like a diffusion term and the causes the dissipation of sharp structures. ∎

9.3 Elementary Soliton Solutions

In this section, we will begin our study of a particular type of solution to nonlinear equations called solitary waves, or solitons. The term 'soliton' was used by Zabusky and Kruskal (1965) to emphasize the localized nature of these solutions and the retention of their form after interactions with other solitons. Although there is no precise definition of a soliton, several properties are generally found for this type of solution to a nonlinear equation: (i) the form is localized in the sense that the solitary wave becomes constant (possibly zero) at infinity, so the derivatives of the solution vanish at infinity, (ii) a single soliton propagates without change in form and, most strikingly, (iii) solitons interacting with other solitons retain their original form after the interaction. In this way, solitons are said to have 'particle-like' properties.

Before developing the theory necessary to systematically construct solutions with successively larger numbers of solitons and to study their interactions, we examine 'elementary' soliton solutions of some nonlinear equations. These are solutions that can be derived from a specific form of trial solution, $\phi(x,t) = u(x - vt) \equiv u(\xi)$, which corresponds to a constant wave form moving with a velocity v. Both the function u and the velocity v are to be determined. This transformation is the usual starting point in attempting to find single-soliton solutions of nonlinear equations because the replacement $\phi(x,t) \to u(x-vt)$ has the important effect of transforming a *partial* differential equation in the two variables x and t into an *ordinary* differential equation in the single variable ξ.

We will first derive the single-soliton solutions for the sine-Gordon equation in (9.10):

$$\frac{\partial^2 \phi}{\partial x^2} - \frac{\partial^2 \phi}{\partial t^2} = \sin \phi \qquad (9.44)$$

Substituting the trial solution $\phi(x,t) = u(x-vt) \equiv u(\xi)$ into (9.44), we obtain an ordinary differential equation for u:

$$\frac{d^2u}{d\xi^2} - v^2\frac{d^2u}{d\xi^2} = (1-v^2)\frac{d^2u}{d\xi^2} = \sin u \qquad (9.45)$$

Dividing both sides of (9.45) by $1-v^2$ and then multiplying both sides by $du/d\xi$ produces

$$\frac{d^2u}{d\xi^2}\frac{du}{d\xi} = \frac{\sin u}{1-v^2}\frac{du}{d\xi} \qquad (9.46)$$

The effect of multiplying by $du/d\xi$ is to enable (9.46) to be written as a total derivative:

$$\frac{d}{d\xi}\left[\frac{1}{2}\left(\frac{du}{d\xi}\right)^2 + \frac{\cos u}{1-v^2}\right] = 0 \qquad (9.47)$$

Equation (9.47) states that the quantity enclosed in brackets is equal to a constant of integration, which we call A:

$$\frac{1}{2}\left(\frac{du}{d\xi}\right)^2 + \frac{\cos u}{1-v^2} = A \qquad (9.48)$$

Solving for the derivative $du/d\xi$, we obtain a first-order ordinary differential equation for u:

$$\frac{du}{d\xi} = \left(2A - \frac{2\cos u}{1-v^2}\right)^{1/2} \qquad (9.49)$$

We can separate the variables in (9.49) and rearrange the equation as:

$$\int_{u_0}^{u}\frac{du'}{\sqrt{B-\cos u'}} = \left(\frac{2}{1-v^2}\right)^{1/2}\int_{\xi_0}^{\xi}d\xi' \qquad (9.50)$$

In obtaining (9.50) from (9.49), we have used B to signify the quantity $B = A(1-v^2)$.

The integral in (9.50) is a function of the two parameters v, the velocity of the soliton and the constant of integration B. Depending upon the sign and magnitude of these quantities, the stable solutions are solitary waves, periodic waves or a monotonically increasing function of ξ (Scott, 1969). By choosing $B = 1$, a solitary wave solution is obtained for *any* velocity $0 < |v| < 1$. In this case, the integrand on the left-hand side of (9.50) can be written in a slightly more suggestive form by noting that $1 - \cos u = 2\sin^2\left(\frac{1}{2}u\right)$. Combined with

$$\frac{d}{du}\ln\tan\left(\tfrac{1}{4}u\right) = \frac{1}{2\sin(\tfrac{1}{2}u)} \qquad (9.51)$$

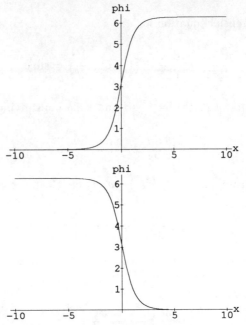

Figure 9.3 The soliton solution (9.54) (top) and the anti-soliton solution (9.55) (bottom) of the sine-Gordon equation. The value $v = \frac{1}{2}$ has been used for both solutions.

both sides of Equation (9.50) can be integrated to obtain

$$\left[\frac{2}{1-v^2}\right]^{1/2} (\xi - \xi_0) = \sqrt{2} \int_{u_0}^{u} \frac{du'}{2\sin(\frac{1}{2}u')}$$

$$= \sqrt{2} \ln \left[\frac{\tan\left(\frac{1}{4}u\right)}{\tan\left(\frac{1}{4}u_0\right)}\right] \qquad (9.52)$$

Solving for u, we obtain the solution

$$u(\xi) = 4\tan^{-1}\left\{\tan(\tfrac{1}{4}u_0)\exp\left[\frac{\xi - \xi_0}{\sqrt{1-v^2}}\right]\right\} \qquad (9.53)$$

The constants of integration u_0 and ξ_0 can be consolidated in a variety of ways. One attractive choice is to write the solution simply as

$$\phi(x,t) = 4\tan^{-1}\left\{\exp\left[\frac{x-vt}{\sqrt{1-v^2}}\right]\right\} \qquad (9.54)$$

The profile of this solution, which is sometimes called a **kink**, is shown in Figure 9.3. The other solution that can be obtained from (9.50) is

$$\phi(x,t) = 4\cot^{-1}\left\{\exp\left[\frac{x-vt}{\sqrt{1-v^2}}\right]\right\} \qquad (9.55)$$

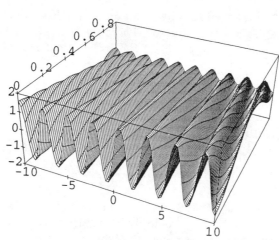

Figure 9.4 The solution (9.57) of the sine-Gordon equation shown as a surface plot to illustrate the effect of varying the integration constant B from 0 to 0.95 and the speed v set equal to 1.05.

This solution, which called an **anti-soliton**, or **anti-kink**, is also shown in Figure 9.3.

If $B \neq 1$, other types of solution of the sine-Gordon are obtained. For example, if $v > 1$ and $|B| < 1$, then the integral in (9.50) can be written as

$$\xi = \sqrt{v^2 - 1} \int_0^{k^{-1}\sin(\phi/2)} \frac{ds}{\sqrt{1-s^2}\sqrt{1-k^2 s^2}} \qquad (9.56)$$

where $k = \sqrt{(1-B)/2}$. This integral can be represented in the form of the solution ϕ as a function of ξ by introducing the Jacobi elliptic function sn: $k^{-1}\sin(\frac{1}{2}\phi) = \operatorname{sn}(\xi/\sqrt{v^2-1}, k)$, or

$$\phi(\xi) = 2\sin^{-1}[k\operatorname{sn}(\xi/\sqrt{v^2-1}, k)] \qquad (9.57)$$

The quantity k is called the **modulus** of the elliptic function. The motivation for notation of the Jacobi elliptic functions, as well as their various properties and interrelationships, is discussed by Whittaker and Watson (1963) and Abramowitz and Stegun (1965).

Mathematica can be used to plot the solution in Equation (9.57) and to investigate the effect of changing B and v by using the in-built function JacobiSN. A surface plot is shown in Figure 9.4 where B is varied from 0 to 0.95 and $v = 1.05$. The solution is seen to be periodic, with the amplitude diminishing and the period decreasing as B increases.

To find the solitary wave solutions of the Korteweg–de Vries equation we again begin with a trial solution of the form $u(x,t) = z(x-vt) \equiv z(\xi)$ and substitute into (9.15) to obtain the ordinary differential equation

$$-v\frac{dz}{d\xi} + 6z\frac{dz}{d\xi} + \frac{d^3 z}{d\xi^3} = 0 \tag{9.58}$$

Equation (9.58) is already in a form of a total derivative and so may be integrated directly:

$$-vz + 3z^2 + \frac{d^2 z}{d\xi^2} = A \tag{9.59}$$

where A is a constant of integration. To integrate this equation to obtain a first-order equation for z, we multiply by $dz/d\xi$ to make every term in the equation a derivative. Integrating, we obtain

$$-\frac{v}{2}z^2 + z^3 + \frac{1}{2}\left(\frac{dz}{d\xi}\right)^2 - Az = B \tag{9.60}$$

where B is a second constant of integration. We now require that z, $dz/d\xi$ and $d^2 z/d\xi^2$ all vanish as $x \to \pm\infty$. Using Equations (9.59) and (9.60), we deduce from these requirements that $A=0$ and $B=0$. Just as in the case of the sine-Gordon equation, more general solutions can be found for other choices of A and B, which can also be represented in terms of elliptic integrals. These solutions are discussed in detail by Drazin (1983) and by Ablowitz and Clarkson (1991). For our choice of integration constants, (9.60) can be rearranged in the form of a first-order ordinary differential equation for z:

$$\left(\frac{dz}{d\xi}\right)^2 = z^2(v - 2z) \tag{9.61}$$

The variables in (9.61) can be separated and the equation can be written in integral form as

$$\int_{z_0}^{z} \frac{dz'}{z'\sqrt{v - 2z'}} = \int_{\xi_0}^{\xi} d\xi' \tag{9.62}$$

The integration on the left-hand side of (9.62) can be carried out by using the trigonometric substitution $s = \frac{1}{2}v\,\text{sech}^2 w$. In terms of the variable w, the integrand and the element of integration become, respectively

$$v - 2z' = v[1 - \text{sech}^2 w] = v\tanh^2 w \tag{9.63}$$

and

$$dz' = -v\frac{\sinh w}{\cosh^3 w}\,dw \tag{9.64}$$

In addition, the upper and lower limits of integration z_0 and z are transformed to
$$w_0 = \text{sech}^{-1}\sqrt{2z_0/v}, \qquad w_1 = \text{sech}^{-1}\sqrt{2z/v} \tag{9.65}$$
Substituting Equations (9.63)–(9.65) into (9.62), the integration can now be done easily, and we obtain
$$\begin{aligned}\xi - \xi_0 &= -\frac{2}{\sqrt{v}}\int_{w_0}^{w_1} \frac{1}{\text{sech}^2 w \, \tanh w} \frac{\sinh w}{\cosh^3 w} dw \\ &= -\frac{2}{\sqrt{v}}\int_{w_0}^{w_1} dw \\ &= -\frac{2}{\sqrt{v}}(w_1 - w_0)\end{aligned} \tag{9.66}$$

Using (9.65) to transform back to the original function z, we obtain the solitary wave solution in the form
$$z(\xi) = \tfrac{1}{2}v\,\text{sech}^2\left[\tfrac{1}{2}\sqrt{v}(\xi - \xi_0) - w_0\right] \tag{9.67}$$

As was the case with Equation (9.52), the constants u_0 and ξ_0 can be chosen in a variety of ways. By requiring that the solution attains its maximum value at the origin at $t = 0$, we obtain the simple form
$$u(x,t) = \tfrac{1}{2}v\,\text{sech}^2\left[\tfrac{1}{2}\sqrt{v}(x - vt)\right] \tag{9.68}$$

The profile of this solution is shown in Figure 9.5.

There are two interesting features of this solution. First, since v enters the argument of the solution as a square root, this quantity must be chosen to be *positive* in order to obtain a solution that satisfies the requirement that the function itself and its first two derivatives vanish as $\xi \to \pm\infty$. In other words, the solitary-wave solutions move only *to the right*. This contrasts with the propagation of the traveling-wave solutions of the linearized Korteweg–de Vries equation, which move only *to the left* (Problem 3). The second point concerning (9.67) is that the amplitude is proportional to the speed. Thus, larger amplitude solitary waves move with a greater speed than smaller amplitude solitary waves. Both aspects of these solitary wave solutions will appear again in the next chapter when we investigate solutions of the Korteweg–de Vries equation that involve several solitons and the remnants of the traveling-wave solutions to the linearized equation.

9.4 Bäcklund Transformations

Bäcklund transformations were originally devised in the late nineteenth century for the study of mappings between surfaces. However, now these transformations provide a very rich arena for investigating properties of partial

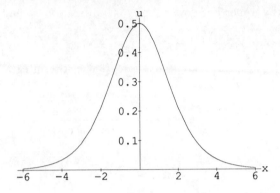

Figure 9.5 The profile at $t=0$ of the soliton solution (9.68) of the Korteweg–de Vries equation. We have set $v = 1$. Notice that in contrast to the sine-Gordon solitons in Figure 9.3, the amplitude of the Korteweg–de Vries soliton is appreciable only in a small region of space at any one time.

differential equations. In these latter applications, Bäcklund transformations may be viewed as transformations among solution surfaces. Thus, Bäcklund transformations can be used to connect solutions of a nonlinear equation either to those of another nonlinear equation whose properties are better understood, or even to those of a linear equation, for which the general solution can be found and then transformed back. The Hopf–Cole transformation of the Burgers equation into the diffusion equation discussed in Section 9.2 provides an example of this. Alternatively, the invariance of an equation under a Bäcklund transformation means that a given solution of the equation is transformed to *another* solution of the same equation. This property can be used to generate hierarchies of solutions to nonlinear equations solely by algebraic methods, so producing *nonlinear superposition principles*.

To construct a transformation between two surfaces $u(x, y)$ and $u'(x', y')$, we must identify the quantities that characterize the surface in the neighborhood of any point. We suppose that these surfaces are solutions to second-order partial differential equations. From our discussion in Chapter 3 of linear second-order equations, we know that the points of the surface and the orientation of the tangent plane at each point are sufficient to determine the solution surface. Thus, the pertinent quantities of the two surfaces are x, y, u, u_x and u_y and x', y', u', $u'_{x'}$ and $u'_{y'}$. If each quantity is regarded as being independent, five equations are required to connect them. However, these quantities are related by the total differentials

$$\mathrm{d}u = u_x \, \mathrm{d}x + u_y \, \mathrm{d}y, \qquad \mathrm{d}u' = u'_{x'} \, \mathrm{d}x' + u'_{y'} \, \mathrm{d}y' \qquad (9.69)$$

so only four independent equations are required to map between the primed and unprimed coordinate systems. We can represent these four equations symbolically as

$$F_j(x,y,u,u_x,u_y; x',y',u',u'_{x'},u'_{y'}) = 0 \qquad (9.70)$$

for $j=1,2,3$ and 4. When the quantities u and u' satisfy second-order partial differential equations, a mapping between these two differential equations is provided by the four equations in (9.70). These four equations are called a **Bäcklund transformation**. The functions F_j as written in (9.70) are the most general form of a Bäcklund transformation. In practice, simplifying assumptions are usually made for deriving these transformations for particular equations (Rogers and Shadwick, 1982; Lamb, 1980).

Although Equations (9.69) supplement the Bäcklund transformation (9.70), the construction of the four relations (9.70) does not *guarantee* that (9.69) can be integrated to obtain u as a function of x and y. To see this, we work not with the general form (9.70), but with a reduced form that is frequently used:

$$\begin{aligned}
u_x &= F_1(x',y',u',u'_{x'},u'_{y'};u) \\
u_y &= F_2(x',y',u',u'_{x'},u'_{y'};u) \\
x &= F_3(x',y',u',u'_{x'},u'_{y'};u) \\
y &= F_4(x',y',u',u'_{x'},u'_{y'};u)
\end{aligned} \qquad (9.71)$$

with similar expressions in the primed coordinate system:

$$\begin{aligned}
u'_{x'} &= F'_1(x,y,u,u_x,u_y;u') \\
u'_{y'} &= F'_2(x,y,u,u_x,u_y;u') \\
x' &= F'_3(x,y,u,u_x,u_y;u') \\
y' &= F'_4(x,y,u,u_x,u_y;u')
\end{aligned} \qquad (9.72)$$

These equations yield the following expressions for du and du':

$$du = u_x\,dx + u_y\,dy = F_1\,dx + F_2\,dy \qquad (9.73a)$$

$$du' = u'_{x'}\,dx' + u'_{y'}\,dy' = F'_1\,dx' + F'_2\,dy' \qquad (9.73b)$$

We require that these differential equations be integrable to obtain $u(x,y)$ and $u'(x',y')$. In other words, Equation (9.73a) associates with every point (x,y,u) the orientation of a plane passing through that point, as discussed in Chapter 3. The equation of this plane can be written as

$$(x-\bar{x})F_1 + (y-\bar{y})F_2 - (u-\bar{u}) = 0 \qquad (9.74)$$

This equation gives the coordinates of points $(\bar{x}, \bar{y}, \bar{u})$ that lie on the plane passing through (x, y, u) with orientations $F_1 = u_x$ and $F_2 = u_y$ with respect to the x and y directions, respectively. An analogous equation holds in the primed coordinate system. The integrability of (9.73) thus rests with determining the surface $u(x, y)$ whose tangent planes are given by (9.74). A necessary and sufficient condition that (9.73) be integrable, i.e. that the two equations (9.73a, b) are **exact differentials**, is the equality of mixed second partial derivatives: $\partial F_1/\partial y = \partial F_2/\partial x$ and $\partial F_1'/\partial y' = \partial F_2'/\partial x'$, or

$$\frac{\partial^2 u}{\partial x \partial y} = \frac{\partial^2 u}{\partial y \partial x}, \quad \frac{\partial^2 u'}{\partial x' \partial y'} = \frac{\partial^2 u'}{\partial y' \partial x'} \qquad (9.75)$$

These are the **integrability conditions** that the Bäcklund transformation must satisfy.

Example 9.2. As an example of a Bäcklund transformation we consider the Hopf–Cole transformation of the Burgers equation derived in Section 9.2. Given a solution $z(x, t)$ of the diffusion equation a solution $u(x, t)$ of the Burgers equation can be constructed using (9.34). This equation can be written as one component of a Bäcklund transformation as

$$z_x = -\tfrac{1}{2} u z \qquad (9.76)$$

Alternatively, given a solution of the Burgers equation, a solution of the diffusion equation can be obtained by integrating (9.76):

$$z(x, t) = \exp\left[-\frac{1}{2} \int^x u(s, t)\, ds\right] \qquad (9.77)$$

This can be expressed in the form of the transformations in (9.71) and (9.72) by differentiating both sides with respect to t and using the fact that u is a solution of the Burgers equation. Carrying out these steps yields

$$z_t = \tfrac{1}{4}(u^2 - 2u_x) z \qquad (9.78)$$

Equations (9.76) and (9.78), together with $x = x'$ and $t = t'$, which state that the independent variables are left invariant by the Hopf–Cole transformation, constitute the Bäcklund transformation relating solutions of the Burgers equation to those of the diffusion equation. The integrability conditions applied to (9.76) and (9.78) show explicitly that $u(x, t)$ is a solution of the Burgers equation. Differentiating (9.76) with respect to x and using both (9.76) and (9.78) shows that z solves the diffusion equation (Problem 15). ∎

An important type of Bäcklund transformation is one that leaves an equation invariant. This is called an **auto-Bäcklund transformation**

Introduction to Nonlinear Partial Differential Equations 375

of that equation. The invariance does not imply that each solution of the equation is transformed into itself, only that the two solution surfaces connected by the Bäcklund transformation each satisfy the same equation. The following is an example of an auto-Bäcklund transformation.

Example 9.3. Consider the following Bäcklund transformation:

$$u'_{x'} = u_x - 2\beta \sin\left[\tfrac{1}{2}(u+u')\right] \qquad (9.79a)$$

$$u'_{y'} = -u_y + 2\beta^{-1} \sin\left[\tfrac{1}{2}(u-u')\right] \qquad (9.79b)$$

$$x' = x \qquad (9.79c)$$

$$y' = y \qquad (9.79d)$$

Notice that only the functions u and u' and their first derivatives appear in the transformation. Furthermore, Equations (9.79c) and (9.79d) mean that we can drop the distinction between primed and unprimed independent variables, just as we did in Example 9.2. The quantity β is a real constant that is called the **Bäcklund parameter** and represents a degree of freedom in the transformation, i.e. a degree of freedom in the solution surfaces that are transformed among one another.

To determine the differential equations implied by this Bäcklund transformation, we apply the integrability condition (9.75). The required second partial derivatives are first calculated by differentiating (9.79a) with respect to y and differentiating (9.79b) with respect to x:

$$u'_{yx} = u_{yx} - \beta(u_y + u'_y)\cos\left[\tfrac{1}{2}(u+u')\right]$$

$$= u_{yx} - 2\sin\left[\tfrac{1}{2}(u-u')\right]\cos\left[\tfrac{1}{2}(u+u')\right] \qquad (9.80a)$$

$$u'_{xy} = -u_{xy} + 2\beta^{-1}(u_x - u'_x)\cos\left[\tfrac{1}{2}(u-u')\right]$$

$$= -u_{xy} + 2\sin\left[\tfrac{1}{2}(u+u')\right]\cos\left[\tfrac{1}{2}(u-u')\right] \qquad (9.80b)$$

In obtaining these equations, we have used (9.79a) to eliminate $\beta(u_y + u'_y)$ in (9.80a) and (9.79b) to eliminate $2\beta^{-1}(u_x - u'_x)$ in (9.80b). The integrability conditions are now applied by equating the mixed partial derivatives in (9.80). We then obtain the differential equations that are connected by the transformation (9.79) by adding and subtracting the two equations in (9.80) and using the trigonometric formulae $\sin(x \pm y) = \sin x \cos y \pm \cos x \sin y$. Adding (9.80a) to (9.80b) yields the differential equation in the primed coordinate system:

$$u'_{xy} = \sin\left[\tfrac{1}{2}(u+u')\right]\cos\left[\tfrac{1}{2}(u-u')\right]$$

$$\quad - \sin\left[\tfrac{1}{2}(u-u')\right]\cos\left[\tfrac{1}{2}(u+u')\right]$$

$$= \sin u' \qquad (9.81)$$

which is the sine-Gordon equation in the form shown in Equation (9.11). Subtracting (9.80b) from (9.80a) to obtain the equation in the unprimed coordinate system, we find

$$u_{xy} = \sin\left[\tfrac{1}{2}(u+u')\right]\cos\left[\tfrac{1}{2}(u-u')\right]$$
$$+ \sin\left[\tfrac{1}{2}(u-u')\right]\cos\left[\tfrac{1}{2}(u+u')\right]$$
$$= \sin u \qquad (9.82)$$

which is again the sine-Gordon equation in Equation (9.11)! Thus, we have shown that the transformation (9.79) is an auto-Bäcklund transformation of the sine-Gordon equation. ∎

The auto-Bäcklund transformation can be used to construct a sequence of solutions of the sine-Gordon equation beginning with any given solution. For example, the single-soliton and anti-soliton solutions in Equations (9.54) and (9.55) can be generated from the trivial solution (Problem 17). The procedure described in Problem 17 provides a way of generating solutions of the sine-Gordon equation by iterating the Bäcklund transformation (9.79). Beginning with a solution obtained after n iterations of this procedure, which we denote by ϕ, the solution obtained upon an additional iteration, denoted by ψ, is determined by solving (9.79a) and (9.79b). We make the identifications $u = \phi$ and $u' = \psi$ to obtain

$$\psi_x = \phi_x - 2\beta\sin[\tfrac{1}{2}(\phi+\psi)]$$
$$= \phi_x - 2\beta\sin(\tfrac{1}{2}\phi)\cos(\tfrac{1}{2}\psi) - 2\beta\cos(\tfrac{1}{2}\phi)\sin(\tfrac{1}{2}\psi) \qquad (9.83a)$$
$$\psi_y = -\phi_y + 2\beta^{-1}\sin[\tfrac{1}{2}(\phi-\psi)]$$
$$= -\phi_y + 2\beta^{-1}\sin(\tfrac{1}{2}\phi)\cos(\tfrac{1}{2}\psi) - 2\beta^{-1}\cos(\tfrac{1}{2}\phi)\sin(\tfrac{1}{2}\psi) \qquad (9.83b)$$

By multiplying (9.83a) by β^{-1}, (9.83b) by β and subtracting the two equations, we obtain

$$\beta^{-1}\phi_x - \beta\phi_y = \beta^{-1}\psi_x + \beta\psi_y + 4\sin(\tfrac{1}{2}\phi)\cos(\tfrac{1}{2}\psi) \qquad (9.84)$$

Since ϕ is a known function, this is a quasi-linear first-order equation for ψ whose general solution can be obtained by the method of characteristics in terms of an arbitrary function, as described in Chapter 2. This arbitrary function can be determined to within an additive constant by substituting into either (9.83a) or (9.83b), as shown in Problem 17.

While this procedure does generate a hierarchy of solutions of the sine-Gordon equation, it also involves an integration at each stage that

Introduction to Nonlinear Partial Differential Equations 377

becomes increasingly cumbersome as we iterate and the algebraic expressions for the solutions become increasingly complex. There is an alternative method to generating solutions that involves only *algebraic* manipulations—in effect, a nonlinear superposition principle. This will be described in the next section.

9.5 Nonlinear Superposition Principles

Nonlinear superposition principles have been derived for a number of equations for which auto-Bäcklund transformations are known. In this section, the nonlinear superposition principle for the Korteweg–de Vries equation will be derived and its use will then be illustrated by generating a hierarchy of soliton solutions. In the problems at the end of the chapter the corresponding steps are carried out for the sine-Gordon equation.

The auto-Bäcklund transformation for the Korteweg–de Vries equation is formulated not in terms of the function v that satisfies the Korteweg–de Vries equation,

$$v_t + 6vv_x + v_{xxx} = 0 \tag{9.85}$$

but in terms of the function u given by $u_x = v$. Upon substitution into (9.85), the equation for u is found to be

$$u_{tx} + 6u_x u_{xx} + u_{xxxx} = \frac{\partial}{\partial x}\left[u_t + 3u_x^2 + u_{xxx}\right] = 0 \tag{9.86}$$

Integrating this equation, we obtain

$$u_t + 3u_x^2 + u_{xxx} = f(t) \tag{9.87}$$

where $f(t)$ is a function of integration. The change of variable

$$\bar{u} = u - \int^t f(s)\,ds \tag{9.88}$$

eliminates f from (9.87). Moreover, since solutions v of (9.85) will be obtained from solutions u of (9.87) by a derivative of u with respect to x, and since $\bar{u}_x = u_x$, we need not make a distinction between u and $\bar{u}$. For our purposes, therefore, we can set $f = 0$ in (9.87) and (9.88) without loss of generality.

With the Korteweg–de Vries equation written as

$$u_t + 3u_x^2 + u_{xxx} = 0 \tag{9.89}$$

the auto-Bäcklund transformation is given by the set of equations

$$u'_{x'} = -u_x + \beta - \tfrac{1}{2}(u - u')^2 \tag{9.90a}$$

$$u'_{t'} = -u_t + (u - u')(u_{xx} - u'_{x'x'})$$
$$- 2(u_x^2 + u_x u'_{x'} + u'^2_{x'}) \tag{9.90b}$$

$$x' = x \tag{9.90c}$$

$$t' = t \tag{9.90d}$$

Notice the appearance of the second-order derivatives in (9.90b), which results from the Korteweg–de Vries equation being of third order. The explicit demonstration that (9.89) is invariant under the transformation (9.90) is left to the problems at the end of the chapter.

We can generate a nontrivial solution of the Korteweg–de Vries equation by applying the transformation (9.90) to the trivial solution. With $u' = 0$, Equations (9.90a) and (9.90b) become differential equations for u:

$$u_x = \beta - \tfrac{1}{2}u^2 \tag{9.91a}$$

$$u_t = u u_{xx} - 2u_x^2 \tag{9.91b}$$

Equation (9.91a) may be readily integrated and the solution expressed in terms of an arbitrary function of t which can be determined by substituting into (9.91b) (Problems 22 and 23). We find that there are two types of solutions, a regular solution

$$u(x,t) = \sqrt{2\beta}\, \tanh\left[\sqrt{\beta/2}(x - 2\beta t)\right] \tag{9.92}$$

and an irregular solution,

$$\bar{u}(x,t) = \sqrt{2\beta}\, \coth\left[\sqrt{\beta/2}(x - 2\beta t)\right] \tag{9.93}$$

which has a singularity where the argument vanishes. Both in Equation (9.93) and below, we will signify irregular solutions by an overbar. The corresponding solutions of the Korteweg–de Vries equation (9.85) are obtained by differentiating these expressions with respect to x, whereupon we obtain

$$v(x,t) = u_x(x,t) = \beta\, \text{sech}^2\left[\sqrt{\beta/2}(x - 2\beta t)\right] \tag{9.94a}$$

$$\bar{v}(x,t) = \bar{u}_x(x,t) = -\beta\, \text{cosech}^2\left[\sqrt{\beta/2}(x - 2\beta t)\right] \tag{9.94b}$$

The solution (9.94a) is seen to be the single-soliton solution (9.67) with β corresponding to the speed of the soliton. We must therefore restrict β to have only positive values in order to associate (9.94) with real solutions of the Korteweg–de Vries equation. Although the irregular solution (9.94b) is not an acceptable physical solution by itself, it will be required to construct regular higher-order solutions using the nonlinear superposition principle that will be derived from (9.90).

Just as we found for the sine-Gordon equation, the procedure used to obtain (9.94) can be iterated and a sequence of solutions can be generated. At each step, an integration of (9.90) must be carried out, which also becomes increasingly cumbersome as successively more complex solutions are generated. What we seek is an alternative approach whereby new solutions can be generated simply by performing algebraic operations on known solutions. Such a procedure would play the role for a particular nonlinear equation that simple linear superposition does for linear equations. This nonlinear superposition is also expected to be less general than the superposition principle for linear equations in that it must be formulated for each nonlinear equation separately—if it exists at all!

We begin by introducing some notation. Let ϕ be any solution of (9.89) and let ϕ_β be a solution obtained from ϕ by applying the Bäcklund transformation with Bäcklund parameter β. For example, if $\phi = 0$, then ϕ_β represents either of the solutions (9.92) or (9.93). The derivation of a nonlinear superposition principle from (9.90) requires only (9.90a); however, if u and u' in (9.90) are solutions of (9.89), then (9.90a) implies (9.90b) and vice versa.

Consider the two solutions obtained by applying the Bäcklund transformation to ϕ with parameters β_1 and β_2:

$$\phi_{\beta_1,x} = -\phi_x + \beta_1 - \tfrac{1}{2}(\phi - \phi_{\beta_1})^2 \qquad (9.95a)$$

$$\phi_{\beta_2,x} = -\phi_x + \beta_2 - \tfrac{1}{2}(\phi - \phi_{\beta_2})^2 \qquad (9.95b)$$

Let $\phi_{\beta_2\beta_1}$ be the solution obtained from ϕ by the successive application of (9.90), first with parameter β_1 and then with parameter β_2:

$$\phi_{\beta_2\beta_1,x} = -\phi_{\beta_1,x} + \beta_2 - \tfrac{1}{2}(\phi_{\beta_1} - \phi_{\beta_2\beta_1})^2 \qquad (9.96)$$

Similarly, let $\phi_{\beta_1\beta_2}$ be the solution obtained from ϕ by the successive application of (9.90), first with parameter β_2 and then with parameter β_1:

$$\phi_{\beta_1\beta_2,x} = -\phi_{\beta_2,x} + \beta_1 - \tfrac{1}{2}(\phi_{\beta_2} - \phi_{\beta_1\beta_2})^2 \qquad (9.97)$$

We now impose the requirement that

$$\phi_{\beta_2\beta_1} = \phi_{\beta_1\beta_2} = \psi \qquad (9.98)$$

and solve (9.96) and (9.97) for ψ. Subtracting (9.95b) from (9.95a), we obtain

$$\phi_{\beta_1,x} - \phi_{\beta_2,x} = \beta_1 - \beta_2 - \tfrac{1}{2}(\phi_{\beta_1} - \phi_{\beta_2})(\phi_{\beta_1} + \phi_{\beta_2} - 2\phi) \qquad (9.99)$$

and subtracting (9.97) from (9.96) produces, after some rearrangement, the expression

$$\phi_{\beta_1,x} - \phi_{\beta_2,x} = \beta_2 - \beta_1 - \tfrac{1}{2}\left(\phi_{\beta_1} - \phi_{\beta_2}\right)\left(\phi_{\beta_1} + \phi_{\beta_2} - 2\psi\right) \qquad (9.100)$$

Eliminating the quantity $\phi_{\beta_1,x} - \phi_{\beta_2,x}$ from (9.99) and (9.100), and solving for ψ produces an expression for ψ:

$$\psi = \phi + 2\frac{\beta_1 - \beta_2}{\phi_{\beta_1} - \phi_{\beta_2}} \qquad (9.101)$$

It remains only to show that (9.101) is indeed a solution of (9.89). This is a straightforward though somewhat lengthy calculation, which can be simplified considerably by using *Mathematica*. We will present only the results of the calculations without reproducing the actual *Mathematica* session. Performing the required differentiations of (9.101) and substituting into (9.89), we obtain

$$\psi_t + 3\psi_x^2 + \psi_{xxx} = \phi_t + \phi_{xxx}$$

$$- 2(\beta_1 - \beta_2)\frac{\phi_{\beta_1,t} - \phi_{\beta_2,t}}{(\phi_{\beta_1} - \phi_{\beta_2})^2} + 3\left[\phi_x - 2(\beta_1 - \beta_2)\frac{\phi_{\beta_1,x} - \phi_{\beta_2,x}}{(\phi_{\beta_1} - \phi_{\beta_2})^2}\right]^2$$

$$- 2(\beta_1 - \beta_2)\frac{\phi_{\beta_1,xxx} - \phi_{\beta_2,xxx}}{(\phi_{\beta_1} - \phi_{\beta_2})^2} - 12(\beta_1 - \beta_2)\frac{(\phi_{\beta_1,x} - \phi_{\beta_2,x})^3}{(\phi_{\beta_1} - \phi_{\beta_2})^4}$$

$$+ 12(\beta_1 - \beta_2)\frac{(\phi_{\beta_1,x} - \phi_{\beta_2,x})(\phi_{\beta_1,xx} - \phi_{\beta_2,xx})}{(\phi_{\beta_1} - \phi_{\beta_2})^3} \qquad (9.102)$$

Differentiating (9.99) with respect to x and substituting the resulting expression into the last term on the right-hand side of (9.102), and using the fact that ϕ, ϕ_{β_1} and ϕ_{β_2} are solutions of (9.89) yields

$$\psi_t + 3\psi_x^2 + \psi_{xxx}$$

$$= 12(\beta_1 - \beta_2)^2 \frac{(\phi_{\beta_1,x} - \phi_{\beta_2,x})^2}{(\phi_{\beta_1} - \phi_{\beta_2})^4} - 12(\beta_1 - \beta_2)\frac{(\phi_{\beta_1,x} - \phi_{\beta_2,x})^3}{(\phi_{\beta_1} - \phi_{\beta_2})^4}$$

$$+ 6(\beta_1 - \beta_2)\frac{(\phi_{\beta_1,x} - \phi_{\beta_2,x})^2(\phi_{\beta_1} + \phi_{\beta_2} - 2\phi)}{(\phi_{\beta_1} - \phi_{\beta_2})^3} \qquad (9.103)$$

> Using (9.99) to substitute for the factor $\phi_{\beta_1}+\phi_{\beta_2}-2\phi$ in (9.103) leads to a cancellation of all the terms on the right-hand side, which shows that ψ is indeed a solution of (9.89).

Having shown that the right-hand side of (9.101) is a solution of (9.89), we can state our result as follows. Given a solution ϕ of (9.89), from which are obtained the two solutions ϕ_{β_1} and ϕ_{β_2} by the application of the Bäcklund transformation in (9.90), the quantity ψ in (9.101) is also a solution. According to the transformation between (9.85) and (9.89), we can then obtain from (9.101) the corresponding solution of the Korteweg–de Vries equation by differentiating ψ with respect to x. This is the nonlinear superposition principle for the Korteweg–de Vries equation.

Example 9.4. Since we have already obtained solutions of the Korteweg–de Vries equation in (9.92) and (9.93), we can use (9.101) to construct another solution. In particular, if we set $\phi=0$ in (9.101), then the ϕ_{β_i} correspond to either of the two solutions in (9.92) and (9.93) obtained by solving (9.91). A glance at the form of (9.101) reveals, however, that a regular solution ψ cannot be obtained from two regular solutions u_{β_i}. This can be seen by observing that the limit of (9.92) as the argument approaches $\pm\infty$ is $\pm(2\beta)^{1/2}$, and since (9.92) is a continuous function, there will always be one value of the argument for which the two solutions in (9.101) are equal, leading to a singularity in ψ. The only way to generate a regular solution ψ from solutions of (9.91) is to use both the regular solution (9.92), u_{β_1}, and the irregular solution (9.93), $\bar{u}_{\beta_2}$ with the Bäcklund parameters chosen such that $\beta_2 > \beta_1$. This insures that the denominator in (9.101) does not vanish for any value of the arguments (Problem 30). In this case, (9.101) yields

$$\psi(x,t) = \frac{2(\beta_1 - \beta_2)}{u_{\beta_1}(x,t) - \bar{u}_{\beta_2}(x,t)}, \qquad \beta_2 > \beta_1 \qquad (9.104)$$

> There are several ways that *Mathematica* can be used to plot soliton solutions. The surface plot in Figure 9.6 represents the entire solution surface, but leaves some interesting aspects of the soliton interaction obscured. The contour plot in the x-t plane in Figure 9.6 shows clearly the shift of the soliton trajectories as a result of the interaction. This phase shift is the only memory of the interaction, since the forms of the solitons are unchanged.
>
> The retention of the wave form of the solitons after interactions with other solitons can be seen most clearly in Figure 9.7. `Plot` has been used to generate a time sequence of (9.104) before, during, and after the interaction of the solitons. These snapshots show how the wave forms are deformed as the two solitons approach and interact, but revert to their original forms after the interaction.

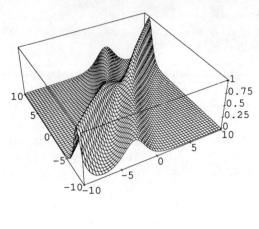

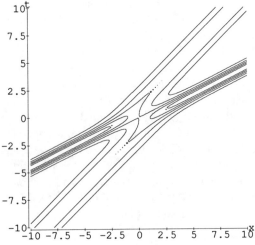

Figure 9.6 (Top) Solution surface of the Korteweg–de Vries equation describing the interaction of two solitons. The amplitudes and speeds of the two solitons are determined by the Bäcklund parameters in (9.101). The values chosen for this solution are $\beta_1 = 0.5$ and $\beta_2 = 1.0$. (Bottom) Contour plot of the same solution as shown at the top. Notice the shift in the trajectories of the two solitons after the interaction.

The animation of the sequence of frames shown in Figure 9.7, which is readily achieved in *Mathematica*, provides yet another way to visualize the soliton solution. Animation, of course, emphasizes the dynamical nature of the solution and provides the most vivid and natural illustration of the interactions among the solitons. Thus, `Plot3D` (Figure 9.6), `ContourPlot` (Figure 9.6), `Plot` (Figure 9.7) and the animation each highlight a different aspect of the soliton solution.

Introduction to Nonlinear Partial Differential Equations   383

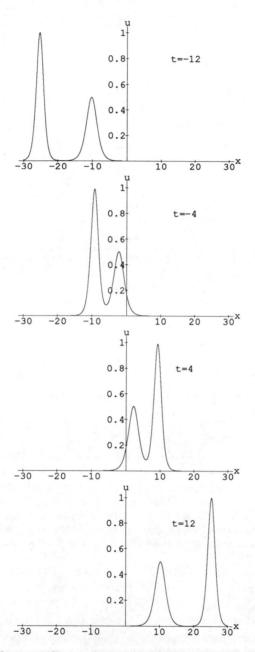

Figure 9.7 The two-soliton solution of the Korteweg–de Vries equation in Figure 9.6 shown at individual times. The time is $t=-12$ in the top plot and increases by 8 units in each successive plot downward. This representation of the solution clearly demonstrates that the profile of each soliton is regained after the interaction.

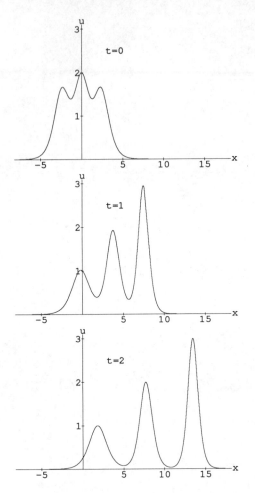

Figure 9.8 The early stages of the time development of the three-soliton solution of the Korteweg–de Vries equation generated from (9.108) with the parameters $\beta = 1$, $\beta_1 = 2$ and $\beta_2 = 3$. The top plot corresponds to $t = 0$ and the time step increases by one unit in each successive panel downward. The emergence of three solitons in the lower two plots can be clearly seen.

The procedure described in Example 9.4 can be iterated to produce a sequence of solutions of the Korteweg–de Vries equation without additional integration. However, we must again take note of the fact that we cannot construct a regular solution ψ from (9.101) if both ϕ_{β_1} and ϕ_{β_2} are themselves regular. We can again choose the limits of the allowable values of the various Bäcklund parameters by examining the asymptotic behavior of the respective solutions. We obtain the required solutions as

$$\bar{\phi}_{\beta_2 \beta}(x,t) = \frac{2(\beta_2 - \beta)}{u_{\beta_2}(x,t) - u_\beta(x,t)} \tag{9.105}$$

$$\phi_{\beta_1\beta}(x,t) = \frac{2(\beta - \beta_1)}{u_\beta(x,t) - \bar{u}_{\beta_1}(x,t)} \tag{9.106}$$

For (9.106) to be regular, the Bäcklund parameters must satisfy the inequality

$$\beta_1 > \beta \tag{9.107}$$

Similarly, for the solution constructed from (9.105) and (9.106) according to (9.101) to be regular, the respective Bäcklund parameters must satisfy the analogous inequality:

$$\beta_2 - \beta > \beta_1 - \beta \tag{9.108}$$

Combining (9.107) and (9.108) shows that a regular three-soliton solution of the Korteweg–de Vries equation must have Bäcklund parameters that satisfy the inequality

$$\beta_2 > \bar{\beta}_1 > \beta \tag{9.109}$$

where the bar has been placed over the Bäcklund parameter corresponding to the irregular solution.

The early stages of evolution of the regular solution

$$\psi(x,t) = u_\beta(x,t) + 2\frac{\beta_2 - \beta_1}{\phi_{\beta_1\beta}(x,t) - \bar{\phi}_{\beta_2\beta}(x,t)} \tag{9.110}$$

are plotted in Figure 9.8 for the parameters $\beta = 1$, $\beta_1 = 2$ and $\beta_2 = 3$. The emergence of three solitons in this solution is evident. Higher-order solutions can be constructed using this procedure, with each iteration producing an additional soliton in the asymptotic solution.

Further Reading

There are several books that provide general introductions to nonlinear equations. A comprehensive discussion of many methods may be found in the two volumes by Ames (1965, 1972). The book by Ablowitz and Segur (1981) includes a chapter on nonlinear equations used to describe physical phenomena. Hasegawa (1990) introduces solitons as optical pulses in dielectric fibers and includes discussions of the physical principles involved in setting up the equations. Drazin (1983), Drazin and Johnson (1989), Ablowitz and Segur (1981) and Ablowitz and Clarkson (1991) provide excellent introductions to solitons and other solutions of nonlinear equations. A detailed account of the Burgers equation and some of the more intriguing features of its solutions is given by Burgers (1974). Whitham (1974) addresses many aspects of nonlinear wave motion, as well as providing complementary discussions on several of the topics included in this chapter. A compilation of solutions to the Burgers equation is provided by Benton and Platzmann (1972). Rogers and Chadwick (1982) give many examples of Bäcklund transformations, and cite abundant references on the applications to various nonlinear equations.

References

Ablowitz M. J. and Clarkson P. A. (1991). *Solitons, Nonlinear Evolution Equations and Inverse Scattering.* Cambridge: Cambridge University Press.

Ablowitz M. J. and Segur H. (1981). *Solitons and the Inverse Scattering Transform.* Philadelphia: Society for Industrial and Applied Mathematics.

Abramowitz M. and Stegun I. (1965). *Handbook of Mathematical Functions.* New York: Dover.

Ames W. F. (1965). *Nonlinear Partial Differential Equations in Engineering* Vol. 1. New York: Academic.

Ames W. F. (1972). *Nonlinear Partial Differential Equations in Engineering* Vol. 2. New York: Academic.

Benton E. R. and Platzmann G. W. (1972). A Table of Solutions of the One-Dimensional Burgers Equation. *Quart. Appl. Math.* **30**, 195–212.

Burgers J. M. (1974). *The Nonlinear Diffusion Equation.* Dordrecht: Reidel.

Cole J. D. (1951). On a Quasi-Linear Parabolic Equation Occurring in Aerodynamics. *Quart. Appl. Math.* **9**, 225–36.

Crandall R. E. (1991). *Mathematica for the Sciences.* Redwood City, CA: Addison-Wesley.

Drazin P. G. (1983). *Solitons.* Cambridge University Press.

Drazin P. G. and Johnson R. (1989). *Solitons: An Introduction.* Cambridge University Press.

Enz U. (1963). Discrete Mass, Elementary Length, and a Topological Invariant as a Consequence of a Relativistic Invariant Variational Principle. *Phys. Rev.* **131**, 1392–94.

Hasegawa A. (1990). *Optical Solitons in Fibers* 2nd edn. Berlin: Springer.

Hopf E. (1950). The Partial Differential Equation $u_t+uu_x=\mu u_{xx}$. *Commun. Pure Appl. Math.* **3**, 201–30.

Josephson B. D. (1965). Supercurrents through Barriers. *Adv. Phys.* **14**, 419–51.

Lamb G. L. Jr. (1971). Analytical Descriptions of Ultrashort Optical Pulse Propagation in a Resonant Medium. *Rev. Mod. Phys.* **49**, 99–124.

Lamb G. L. Jr. (1980). *Elements of Soliton Theory.* New York: Wiley.

Rogers C. and Shadwick W.F. (1982). *Bäcklund Transformations and Their Applications.* New York: Academic.

Scott A. C. (1969). A Nonlinear Klein-Gordon Equation. *Am. J. Phys.* **37**, 52–61.
Skyrme T. H. R. (1958). A Nonlinear Theory of Strong Interactions. *Proc. Roy. Soc. London.* **A247**, 260–78.
Skyrme T. H. R. (1961). Particle States of a Quantized Meson Field. *Proc. Roy. Soc. London.* **A262**, 237–45.
Whitham G. B. (1974). *Linear and Nonlinear Waves.* New York: Wiley.
Whittaker E. T. and Watson G. N. (1963). *A Course of Modern Analysis* 4th edn. Cambridge: Cambridge University Press.
Zabusky N. J. and Kruskal M. D. (1965). Interactions of "Solitons" in a Collisionless Plasma and the Recurrence of Initial States. *Phys. Rev. Lett.* **15**, 240–3.
Zakharov V. E. (1974). On Stochastization of One-Dimensional Chains of Nonlinear Oscillators. *Sov. Phys. JETP* **38**, 108–10.

Problems

1. Consider the problem of heat conduction within a volume $\mathcal{V}$ bounded by a surface $\mathcal{S}$. The temperature distribution is given by $T(\mathbf{x},t)$ and the heat flux vector is $\mathbf{q}(\mathbf{x},t)$. The heat energy contained within $\mathcal{V}$ is due to the internal energy of the material within $\mathcal{V}$, which is assumed to have density ρ and specific heat C:

$$\int_{\mathcal{V}} C(\mathbf{x})\rho(\mathbf{x})T(\mathbf{x},t)\,d\mathcal{V}$$

where it is assumed that both C and ρ may depend upon the position but not the temperature. The rate of change of the internal energy is due to the gain or loss of heat because of the flux across $\mathcal{S}$:

$$\int_{\mathcal{S}} \mathbf{q}\cdot d\mathbf{S}$$

where the sign is *positive* for heat *loss*, i.e. the vector $\mathbf{q}$ points along the outward normal of $\mathcal{S}$. If there are no sources or sinks of heat within $\mathcal{V}$, such as chemical reactions, which can generate or consume heat, the rate of change of internal energy must equal the negative of the flux across the bounding surface. Apply the divergence theorem to obtain this conservation law in differential form:

$$C\rho\frac{\partial T}{\partial t} + \nabla\cdot\mathbf{q} = 0$$

Although this equation is linear in both T and $\mathbf{q}$, a *constitutive equation* is still required to establish a relationship between T and $\mathbf{q}$ in order to obtain a closed equation for T. The simplest form of constitutive equation is to assume a linear relationship between $\mathbf{q}$ and T which, in the context of heat conduction, is known as **Fourier's law**:

$$\mathbf{q} = -K\nabla T$$

where K is the thermal conductivity. This states that the flow of heat is in the direction of decreasing temperature. If there is a strictly linear relation between T and $\mathbf{q}$, then K must be a constant, independent of T. However, in general, K depends upon both the position and the temperature: $K=K(\mathbf{x},T)$. Substituting this equation into the conservation equation, obtain an equation involving T alone:

$$C\rho\frac{\partial T}{\partial t} = \nabla\cdot[K\nabla T]$$
$$= \nabla K\cdot\nabla T + K\nabla^2 T$$

If K depends upon the temperature, then the equation of motion obtained for T is seen to be nonlinear. On the other hand, if Fourier's law is obeyed strictly and K is a constant, independent of the position and the temperature, show that reduces to the familiar diffusion equation

$$\frac{\partial T}{\partial t} = \mathcal{K}\nabla^2 T$$

where $\mathcal{K} = K/C\rho$ is the thermal diffusivity.

2. The Korteweg–de Vries equation can be derived from a one-dimensional anharmonic lattice that was studied by Fermi, Pasta and Ulam. Numerical experiments were performed on a one-dimensional chain of nonlinear oscillators to ascertain how the energy distribution is randomized in a system with many degrees of freedom. Instead of a rapid transition to a uniform energy distribution, an anomalously slow randomization was observed. This has become known as the 'Fermi-Pasta-Ulam' problem. A discussion of this problem in light of modern developments in the theory of nonlinear partial differential equations (which are the subject of Chapter 10) has been given by Zakharov (1974).

To draw a connection between an anharmonic lattice and the Korteweg–de Vries equation, consider a chain of masses m that are connected with *nonlinear* springs whose force F as a function of displacement x is given by

$$F(x) = -kx - \alpha k x^2$$

Assuming only interactions between nearest neighbors, obtain the equation of motion of the jth spring:

$$m\frac{d^2 y_j}{dt^2} = k(y_{j+1} + y_{j-1} - 2y_j) + k\alpha\left[(y_{j+1} - y_j)^2 - (y_j - y_{j-1})^2\right]$$

To take the continuum limit of this equation, let a be the nearest-neighbor spacing. Then expand $y_{j\pm 1}$ in a Taylor series about y_j and show that, up to terms of third order in a, that the equation of motion is given by

$$\omega_0^{-2} y_{tt} = y_{xx} + \epsilon y_x y_{xx} + \frac{h^2}{12} y_{xxxx}$$

where $\epsilon = 2\alpha a$. By looking for solutions of the form

$$y(x,t) = \psi(\xi,\tau), \qquad \xi = x - t, \qquad \tau = \tfrac{1}{2}\epsilon t$$

and defining $u = \psi_x$, show that the equation for u takes the form of a Korteweg–de Vries equation:

$$u_\tau + u u_\xi + \delta^2 u_{\xi\xi\xi} = 0$$

where $\delta^2 = a^2/12\epsilon$.

3. Beginning with a function $u(x,t)$ that satisfies the wave equation

$$\frac{\partial^2 u}{\partial x^2} - \frac{\partial^2 u}{\partial t^2} = 0$$

perform Fourier transforms in space and time

$$u(k,\omega) = \int_{-\infty}^{\infty}\int_{-\infty}^{\infty} u(x,t)\, e^{ikx-\omega t}\, dx\, dt$$

to obtain the dispersion relation $\omega^2 = k^2$, or $\omega = \pm k$. Thus, deduce that waves travel both to the left and right.

While the Fourier transform cannot be applied to the Korteweg–de Vries equation, it can be applied to the *linearized* Korteweg–de Vries equation. If the amplitude is assumed to be sufficiently small to justify the retention only of the linear terms this equation is:

$$\frac{\partial u}{\partial t} + \frac{\partial^3 u}{\partial x^3} = 0$$

Obtain the dispersion relation for this equation. Can waves travel in both directions?

4. Using the dispersion relation obtained in Problem 3, determine the solution to the initial-value problem for the linearized Korteweg–de Vries equation in the form

$$u(x,t) = \frac{1}{2\pi} \int_{-\infty}^{\infty} u_0(k) \exp\left[i(kx + k^3 t)\right] dk$$

where the $u_0(k)$ are the Fourier components of the value of u at $t=0$:

$$u(x,0) = u_0(x) = \frac{1}{2\pi} \int_{-\infty}^{\infty} u_0(k) e^{ikx} \, dk$$

Specialize this result to the case where the initial condition is a delta function

$$u_0(x) = \delta(x)$$

and express your solution in the form of an Airy function:

$$Ai(x) = \frac{1}{\sqrt{\pi}} \int_0^{\infty} \cos\left(\tfrac{1}{3} k^3 + kx\right) dx$$

5. Use the function AiryAi of *Mathematica* to plot the evolution of the solution obtained in Problem 4.

Another way to see how solutions of the linearized Korteweg–de Vries equation evolve is to integrate the equation directly using, for example, the integrator given by Crandall (1991). This permits a wider variety of initial conditions to be examined than if we restrict ourselves to analytic constructions. In Figure 9.9 is shown the evolution from the initial condition

$$u_0(x) = \exp(-x^2)$$

Notice how the solution changes into a decaying wave train moving *to the left*. This feature, which is a direct result of the dispersion relation obtained in Problem 3, will surface again in the next chapter when we discuss the solution of the Korteweg–de Vries equation, including the nonlinear term, by the method of inverse scattering.

Introduction to Nonlinear Partial Differential Equations 391

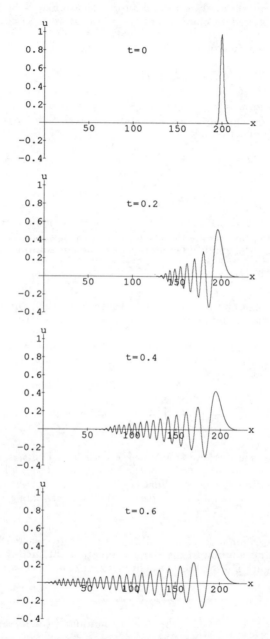

Figure 9.9 The development of the solution of the linearized Korteweg–de Vries equation from the initial condition $u(x,0) = \exp(-x^2)$ obtained by numerical integration. Note the decaying wave train moving to the left. This results from the dispersion relation found in Problem 3 showing that the wave solutions of this equation travel only from right to left. The time sequence is downward.

6. The relationship between the diffusion equation and the Burgers equation can be exploited to obtain solutions of the Burgers equation that have some intriguing properties. Beginning with the diffusion equation, $u_t = u_{xx}$, find a solution as a function of the variable $\xi = x - vt$, i.e.

$$u(x,t) = z(x - vt)$$

Substitute this form into the diffusion equation and perform the integration with respect to ξ. This produces a constant of integration A. By choosing the initial condition of z to be

$$z(0) = \frac{2A}{v}$$

obtain the solution

$$z(\xi) = \frac{A}{v}\left[1 + e^{-v(\xi - \xi_0)}\right]$$

Use the Hopf–Cole transformation to find the corresponding solution of the Burgers equation, and sketch the evolution of this solution. What is the relationship between the amplitude and the speed of this solution?

7. The solution of the Burgers equation derived in Problem 6 exhibits two characteristics generally associated with solitons, namely, the wave form approaches a constant at infinity, and there are appreciable variations of the wave form only in a small region of space. To examine the behavior of these solutions with respect to one another, we can exploit the superposition principle for the diffusion equation to write a solution composed of N elementary solution of the type obtained in Problem 6:

$$z_N(\xi) = \sum_{j=1}^{N} \frac{A_j}{v_j}\left[1 + e^{-v_j(\xi - \xi_j)}\right] \tag{3}$$

where the velocities of these solutions are chosen as $0 < c_1 < c_2 < \cdots < c_N$. Obtain the corresponding solution of the Burgers equation and determine the limit of this solution as $t \to \infty$. What can you conclude about the behavior of the solutions in Problem 6 with respect to interactions among one another?

8. Use the commands Plot and ContourPlot to construct profiles and trajectories of the solution to the Burgers equation obtained in Problem 7, as is shown in Figure 9.6 for the two-soliton solution of the Korteweg–de Vries equation.

9. Consider the nonlinear Schrödinger equation in the dimensionless form

$$iu_t + u_{xx} + |u|^2 u = 0$$

Attempt to find a soliton solution of the form

$$u(\xi) = A(\xi)\, e^{i\phi(\xi)}$$

where the amplitude A and phase ϕ are real functions, $\xi = x - vt$ and v is a velocity. By substituting this form into the nonlinear equation and equating real and imaginary parts, obtain the two ordinary differential equations that determine A and ϕ:

$$v\phi' A + A'' - (\phi')^2 A + A^3 = 0$$
$$-vA' + 2\phi' A' + \phi'' A = 0$$

By requiring that A and its derivatives vanish as $\xi \to \pm\infty$, solve for ϕ and obtain

$$\phi(\xi) = \tfrac{1}{2}v\xi + \text{constant}$$

and that the equation for $\mathcal{A} = A^2$ becomes

$$(\mathcal{A}')^2 = -v^2 \mathcal{A}^2 - 2\mathcal{A}^3 = -\mathcal{A}^2[v^2 + 2\mathcal{A}]$$

Deduce that $\mathcal{A}$, and therefore A, must be complex, contrary to our initial assumption.

To rectify this situation, modify the form of the solitary wave to

$$u(\xi) = A(\xi)\, e^{i[\phi(\xi) + \alpha t]}$$

where α is a real constant. By substituting this into the nonlinear Schrödinger equation, obtain the solitary wave solution

$$u(x,t) = a\,\text{sech}[\tfrac{1}{\sqrt{2}}a(x - vt)]\exp\{i[\tfrac{1}{2}v(x - vt) + \alpha t]\}$$

where

$$a^2 = 2(\alpha - \tfrac{1}{4}v^2)$$

10. The following expressions are solutions of the sine-Gordon equation in the form $u_{xx} - u_{tt} = \sin u$:

(a) $u(x,t) = 4\tan^{-1}\left\{\exp\left[\dfrac{x - vt}{\sqrt{1 - v^2}}\right]\right\}$ (single soliton)

(b) $u(x,t) = 4\tan^{-1}\left[\dfrac{v\sinh(x/\sqrt{1 - v^2})}{\cosh(vt/\sqrt{1 - v^2})}\right]$ (two solitons)

(c) $u(x,t) = 4\tan^{-1}\left[\dfrac{\sinh(vt/\sqrt{1 - v^2})}{v\cosh(x/\sqrt{1 - v^2})}\right]$ (soliton and an anti-soliton)

(d) $u(x,t) = 4\tan^{-1}\left[\dfrac{\sqrt{1 - v^2}}{v}\dfrac{\sin(vt)}{\cosh(x\sqrt{1 - v^2})}\right]$ ("breather")

Use *Mathematica* to show that these expressions are indeed solutions of the sine-Gordon equation. This is most easily done by determining the sine of u

before inputting into *Mathematica*. The appropriate commands to show, for example, that (a) solves the sine-Gordon equation are

```
f[x_,t_]:=Exp[(x-c t)/Sqrt[1-c^2]];
u[x_,t_]:=4 ArcTan[f[x,t]];
Sine[x_,t_]:=4 (f[x,t](1-(f[x,t])^2)/(1+(f[x,t])^2)^2);
Together[D[u[x,t],{x,2}]-D[u[x,t],{t,2}]-Sine[x,t]]
```

Plot these solutions for various times using the `Plot` command and animate a series of frames to observe the time development of each solution. Then for each solution use `Plot3D` to plot the solution surfaces and, for the solutions in (b) and (c), use `ContourPlot` to display the trajectories of the solitons. In all of these applications, it may be more instructive to display the derivative of each solution with respect to x, rather than the solution itself.

11. Find the solitary wave solution of the modified Korteweg–de Vries equation

$$v_t - 6v^2 v_x + v_{xxx} = 0$$

by assuming a solution of the form $v(x,t) = v(x-ct)$. Obtain the solution in the form

$$v(x,t) = -\sqrt{c}\,\text{cosech}[\sqrt{c}(x-ct)]$$

Show that if v is a solution of the modified Korteweg–de Vries equation, then

$$u = v^2 + v_x$$

is a solution of the Korteweg–de Vries equation:

$$u_t - 6uu_x + u_{xxx} = 0$$

Transform the solitary wave solution of the modified Korteweg–de Vries equation to obtain a solution of the Korteweg–de Vries equation.

12. By taking $\xi < \xi_0$ in Equation (9.50) obtain the anti-soliton solution of the sine-Gordon equation:

$$\phi(x,t) = 4\cot^{-1}\left\{\exp\left[\frac{x-vt}{\sqrt{1-v^2}}\right]\right\}$$

13. Consider the nonlinear equation obtained from the anharmonic lattice in Problem 2. By a suitable rescaling of the variables, this equation can be written as

$$u_{tt} = u_{xx} + u_x u_{xx} + u_{xxxx}$$

Find the soliton solutions of the form $\phi(\xi) \equiv \phi(x-ct)$ such that all the derivatives of ϕ vanish at infinity, but ϕ itself need not.

Proceed by substituting ϕ into the equation, integrating and invoking the boundary conditions. Then multiplying by the appropriate quantity, integrate the equation again and use the boundary conditions to obtain

$$\tfrac{1}{2}(1-c^2)(\phi')^2 + \tfrac{1}{6}(\phi')^3 + \tfrac{1}{2}(\phi'')^2 = 0$$

where the primes signify derivatives with respect to the argument of ϕ. Interpret this as an equation for $w = \phi'$, separate variables and integrate. Finally, integrate the equation for w to obtain the soliton solution ϕ.

Are there any restrictions on the velocity c?

14. The Cauchy–Riemann equations

$$v_x = -u_y, \qquad u_x = v_y$$

provide one of the simplest examples of a Bäcklund transformation. Determine the equations satisfied by u and v by applying the integrability condition.

15. Show that the Burgers equation for u and the diffusion equation for v can be obtained from the Bäcklund transformation derived in Example 9.2:

$$v_x = -\tfrac{1}{2}uv, \qquad v_t = \tfrac{1}{4}(u^2 - 2u_x)v$$

In particular, use the integrability condition to obtain Burgers equation, but show that the diffusion equation can be obtained simply by differentiating the first of these equations.

16. Use the Bäcklund transformation

$$v_x = u - v^2, \qquad v_t = -u_{xx} + 2(uv_x + u_x v)$$

to deduce that u satisfies the Korteweg–de Vries equation,

$$u_t - 6uu_x + u_{xxx} = 0$$

and that v satisfies the modified Korteweg–de Vries equation:

$$v_t - 6v^2 v_x + v_{xxx} = 0$$

17. The auto-Bäcklund transformation of the sine-Gordon equation in the form $u_{xy} = \sin u$ can be used to generate a nontrivial solution beginning with the solution $u' = 0$. By substituting $u' = 0$ into (9.81) and (9.82), obtain

$$u_x = -2\beta \sin(\tfrac{1}{2}u), \qquad u_y = 2\beta^{-1} \sin(\tfrac{1}{2}u)$$

Construct from these equations the first-order linear equation

$$\frac{1}{\beta}\frac{\partial u}{\partial x} + \beta\frac{\partial u}{\partial y} = 0$$

and show that the general solution is

$$u(x,y) = F(\beta x - \beta^{-1} y)$$

where F is a once-differentiable but otherwise arbitrary function that is to be determined. Substitute this expression into either of the equations of the Bäcklund transformation, determine F and show that the nontrivial solution generated by the auto-Bäcklund transformation applied to the trivial solution is

$$u(x,y) = 4\tan^{-1}[\exp(\beta x - \beta^{-1} y + A)]$$

where A is a constant of integration.

18. In order to compare the solution in Problem 17 with that in Equation (9.54), transform the 'normal' form of the sine-Gordon equation, $u_{xy} = \sin u$, into the form (9.44) by the change of variables,

$$x = \tfrac{1}{2}(z+t), \qquad y = \tfrac{1}{2}(z-t)$$

as discussed in Section 3.1. In the transformed coordinate system, show that the argument of the exponential can be written as

$$\beta x - \beta^{-1} y = \frac{1+\beta^2}{2\beta}\left[z - \frac{\beta^2 - 1}{\beta^2 + 1}t\right]$$

Thus, defining

$$v = \frac{\beta^2 - 1}{\beta^2 + 1}$$

obtain the solution in the form

$$u(z,t) = 4\tan^{-1}\{\exp[(z-vt)/\sqrt{1-v^2}]\}$$

for $\beta > 0$, which is precisely the soliton solution (9.54). On the other hand, if $\beta < 0$, obtain the *anti*-soliton solution

$$u(z,t) = 4\cot^{-1}\{\exp[(z-vt)/\sqrt{1-v^2}]\}$$

Note that since β is a real number, $v < 1$.

19. Consider the Bäcklund transformation:

$$u'_{x'} = u_x + \beta\, e^{(u+u')/2}$$

$$u'_{y'} = -u_y - 2\beta^{-1}\, e^{(u-u')/2}$$

$$x' = x$$

$$y' = y$$

Introduction to Nonlinear Partial Differential Equations 397

By applying the integrability condition to the first two of these equations, show that the differential equations implied by this transformation are

$$u'_{xy} = 0$$

in the primed coordinate system, and

$$u_{xy} = e^u$$

in the unprimed coordinate system. The latter equation is known as **Liouville's equation**.

20. The Bäcklund transformation in Problem 19 provides a relation between the solution surfaces of a linear equation, $u'_{xy} = 0$, and a nonlinear equation, $u_{xy} = e^u$. Thus, by finding the general solution of the linear equation and applying the Bäcklund transformation, we can obtain the general solution of the nonlinear equation by following a procedure similar to that in Problem 17.

The equation in the primed coordinate system is in the normal form for a hyperbolic equation. Thus, from Section 3.1, the general solution is

$$u'(x, y) = \mathcal{F}(x) + \mathcal{G}(y)$$

where, apart from the requirement of being twice differentiable, the functions $\mathcal{F}$ and $\mathcal{G}$ are arbitrary. By substituting this expression into the Bäcklund transformation in Problem 19, obtain the following first-order quasi-linear equation for u:

$$\frac{1}{\beta} e^{-(\mathcal{F}+\mathcal{G})/2} \frac{\partial u}{\partial x} - \frac{\beta}{2} e^{(\mathcal{F}+\mathcal{G})/2} \frac{\partial u}{\partial y} = \frac{1}{\beta} \mathcal{F}' e^{-(\mathcal{F}+\mathcal{G})/2} + \frac{\beta}{2} \mathcal{G}' e^{(\mathcal{F}+\mathcal{G})/2}$$

The primes in this equation signify differentiation with respect to the argument of the function. Determine the general solution of this equation using the method of characteristics (Section 2.1) and so obtain the general solution of the Liouville equation:

$$u(x, y) = \mathcal{F}(x) - \mathcal{G}(y) - 2 \ln \left[\frac{\beta}{2} \int_{x_0}^{x} e^{\mathcal{F}(x')} \, dx' + \frac{1}{\beta} \int_{y_0}^{y} e^{-\mathcal{G}(y')} \, dy' \right]$$

21. Show that the Korteweg–de Vries equation,

$$v_t + 6vv_x + v_{xxx} = 0$$

is invariant under the Bäcklund transformation

$$u'_x = -u_x + \beta - \tfrac{1}{2}(u - u')^2$$

$$u'_t = -u_t + (u - u')(u_{xx} - u'_{xx}) - 2(u_x^2 + u_x u'_x + u'^2_x)$$

$$x' = x$$

$$t' = t$$

where β is a real number and $u_x = v$. First derive the equation satisfied by u:

$$u_t + 3u_x^2 + u_{xxx} = 0$$

Then show that the equations defining the Bäcklund transformation can be manipulated to produce the two equations:

$$(u_t + 3u_x^2 + u_{xxx}) + (u'_t + 3u'^2_x + u'_{xxx}) = 0$$
$$(u - u')\left[(u_t + 3u_x^2 + u_{xxx}) - (u'_t + 3u'^2_x + u'_{xxx})\right] = 0$$

Hence, deduce that v and $v' = u'_x$ both satisfy the Korteweg–de Vries equation.

22. Beginning with the trivial solution $u' = 0$, show that the solution of the Bäcklund transformation in Problem 21 yields

$$u'(x,t) = (2\beta)^{1/2} \tanh\left[(\beta/2)^{1/2}(x - 2\beta t)\right]$$

and so obtain the single-soliton solution to the Korteweg–de Vries equation:

$$v'(x,t) = \beta \operatorname{sech}^2\left[(\beta/2)^{1/2}(x - 2\beta t)\right]$$

These results are most easily obtained by integrating the first of the Bäcklund equations and then substituting into the second to determine the arbitrary function of integration.

23. Show that the auto-Bäcklund transformation of the Korteweg–de Vries equation in Problem 21 is invariant under the transformation $u' \to 2\beta/u'$. Deduce that in addition to the solution obtained in Problem 21, there is a singular solution, u^*, that can be obtained from the Bäcklund transformation beginning with the trivial solution

$$u^* = (2\beta)^{1/2} \coth\left[(\beta/2)^{1/2}(x - 2\beta t)\right]$$

with the associated solution of the Korteweg–de Vries equation given by

$$v^*(x,t) = -\beta \operatorname{cosech}^2\left[(\beta/2)^{1/2}(x - 2\beta t)\right]$$

24. The first step in obtaining the two-soliton solution of the Korteweg–de Vries equation was shown in Section 9.5 to require the construction of

$$\psi(x,t) = \frac{2(\beta_1 - \beta_2)}{u_{\beta_1}(x,t) - \bar{u}_{\beta_2}(x,t)}$$

with the requirement that $\beta_2 > \beta_1$ and where

$$u_{\beta_1}(x,t) = (2\beta_1)^{1/2} \tanh\left[(\beta_1/2)^{1/2}(x - 2\beta_1 t)\right]$$

and

$$\bar{u}_{\beta_2}(x,t) = (2\beta_2)^{1/2} \coth\left[(\beta_2/2)^{1/2}(x - 2\beta_2 t)\right]$$

Using these formulae, obtain the two-soliton solution, $v(x,t)$, of the Korteweg–de Vries equation

$$v(x,t) = 2(\beta_1 - \beta_2) \frac{\beta_1 \operatorname{sech}^2 \xi_1 + \beta_2 \operatorname{cosech}^2 \xi_2}{(\sqrt{2\beta_2} \coth \xi_2 - \sqrt{2\beta_1} \tanh \xi_1)^2}$$

where

$$\xi_1 = \sqrt{\beta_1/2}\,(x - 2\beta_1 t), \qquad \xi_2 = \sqrt{\beta_2/2}\,(x - 2\beta_2 t)$$

By choosing $\beta_1 = 2$ and $\beta_2 = 8$, show, by using standard formulae for $\cosh\tfrac{1}{2}z$, $\sinh\tfrac{1}{2}z$, and $\cosh(z_1 + z_2)$, that the two-soliton solution can be written as

$$v(x,t) = -12 \frac{3 + 4\cosh(2x - 8t) + \cosh(4x - 64t)}{[3\cosh(x - 28t) + \cosh(3x - 36t)]^2}$$

25. Beginning with the two-soliton solution of Problem 24 (with $\beta_1 = 2$ and $\beta_2 = 8$, transform to the coordinates to a frame moving with velocity $2\beta_1 = 4$ by introducing the variable $\xi_1 = x - 4t$. Show that in terms of the variables ξ and t, the two-soliton solution becomes

$$v(\xi_1,t) = -12 \frac{3 + 4\cosh(\xi_1) + \cosh(4\xi_1 - 48t)}{[3\cosh(\xi_1 - 24t) + \cosh(3\xi_1 - 24t)]^2}$$

Take the limit as $t \to \pm\infty$ of this expression with ξ_1 held fixed to obtain

$$\lim_{t\to\pm\infty} v(\xi_1,t) = -2\operatorname{sech}^2(\xi_1 \pm \tfrac{1}{2}\ln 3)$$

Similarly, by transforming to a frame moving with velocity $2\beta_2 = 16$ and introducing the variable $\xi_2 = x - 16t$, show that the asymptotic form of the two-soliton solution is

$$\lim_{t\to\pm\infty} v(\xi_2,t) = -8\operatorname{sech}^2(2\xi_2 \mp \tfrac{1}{2}\ln 3)$$

Provide an interpretation of these solutions and address specifically the role of the quantity

$$\delta = \tfrac{1}{2}\ln 3$$

26. Use *Mathematica* to verify that the two-soliton solutions in Problems 24 and 25 are solutions of the Korteweg–de Vries equation. Then use the `ContourPlot` command to plot the solution

$$v(\xi_1, t) = -12 \frac{3 + 4\cosh(\xi_1) + \cosh(4\xi_1 - 48t)}{[3\cosh(\xi_1 - 24t) + \cosh(3\xi_1 - 24t)]^2}$$

in (i) the rest frame, (ii) the moving frame $\xi_1 = x - 4t$ and (iii) the moving frame $\xi_2 = x - 16t$.

27. The auto-Bäcklund transformation of the sine-Gordon equation in Equation (9.79) can be used to derive a nonlinear superposition rule by following steps analogous to those used in Section 9.5 for the Korteweg–de Vries equation (Rogers and Shadwick, 1982). Let ϕ be any solution of the sine-Gordon equation. Then signify by ϕ_1 and ϕ_2 the solutions obtained from ϕ by applying the Bäcklund transformation with Bäcklund parameters β_1 and β_2, respectively. Similarly, signify by ϕ_{21} the solution obtained from ϕ by the successive application of (9.79) first with parameter β_1 and then with parameter β_2, and by ϕ_{12} the solution obtained from ϕ by the successive application of (9.79) first with parameter β_2 and then with parameter β_1.

Use Equations (9.79a, b) to obtain the eight equations for ϕ_1, ϕ_2, ϕ_{21} and ϕ_{12}. By taking appropriate sums and differences of pairs of these equations obtain the following four equations:

$$\phi_{21,x} = \phi_x + 2\beta_1 \sin[\tfrac{1}{2}(\phi_1 + \phi)] + 2\beta_2 \sin[\tfrac{1}{2}(\phi_{21} + \phi_1)]$$

$$\phi_{21,y} = \phi_y - 2\beta_1^{-1} \sin[\tfrac{1}{2}(\phi_1 - \phi)] + 2\beta_2^{-1} \sin[\tfrac{1}{2}(\phi_{21} - \phi_1)]$$

$$\phi_{12,x} = \phi_x + 2\beta_1 \sin[\tfrac{1}{2}(\phi_{12} + \phi_2)] + 2\beta_2 \sin[\tfrac{1}{2}(\phi_2 + \phi)]$$

$$\phi_{12,y} = \phi_y + 2\beta_1^{-1} \sin[\tfrac{1}{2}(\phi_{12} - \phi_2)] - 2\beta_2^{-1} \sin[\tfrac{1}{2}(\phi_2 + \phi)]$$

By requiring that $\phi_{21} = \phi_{12} \equiv \psi$, and introducing the abbreviations

$$\Sigma_\pm \equiv \tfrac{1}{4}[\psi + \phi \pm (\phi_2 + \phi_1)]$$

$$\Delta_\pm \equiv \tfrac{1}{4}[\psi - \phi \pm (\phi_2 - \phi_1)]$$

show that the four equations above imply

$$\beta_1 \sin[\Sigma_+ - \Delta_+] + \beta_2 \sin[\Sigma_+ + \Delta_-]$$
$$= \beta_1 \sin[\Sigma_+ + \Delta_+] + \beta_2 \sin[\Sigma_- - \Delta_-]$$

$$\beta_1 \sin[\Sigma_- - \Delta_+] + \beta_2 \sin[\Sigma_- - \Delta_-]$$
$$= \beta_1 \sin[\Sigma_- - \Delta_+] + \beta_2 \sin[\Sigma_- + \Delta_-]$$

By using the standard trigonometric formulae for the sine of the sum and difference of two angles, show that these equations can be written as

$$[\beta_1 \sin \Delta_+ - \beta_2 \sin \Delta_-] \cos \Sigma_+ = 0$$

$$[\beta_1 \sin \Delta_+ - \beta_2 \sin \Delta_-] \cos \Sigma_- = 0$$

Deduce from these equations that $\beta_1 \sin \Delta_+ = \beta_2 \sin \Delta_-$, i.e.,

$$\beta_1 \sin\{\tfrac{1}{4}[\psi - \phi + (\phi_2 - \phi_1)]\} = \beta_1 \sin\{\tfrac{1}{4}[\psi - \phi - (\phi_2 - \phi_1)]\}$$

Use standard trigonometric identities again to obtain the following expression for ψ:

$$\psi = 4\tan^{-1}\left\{\frac{\beta_1 + \beta_2}{\beta_1 - \beta_2}\tan[\tfrac{1}{4}(\phi_1 - \phi_2)]\right\} + \phi$$

This is the nonlinear superposition rule for the sine-Gordon equation.

28. As an illustration of the use of the nonlinear superposition rule for the sine-Gordon equation derived in Problem 27, let ϕ be the trivial solution $\phi = 0$. Then, using the calculation in Problem 17, choose ϕ_1 and ϕ_2 to be

$$\phi_1 = 4\tan^{-1}[\exp(\beta_1 x - \beta_1^{-1} y)] \equiv 4\tan^{-1} e^{\alpha_1}$$
$$\phi_2 = 4\tan^{-1}[\exp(\beta_2 x - \beta_2^{-1} y)] \equiv 4\tan^{-1} e^{\alpha_1}$$

The abbreviations $\alpha_{1,2} = \beta_{1,2} x - \beta_{1,2}^{-1} y$ have been used and the integration constants of both solutions have been set to zero.

Substitute these solutions into the nonlinear superposition rule in Problem 27 and, using standard trigonometric identities, obtain the following solution of the sine-Gordon equation:

$$\psi = 4\tan^{-1}\left\{\frac{\beta_1 + \beta_2}{\beta_1 - \beta_2}\frac{\sinh[\tfrac{1}{2}(\alpha_1 - \alpha_2)]}{\cosh[\tfrac{1}{2}(\alpha_1 + \alpha_2)]}\right\}$$

29. With suitable modifications to the procedure in Problem 10, show that the expression derived in Problem 28 from the nonlinear superposition is a solution of the sine-Gordon equation.

Use *Mathematica* to display this solution using the `Plot` and `ContourPlot` commands for several values of β_1 and β_2 and deduce that this solution corresponds to a two-soliton solution.

30. Use the `Plot` command of *Mathematica* to display the functions

$$u(\xi) = A\tanh\xi, \qquad \bar{u}(\xi) = B\coth\xi$$

on the same graph for $A > B$, $A = B$ and $A < B$. Deduce that the nonlinear superposition rule for the Korteweg–de Vries equation can be used to construct a nonsingular two-soliton solution only from two single-soliton solutions composed from the regular solution u_{β_1} in (9.92) and the irregular solution $\bar{u}_{\beta_2}$ in (9.93) provided $\beta_2 > \beta_1$.

Chapter 10

The Method of Inverse Scattering

The most general technique currently available for solving initial-value problems for certain types of nonlinear equation is the inverse scattering method developed by Gardner, Greene, Kruskal and Miura. The method was initially devised for the Korteweg–de Vries equation, but the general approach was later shown to be applicable to several other nonlinear equations. In this chapter we will describe the inverse scattering method as applied to the Korteweg–de Vries equation, and then discuss the applicability of this method to other equations.

The method of inverse scattering as formulated for the Korteweg–de Vries equation is an ingenious way of solving the initial-value problem by associating the solution of the nonlinear equation with the potential of a time-independent Schrödinger equation. This replaces the problem of finding a solution of the Korteweg–de Vries equation—a *nonlinear* equation—with the less onerous task of solving a *linear* quantum mechanical scattering problem. The basis of the method is contained in a series of remarkable results concerning the simple ways in which the scattering properties of the quantum mechanical problem change as the potential 'evolves' as a solution of the Korteweg–de Vries equation. Thus, once the scattering problem is solved with the initial value of the Korteweg–de Vries equation as the potential, the solution of the scattering problem for potentials corresponding to later 'times' may be obtained directly without explicit reference to that potential. The solution of the scattering problem can then be inverted to find the corresponding potential. This potential is the solution of the Korteweg–de Vries equation at that time.

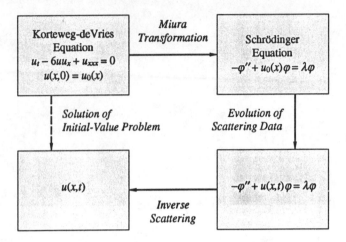

Figure 10.1 Schematic illustration of the method of inverse scattering for solving the initial-value problem for the Korteweg–de Vries equation. The motivation for associating the solution of this equation with the potential of a time-independent Schrödinger equation is provided in the problems at the end of the chapter. The evolution of the eigenfunctions and eigenvalues as the potential changes according to the Korteweg–de Vries equation is discussed in Section 10.1. The determination of the potential from the scattering information with the inverse scattering method is derived in Section 10.2.

It may help to think of the inverse scattering transformation as a generalization of the Fourier and Laplace transforms. For example, by solving an initial-value problem using Fourier and Laplace transforms, differential operators are replaced by multiplicative operations and the transform of the solution can be obtained by a few simple algebraic operations. The last step of the process, the solution of the initial-value problem in terms of the original variables, is then carried out by inverting the transforms. The analogy with the inverse scattering transform is not exact, but the philosophy is similar.

A flow diagram of the inverse scattering method is shown in Figure 10.1. The first step of the method is the association of the solution of the Korteweg–de Vries equation with a potential of a time-independent Schrödinger equation. In light of the results produced by the inverse scattering method and of more recent theoretical developments, this step can be easily justified. However, without these, there are only suggestive arguments based on some general properties of the Korteweg–de Vries equation. These are outlined in the problems at the end of the chapter.

The next step is the solution of the quantum mechanical problem with the initial value for the Korteweg–de Vries equation taken as the potential. This involves calculating the discrete, or bound, eigenfunctions, their normalization constants and eigenvalues and the transmission and reflection coefficients of the continuum, or unbound, states. These quantities are collectively referred to as the **scattering data** of the Schrödinger equation. In Section 10.1, the evolution of the scattering data is calculated for any potential that evolves according to the Korteweg–de Vries equation. Both the discrete and continuous eigenvalues will be shown to be invariant under such changes, and the normalization constants and the transmission and reflection coefficients will be found to evolve according to simple exponential laws.

The heart of the inverse scattering method lies both in the simplicity of the evolution of the scattering data and in the ability to determine the potential from the scattering data. The final step of the method, the inversion of the scattering data to determine the potential, is carried out in Section 10.2. We confine ourselves to heuristic arguments that illustrate the important features of the method, and refer the reader to several more rigorous discussions that may be found elsewhere. Examples of solutions to initial-value problems of the Korteweg–de Vries equation are provided in Section 10.3.

If only the Korteweg–de Vries equation was amenable to this type of solution, then the inverse scattering method would be only an interesting mathematical laboratory which could be used to study various properties of solutions of a particular nonlinear equation, but not of much use as a general method. However, as often happens, once a method for solving a problem has been discovered by a combination of inspired guesswork and *tour de force* calculations, a more concise way of formulating the method is discovered and the generalization to other equations becomes apparent. This is the case with the formulation of inverse scattering due to Lax, which will be described in Section 10.4.

10.1 Evolution of Eigenfunctions and Eigenvalues

Given that one of the underlying reasons for the success of the inverse scattering method is the relationship between the Korteweg–de Vries equation and the time-independent Schrödinger equation (Problem 1), the first task is to determine how the eigenfunctions and eigenvalues change as the potential evolves according to the Korteweg–de Vries equation from a specified initial function.

We first write the time-independent Schrödinger equation with the dependencies upon x and t of the various quantities displayed explicitly:

$$-\psi_{xx}(x,t) + u(x,t)\psi(x,t) = \lambda(t)\psi(x,t) \qquad (10.1)$$

Since the potential depends upon the parameter t, the eigenfunctions and eigenvalues also depend upon t. It must be stressed that the quantity t is not to be regarded as the 'time' in this equation, but simply as a parameter that labels the potentials $u(x,t)$. Thus, the Schrödinger equation (10.1) is to be interpreted as a sequence of time-*independent* equations for the family of potentials $u(x,t)$. The change of the potentials with respect to t from an initial value $u(x,0)$ is governed by the Korteweg–de Vries equation:

$$u_t - 6uu_x + u_{xxx} = 0 \tag{10.2}$$

with the initial condition given by

$$u(x,0) = u_0(x) \tag{10.3}$$

The problem is first to solve the one-dimensional quantum mechanical problem for the *known* potential $u_0(x)$,

$$-\psi_{xx}(x,t) + u_0(x)\psi(x,t) = \lambda(t)\psi(x,t) \tag{10.4}$$

and then determine how the eigenfunctions and eigenvalues evolve as the potential changes from that given in (10.3) according to the equation of motion (10.2). The potential for any $t>0$, $u(x,t)$, is then constructed from the scattering data at time t by the inverse-scattering transformation.

This way of solving the Korteweg–de Vries equation also shows that the class of allowed initial conditions is not unrestricted. The domain of the solution is the entire real line $-\infty < x < \infty$ and the solutions are assumed to vanish sufficiently rapidly as $|x| \to \infty$ to insure that the scattering problem is well-defined. These restrictions should be borne in mind in the following sections.

To determine how the eigenvalues $\lambda(t)$ of the Schrödinger equation change with 'time', we first differentiate (10.1) with respect to t:

$$\psi_{xxt} - (u_t - \lambda_t)\psi - (u - \lambda)\psi_t = 0 \tag{10.5}$$

The Korteweg–de Vries equation (10.2) is now used to eliminate u_t from (10.5):

$$\left[\frac{\partial^2}{\partial x^2} - (u - \lambda)\right]\psi_t + (u_{xxx} - 6uu_x)\psi + \lambda_t\psi = 0 \tag{10.6}$$

The term $u_{xxx}\psi$ can be written in a more useful form by using the identity

$$(u_x\psi)_{xx} = u_{xxx}\psi + u_x\psi_{xx} + 2u_{xx}\psi_x \tag{10.7}$$

Then, solving this equation for $u_{xxx}\psi$, the Schrödinger equation enables the resulting expression to be written as

$$\begin{aligned} u_{xxx}\psi &= \frac{\partial^2}{\partial x^2}(u_x\psi) - u_x\psi_{xx} - 2u_{xx}\psi_x \\ &= \left[\frac{\partial^2}{\partial x^2} - (u - \lambda)\right](u_x\psi) - 2u_{xx}\psi_x \end{aligned} \tag{10.8}$$

Substituting (10.8) into (10.6), we obtain

$$\left[\frac{\partial^2}{\partial x^2} - (u - \lambda)\right](\psi_t + u_x\psi) - 2(3uu_x\psi + u_{xx}\psi_x) + \lambda_t\psi = 0 \quad (10.9)$$

We now focus our attention on the quantity $3uu_x\psi + u_{xx}\psi_x$. The term $u_{xx}\psi_x$ can be rewritten by first using the identity

$$(u\psi_x)_{xx} = u_{xx}\psi_x + 2u_x\psi_{xx} + u\psi_{xxx} \quad (10.10)$$

The Schrödinger equation is then used to eliminate ψ_{xx} in the second term on the right-hand side, and the derivative of the Schrödinger equation with respect to x is used to eliminate ψ_{xxx}:

$$(u\psi_x)_{xx} = u_{xx}\psi_x + 2u_x(u - \lambda)\psi + u[u_x\psi + (u - \lambda)\psi_x]$$
$$= u_{xx}\psi_x - 2\lambda u_x\psi + 3uu_x\psi + (u - \lambda)u\psi_x \quad (10.11)$$

Solving (10.11) for $3uu_x\psi + u_{xx}\psi_x$ and again using Schrödinger's equation and its derivative with respect to x, we obtain

$$3uu_x\psi + u_{xx}\psi_x = 3uu_x\psi + (u\psi_x)_{xx} + 2\lambda u_x\psi - 3uu_x\psi - (u - \lambda)u\psi_x$$
$$= \left[\frac{\partial^2}{\partial x^2} - (u - \lambda)\right](u\psi_x) + 2\lambda u_x\psi$$
$$= \left[\frac{\partial^2}{\partial x^2} - (u - \lambda)\right](u\psi_x) + 2\lambda[\psi_{xxx} - (u - \lambda)\psi_x]$$
$$= \left[\frac{\partial^2}{\partial x^2} - (u - \lambda)\right](u\psi_x + 2\lambda\psi_x) \quad (10.12)$$

Substitution of (10.12) into (10.9) yields

$$\left[\frac{\partial^2}{\partial x^2} - (u - \lambda)\right](\psi_t + u_x\psi - 2u\psi_x - 4\lambda\psi_x) + \lambda_t\psi = 0 \quad (10.13)$$

This equation can be written in a more concise form by introducing the quantity Ψ defined by

$$\Psi = \psi_t + u_x\psi - 2(u + 2\lambda)\psi_x \quad (10.14a)$$
$$= \psi_t + \psi_{xxx} - 3(u + \lambda)\psi_x \quad (10.14b)$$

where in writing (10.14b) we have taken the derivative of the Schrödinger equation (10.1) to substitute for the quantity $u_x\psi$ in (10.14a). Equation (10.13) can now be written as

$$-\lambda_t\psi = \left[\frac{\partial^2}{\partial x^2} - (u - \lambda)\right]\Psi \quad (10.15)$$

Equations (10.14) and (10.15) are the central results from which we will obtain the behavior of all of the required quantities in the scattering problem as the potential changes according to the Korteweg–de Vries equation.

We will consider the bound states first. The eigenfunctions and eigenvalues for the bound states will labeled by a subscript 'n'. The wavefunctions for the bound states can be chosen to be real, since the Schrödinger equation is real. To determine the evolution for the eigenvalues, we multiply (10.15) by ψ_n and again use the Schrödinger equation to simplify the result:

$$-\lambda_{n,t}\psi_n^2 = \psi_n \frac{\partial^2 \Psi}{\partial x^2} - (u - \lambda_n)\psi_n \Psi$$

$$= \psi_n \frac{\partial^2 \Psi}{\partial x^2} - \Psi \frac{\partial^2 \psi_n}{\partial x^2}$$

$$= \frac{\partial}{\partial x}\left[\psi_n \frac{\partial \Psi}{\partial x} - \Psi \frac{\partial \psi_n}{\partial x}\right] \quad (10.16)$$

Integrating (10.16) from $x = -\infty$ to $x = \infty$ yields

$$-\lambda_{n,t}\int_{-\infty}^{\infty}\psi_n^2(x)\,dx = \left[\psi_n \frac{\partial \Psi}{\partial x} - \Psi \frac{\partial \psi_n}{\partial x}\right]\Bigg|_{-\infty}^{\infty} \quad (10.17)$$

The eigenfunctions ψ_n of the bound states together with the corresponding spatial derivatives must vanish as $x \to \pm\infty$. Examining each of the terms in Equation (10.14), and recalling our restrictions on the potential u, we conclude that the right-hand side of (10.17) must vanish. We next observe that the integral on the left-hand side of (10.17) is simply the normalization integral for ψ_n. This quantity must be a positive number since ψ_n is a real function corresponding to a nontrivial solution of the Schrödinger equation. Thus, for the left-hand side of Equation (10.17) to vanish the coefficient of the normalization integral must vanish:

$$\frac{d\lambda_n}{dt} = 0 \quad (10.18)$$

that is, for a potential $u(x,t)$ that changes with t according to the Korteweg–de Vries equation, *the eigenvalues of the bound states are invariant with respect to changes in t*: $\lambda_n(t) = \lambda_n(0)$. This remarkable and unexpected result provides an important clue as to why the Korteweg–de Vries equation is amenable to solution by the inverse scattering method. It also provides a fingerprint for identifying other equations that can be solved using similar approaches. These points will be clarified in Section 10.4.

The invariance of the eigenvalues expressed in Equation (10.18) has several useful consequences. First, we see immediately from Equation (10.15) that $\lambda_{n,t} = 0$ implies that Ψ_n is an eigenfunction of Schrödinger's equation with the eigenvalue λ_n. Thus, this eigenfunction must be a linear combination of ψ_n and a second solution ϕ_n independent of ψ_n,

$$\Psi_n(x,t) = A_n(t)\psi_n(x,t) + B_n(t)\phi_n(x,t) \tag{10.19}$$

The expansion coefficients $A_n(t)$ and $B_n(t)$ are, for the moment, assumed to be t-dependent. Since the potential is assumed to decay to zero as $x \to \pm\infty$, the solutions of the Schrödinger equation in this regime are proportional to exponential functions, one growing and one decaying:

$$\lim_{x \to \pm\infty} \psi_n(x,t) \propto e^{\mp \kappa_n x}$$
$$\lim_{x \to \pm\infty} \phi_n(x,t) \propto e^{\pm \kappa_n x} \tag{10.20}$$

where $\kappa_n = -\lambda_n^2$. Thus, if exponentially unbounded behavior for large values of x is to be avoided in Ψ_n, we must set $B_n(t) = 0$. Thus, Ψ_n and ψ_n are proportional:

$$\Psi_n = \psi_{n,t} + \psi_{n,xxx} - 3(u + \lambda_n)\psi_{n,x} = A_n \psi_n \tag{10.21}$$

To evaluate the proportionality constant $A_n(t)$, we introduce the normalization constant, $\mathcal{N}_n(t)$, for the state $\psi_n(x,t)$ through the usual definition:

$$\mathcal{N}_n^2(t) \int_{-\infty}^{\infty} \psi_n^2(x,t)\,dx = 1 \tag{10.22}$$

Then, by taking the derivative of this equation with respect to t and using Equation (10.21), we obtain the result that if ψ_n is assumed to be normalized to unity, the constant $A_n(t)$ is equal to zero:

$$(\mathcal{N}_n \psi_n)_t + \mathcal{N}_n \psi_{n,xxx} - 3(u + \lambda_n)\mathcal{N}_n \psi_{n,x} = 0 \tag{10.23}$$

We will evaluate this equation in the limit as $x \to \infty$. Using the expression in Equation (10.20), the general asymptotic form of $\mathcal{N}_n \psi_n$ in this limit is given by

$$\lim_{x \to \infty} [\mathcal{N}_n(t)\psi_n(x,t)] \equiv c_n(t)\,e^{-\kappa x} \tag{10.24}$$

It is important to observe that the prefactor of the exponential in this equation is *not* necessarily equal to the normalization constant in Equation (10.22). Substituting this asymptotic form into Equation (10.23) yields an equation for c_n alone, $c_{n,t} - 4\kappa_n^3 c_n = 0$, whose solution is given by

$$c_n(t) = c_n(0)\,e^{4\kappa_n^3 t} \tag{10.25}$$

Thus, the proportionality constant for the asymptotic decay of the bound-state wavefunction exhibits a simple exponential dependence on t.

The equations in (10.18), (10.23) and (10.25) specify completely the changes in the eigenfunctions and the eigenvalues of the bound states of the Schrödinger equation in (10.1) as the potential evolves according to the Korteweg–de Vries equation. We now consider the corresponding changes in the continuum eigenfunctions and eigenvalues.

The important quantities that characterize the changes in the continuum states are identified by imposing the standard boundary conditions used in scattering theory (Landau and Lifshitz, 1965). A traveling wave e^{-ikx} from $x=+\infty$ is incident on the potential. The quantity k is related to the eigenvalue λ by $\lambda=k^2>0$. Upon interaction with the potential the wave has two components: a transmitted wave e^{-ikx} at $x=-\infty$ of amplitude $a(k)$ and a reflected wave e^{ikx} at $x=+\infty$ of amplitude $b(k)$:

$$\psi(x) \to \begin{cases} e^{-ikx} + b(k)\,e^{ikx} & x \to +\infty \\ a(k)\,e^{-ikx} & x \to -\infty \end{cases} \qquad (10.26)$$

The quantity $a(k)$ is called the **transmission coefficient** and the quantity $b(k)$ is called the **reflection coefficient**. Both are in general complex quantities. We could also have defined the scattering problem in terms of an incident wave e^{ikx} incident from $x=-\infty$ which, upon interaction with the potential, is transformed into a reflected wave e^{-ikx} at $x=-\infty$ and a transmitted wave e^{ikx} at $x=+\infty$. Then, the corresponding equations in (10.26) for this case can be obtained by observing that since Schrödinger's equation is real, both ψ and the complex conjugate ψ^* are solutions. Thus,

$$\psi(x) \to \begin{cases} e^{ikx} + b^*(k)\,e^{-ikx} & x \to -\infty \\ a^*(k)\,e^{ikx} & x \to +\infty \end{cases} \qquad (10.27)$$

where $a^*(k)$ and $b^*(k)$ are the complex conjugates of the transmission and reflection coefficients in (10.26). Equations (10.26) and (10.27) are entirely equivalent and we will use both in the following sections.

The equations describing the evolution of the continuum states are derived by first observing that the set of continuum eigenvalues is invariant in the sense of forming part of the eigenvalue spectrum of all potentials for which the scattering problem is well-defined. Thus, in Equation (10.15), we can fix the value of λ_k and calculate the evolution of eigenstates at this value as the potential changes. The pertinent equation is therefore (10.15) with $\lambda_{k,t}$ set equal to zero, so Ψ_k is again an eigenstate of the Schrödinger equation with eigenvalue λ_k. We may then represent Ψ_k as in Equation (10.19) as a linear combination of ψ_k and second solution, ϕ_k, independent of ψ_k:

$$\Psi_k(x,t) = A_k(t)\psi_k(x,t) + B_k(t)\phi_k(x,t) \qquad (10.28)$$

Since Ψ_k and ψ_k are both eigenfunctions of the Schrödinger equation with the same eigenvalue, the Wronskian of these functions,

$$\mathcal{W}(\psi_k, \Psi_k) = \psi_k \Psi_{k,x} - \Psi_k \psi_{k,x} \qquad (10.29)$$

must be a function of t only (Problem 2), i.e.

$$\frac{\partial}{\partial x}(\psi_k \Psi_{k,x} - \Psi_k \psi_{k,x}) = 0 \tag{10.30}$$

Evaluation of W in either of the limits $x \to \pm\infty$ produces (Problem 3)

$$\psi_k \Psi_{k,x} - \Psi_k \psi_{k,x} = 0 \tag{10.31}$$

which implies that $\Psi_k(x,t)$ and $\psi_k(x,t)$ are the same to within a proportionality constant that may depend upon t: $\Psi_k(x,t) = A_k(t)\psi_k(x,t)$. This proportionality constant can be determined by evaluating the limit of Ψ_k/ψ_k as $x \to \infty$:

$$\lim_{x \to \infty}(\Psi_k/\psi_k) = \frac{(b_t - 4ik\lambda_k b)e^{ikx} + 4ik\lambda_k e^{-ikx}}{be^{ikx} + e^{-ikx}} \tag{10.32}$$

For the right-hand side of (10.32) to be independent of x, the coefficients of the same exponential factors in the numerator and denominator must be separately equal. This can be seen by writing (10.32) in the general form

$$A_k = \frac{Ae^{ikx} + Be^{-ikx}}{Ce^{ikx} + De^{-ikx}} \tag{10.33}$$

where $A = b_t - 4ik\lambda_k b$, $B = 4ik\lambda_k$, $C = b$ and $D = 1$. Equation (10.33) can be rearranged as

$$(A - A_k C)e^{ikx} + (B - A_k D)e^{-ikx} = 0 \tag{10.34}$$

Since e^{ikx} and e^{-ikx} are linearly independent functions, the factors multiplying these functions in (10.34) must separately vanish. Thus, $B - A_k D = 0$ yields

$$A_k = 4ik\lambda = 4ik^3 \tag{10.35}$$

so A_k is independent of t. Similarly, $A - A_k C = 0$ provides the evolution equation for the reflection coefficient:

$$b_t = 8ik^3 b \tag{10.36}$$

Integrating this equation, the t-dependence of b is seen to be given by a simple exponential law:

$$b(k,t) = b(k,0)e^{8ik^3 t} \tag{10.37}$$

The evolution of the transmission coefficient is obtained in a similar manner by evaluating the quantity Ψ_k/ψ_k in the limit $x \to -\infty$:

$$\lim_{x \to -\infty}(\Psi_k/\psi_k) = \frac{a_t e^{-ikx} - 4\lambda_k(-ika e^{-ikx})}{ae^{-ikx}}$$

$$= \frac{a_t + 4iak^3}{a} \tag{10.38}$$

Since the ratio Ψ_k/ψ_k is independent of x, the value of the left-hand side of this equation is equal to A_k which, according to (10.35), is equal to $4ik^3$. This immediately implies that $a_t = 0$, i.e. we obtain the result that the transmission coefficient is invariant with respect to changes to t:

$$a(k,t) = a(k,0) \tag{10.39}$$

The results obtained in this section are very suggestive of an underlying relationship between the Korteweg–de Vries equation and the Schrödinger equation. We have found that for a family of potentials of a time-independent Schrödinger equation that change according to the Korteweg–de Vries equation from a specified initial function, the corresponding changes in the quantities that characterize the bound and continuum states are governed by very simple evolution equations. The eigenvalues and normalization constants of the bound states change according to

$$\lambda_n(t) = \lambda_n(0), \quad c_n(t) = c_n(0)\,e^{4\kappa_n^3 t} \tag{10.40}$$

For the continuum states, the evolution equations for the transmission coefficient, $a(k,t)$, and the reflection coefficient, $b(k,t)$, are

$$b(k,t) = b(k,0)\,e^{8ik^3 t}, \quad a(k,t) = a(k,0) \tag{10.41}$$

Equations (10.40) and (10.41) characterize completely the *asymptotic* behavior of both the bound and the continuum states of the potential. We will see in the next section that these equations contain all the necessary quantities that are needed for obtaining the potential from the solution of the Schrödinger equation. Equations such as (10.21), which contain information about the changes in the wavefunction over the *entire* real line, will not be needed.

10.2 The Inverse Scattering Transform

In most treatments of scattering in elementary quantum mechanics, a potential is specified and the bound states and the scattering properties are calculated by solving Schrödinger's equation. The classical analogue of this problem is determining the trajectories of particles given the forces acting on the particles by solving Newton's equations of motion. In both of these examples, a force or a potential is given and the dynamics of the particles is calculated by solving a differential equation. The question now arises as to whether a quantum mechanical potential can be determined from scattering experiments, and whether the forces acting on particles can be identified from the motion of those particles. These are two examples of **inverse problems**.

We have seen in the preceding section that the scattering data follow simple evolution laws. Thus, once the scattering problem is solved with the potential given by the initial condition of the Korteweg–de Vries equation $u_0(x)$ at $t=0$, the scattering data are easily determined for $t>0$. In this section we will derive the inverse scattering transform that will enable the potential $u(x,t)$ to be determined for $t>0$ from the scattering data. This is the final step in the solution of the Korteweg–de Vries equation by the inverse scattering method.

The simplest approach to solving the inverse scattering problem involves integral representations of solutions to Schrödinger's equation. Taking $\lambda = k^2$ in (10.1),

$$-\frac{d^2\psi}{dx^2} + u\psi = k^2\psi \tag{10.42}$$

we introduce the following integral representations of the solutions to this equation (Problem 6):

$$\varphi_k(x) = e^{ikx} + \int_x^\infty \mathcal{K}(x,s)\, e^{iks}\, ds \tag{10.43a}$$

$$\varphi_{-k}(x) = e^{-ikx} + \int_x^\infty \mathcal{K}(x,s)\, e^{-iks}\, ds \tag{10.43b}$$

These solutions, which are called **Jost solutions**, have the important property that

$$\lim_{x\to\infty} \varphi_{\pm k}(x) = e^{\pm ikx} \tag{10.44}$$

If k is a real quantity, then eigenfunctions correspond to continuum states, since $\lambda > 0$. On the other hand if k is a purely imaginary quantity, $k = i\kappa$, where κ is real, then $\lambda < 0$, and the eigenfunctions are bound states. In the following discussion, we will use the label k for continuum states and the label $i\kappa$ for bound states.

The utility of the Jost solutions for the inverse scattering problem becomes evident by the substitution of these expressions into the Schrödinger equation (Problem 7). The kernel $\mathcal{K}$ is then seen to satisfy an inhomogeneous wave equation,

$$\frac{\partial^2 \mathcal{K}}{\partial x^2} - \frac{\partial^2 \mathcal{K}}{\partial s^2} - u\mathcal{K} = 0 \tag{10.45}$$

together with the auxiliary condition

$$u(x,t) = -2\frac{\partial}{\partial x}\mathcal{K}(x,x,t) \tag{10.46}$$

This relationship is of central importance to the method of inverse scattering, since we will derive an integral equation for $\mathcal{K}$ that involves only the scattering data for the potential u. The solution of this integral equation as a function of the parameter t then yields the potential $u(x,t)$ by applying Equation (10.46). This potential is the solution of the Korteweg–de Vries equation at time t.

Since the solutions φ_k and φ_{-k} are linearly independent functions, they may be regarded as fundamental solutions of Equation (10.42). Thus, any solution corresponding to the eigenvalue k can be expressed as a linear combination of these two solutions. In particular, consider the solution ψ_k of Schrödinger's equation characterized by the asymptotic behavior $\psi_k \to e^{-ikx}$ as $x \to -\infty$. This solution can be expressed as a linear combination of φ_k and φ_{-k}:

$$\psi_k(x) = A(k)\varphi_{-k}(x) + B(k)\varphi_k(x) \tag{10.47}$$

This equation states that a wavefunction φ_{-k} travelling to the left and incident on the potential u from $x=\infty$ is transmitted into the wavefunction ψ_k and reflected into the wavefunction φ_k. In fact, by taking the limits of Equation (10.47) as $x \to \pm\infty$, we obtain

$$\lim_{x \to \infty} \psi_k(x) = A(k)e^{-ikx} + B(k)e^{ikx} \tag{10.48a}$$

$$\lim_{x \to -\infty} \psi_k(x) = e^{-ikx} \tag{10.48b}$$

The limit in (10.48a) is obtained from the property of the Jost solutions, while the limit in (10.48b) results from the definition of $\psi_k(x)$. By dividing both sides of Equation (10.47) by $A(k)$, we obtain

$$\frac{1}{A(k)}\psi_k(x) = \varphi_{-k}(x) + \frac{B(k)}{A(k)}\varphi_k(x) \tag{10.49}$$

and the corresponding changes in the limits in Equation (10.48) are

$$\lim_{x \to \infty}\left[\frac{1}{A(k)}\psi_k(x)\right] = e^{-ikx} + \frac{B(k)}{A(k)}e^{ikx} \tag{10.50a}$$

$$\lim_{x \to -\infty}\left[\frac{1}{A(k)}\psi_k(x)\right] = \frac{1}{A(k)}e^{-ikx} \tag{10.50b}$$

Comparing these limits with those in (10.26) allows us to identify the quantity $1/A(k)$ as the transmission coefficient $a(k)$ and the quantity $B(k)/A(k)$ as the reflection coefficient $b(k)$:

$$a(k) = \frac{1}{A(k)}, \qquad b(k) = \frac{B(k)}{A(k)} \tag{10.51}$$

Equations (10.47), (10.49) and (10.51) will now be used to obtain an integral equation for $\mathcal{K}$. We first substitute the expressions for φ_k and

φ_{-k} in (10.43) into Equation (10.49) and make use of the identifications in Equation (10.51):

$$a(k)\psi_k(x) = e^{-ikx} + \int_x^\infty \mathcal{K}(x,s)\,e^{-iks}\,ds$$
$$+ b(k)\left[e^{ikx} + \int_x^\infty \mathcal{K}(x,s)\,e^{iks}\,ds\right] \qquad (10.52)$$

We now multiply both sides of this equation by $(2\pi)^{-1}e^{iky}$, with y *strictly* greater than x, and integrate over k:

$$\frac{1}{2\pi}\int_{-\infty}^\infty a(k)\psi_k(x)\,e^{iky}\,dk$$
$$= \frac{1}{2\pi}\int_{-\infty}^\infty e^{ik(y-x)}\,dk + \int_x^\infty \mathcal{K}(x,s)\left[\frac{1}{2\pi}\int_{-\infty}^\infty e^{ik(y-s)}\,dk\right]ds$$
$$+ \frac{1}{2\pi}\int_{-\infty}^\infty b(k)e^{ik(x+y)}\,dk + \int_x^\infty \mathcal{K}(x,s)\left[\frac{1}{2\pi}\int_{-\infty}^\infty b(k)\,e^{ik(y+s)}\,dk\right]ds$$
$$= \mathcal{K}(x,y) + \mathcal{B}_0(x+y) + \int_x^\infty \mathcal{K}(x,s)\mathcal{B}_0(s+y)\,ds \qquad (10.53)$$

In writing the second equation in (10.53) we have made use of the fact that

$$\frac{1}{2\pi}\int_{-\infty}^\infty e^{ik(y-x)}\,dk = \delta(y-x) \qquad (10.54)$$

where $\delta(x-y)$ is the Dirac delta function (Problems 7.16–7.20). In particular, if $x \neq y$, this quantity vanishes. We have also introduced the quantity

$$\mathcal{B}_0(z) = \frac{1}{2\pi}\int_{-\infty}^\infty b(k)\,e^{ikz}\,dk \qquad (10.55)$$

which is the Fourier transform of the reflection coefficient.

The right-hand side of (10.53) contains only terms involving $\mathcal{K}$, which is the quantity to be determined, and $\mathcal{B}_0$, which is given in terms of a known quantity, the reflection coefficient of the potential. To simplify the left-hand side of (10.53), the integral must be evaluated. This will be carried out using contour integral methods. In order to do this, we need to examine the analytic structure of the integrand. Equation (10.20) shows that a bound state is characterized by the exponential decay of the wavefunction sufficiently far away from the potential, with the rate of decay determined by κ_n. In terms of the quantities in Equation (10.47), this behavior can be reproduced by identifying the presence of a bound state as a purely imaginary value of k and thereby setting $k = i\kappa$. Then, by analogy with Equation (10.48), the asymptotic properties of ψ_k, φ_k, and φ_{-k} yield

$$\lim_{x \to \infty} \psi_{i\kappa}(x) = A(i\kappa) e^{\kappa x} + B(i\kappa) e^{-\kappa x} \qquad (10.56a)$$

$$\lim_{x \to -\infty} \psi_{i\kappa}(x) = e^{\kappa x} \qquad (10.56b)$$

To avoid having unbounded exponential growth in $\psi_{i\kappa}(x)$ as x becomes large, the quantity $A(i\kappa)$ must be required to vanish. Since $a(k) = 1/A(k)$, a zero of $A(k)$ corresponds to a pole of $a(k)$, i.e. *the transmission coefficient has poles in the complex k plane corresponding to the bound states of the potential.*

Example 10.1. *Mathematica* can be used to examine the analytic structure of the transmission coefficient in a simple example that is well-known to students of introductory quantum mechanics courses, namely, the square-well potential:

$$u(x) = \begin{cases} 0 & \text{if } |x| > a \\ -V & \text{if } |x| \leq a \end{cases} \qquad (10.57)$$

The potential well has width a, depth V, and is centered at the origin. The calculation of the transmission coefficient for this potential is discussed, for example, by Landau and Lifshitz (1965).

In Figure 10.2 the transmission coefficient $a(k)$ is shown in the upper half of the complex k plane as both surface and contour plots, with the well width and well depth chosen to support two bound states. The singularities in the transmission coefficient for points on the imaginary k axis are evident in both the surface and the contour representation. These singularities occur at values of $i\kappa$ that correspond to the bound states of the potential.

To determine the order of the poles of the transmission coefficient, we differentiate the Schrödinger equation for ψ_k with respect to k. Then, using both the original and differentiated Schrödinger equations to eliminate the potential and signifying the differentiation of ψ_k with respect to k with a prime, we obtain

$$\left[\psi_k \frac{d\psi'_k}{dx} - \psi'_k \frac{d\psi_k}{dx} \right]\bigg|_{-\infty}^{\infty} = -2k \int_{-\infty}^{\infty} \psi_k^2 \, dx \qquad (10.58)$$

The Jost solutions in (10.43) and their asymptotic behavior can be used to calculate the limits in (10.58) which, together with the fact that A vanishes

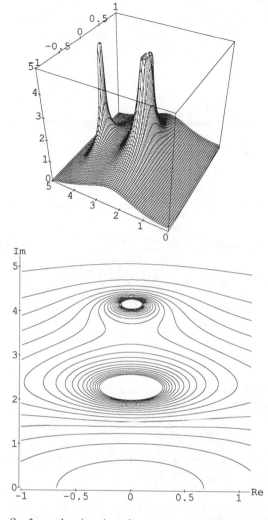

Figure 10.2 Surface plot (top) and contour plot (bottom) of the transmission coefficient $a(k)$ for the square-well potential in Equation (10.57). The poles of $a(k)$, which correspond to the bound states of the potential, cause the sharp singular behavior seen in the two plots.

for bound states, yield (Problem 8)

$$iA'(i\kappa) = \frac{1}{B(i\kappa)} \int_{-\infty}^{\infty} \psi_{i\kappa}^2 \, dx \qquad (10.59)$$

Since $\psi_{i\kappa}(x)$ can be chosen to be a real function for bound states, the right-hand side of (10.58) is real and non-zero. Therefore, since $A'(i\kappa) \neq 0$, the zeros of A are simple or, in other words, *the poles of $a(k) = A^{-1}(k)$ are simple*. Thus, it is natural to consider the evaluation of the left-hand side

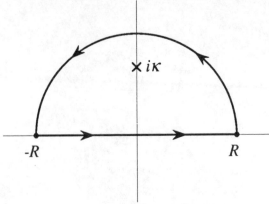

Figure 10.3 The contour of integration used to evaluate the integral in Equation (10.60). The bound state at $i\kappa$ is indicated by a cross.

of (10.53) along the contour $\mathcal{C}$ in the complex plane shown in Figure 10.3:

$$\lim_{R\to\infty}\left[\frac{1}{2\pi}\int_\mathcal{C} a(k)\psi_k(x)\,e^{iky}\,dk\right] \tag{10.60}$$

Before we set this integral equal to the left-hand side of (10.53) we need to show that the contribution from the semi-circular part of the contour vanishes in the limit $R \to \infty$. The asymptotic form of the integral over $\mathcal{C}$ can be determined by examining the behavior of $\psi_k(x)$ in the limit of large imaginary values of k. In this limit, the term $u\psi$ in the Schrödinger equation in (10.42) can be neglected in comparison with the term $k^2\psi$, and the general solution $\psi_k(x)$ can be written as

$$\lim_{k\to i\infty}\psi_k(x) = c_1\,e^{(\mathrm{Im}\,k)x} + c_2\,e^{-(\mathrm{Im}\,k)x} \tag{10.61}$$

where c_1 and c_2 are constants and where the imaginary part of k has been written as $\mathrm{Im}\,k$. A comparison of (10.61) with (10.43) and (10.47) indicates that neither $A(k)$ nor $B(k)$ vanishes in this limit. In particular, the transmission coefficient $a(k)$ remains finite. Thus, substituting (10.61) into the integral in (10.60) and recalling that $y > x$ shows that the contribution from the semi-circular contour vanishes as $R \to \infty$, so we can use the residue theorem to write

$$\frac{1}{2\pi}\int_{-\infty}^{\infty} a(k)\psi_k(x)\,e^{iky}\,dk = \lim_{R\to\infty}\left[\frac{1}{2\pi}\int_\mathcal{C} a(k)\psi_k(x)\,e^{iky}\,dk\right]$$
$$= i\,\mathrm{Res}\left[\frac{\psi_k(x)\,e^{iky}}{A(k)}\right]\bigg|_{k=i\kappa} \tag{10.62}$$

The residue in (10.62) is calculated by applying L'Hospital's rule and using the fact that the poles of $A(k)$ are simple:

$$i\,\mathrm{Res}\left[\frac{\psi_k(x)\,e^{iky}}{A(k)}\right]\bigg|_{k=i\kappa} = i\lim_{k\to i\kappa}\left[(k-i\kappa)\frac{\psi_k(x)\,e^{iky}}{A(k)}\right]$$

$$= i\frac{\psi_{i\kappa}(x)\,e^{-\kappa y}}{A'(i\kappa)} \tag{10.63}$$

Using the value obtained in Equation (10.59) for $A'(k)$ at a zero of $A(k)$ and the fact that at a bound state the wavefunction is given by

$$\psi_{i\kappa}(x) = B(i\kappa)\varphi_{i\kappa}(x) \tag{10.64}$$

we obtain

$$i\,\mathrm{Res}\left[\frac{\psi_k(x)\,e^{iky}}{A(k)}\right]\bigg|_{k=i\kappa} = -B^2(i\kappa)\left[\int_{-\infty}^{\infty}\psi_{i\kappa}^2(x)\,dx\right]^{-1}\varphi_{i\kappa}(x)\,e^{-\kappa y} \tag{10.65}$$

Since the wavefunction $\psi_{i\kappa}(x)$ is assumed to be normalized, the integral in this expression is equal to unity. Furthermore, the limiting form of $\varphi_{i\kappa}(x)$, $\lim_{x\to\infty}\varphi_{i\kappa}(x) = e^{-\kappa x}$, together with a comparison between Equation (10.64) and Equation (10.24), shows that we can make the identification $B(i\kappa) = c_\kappa$. Thus, the residue in Equation (10.65) becomes simply

$$i\,\mathrm{Res}\left[\frac{\psi_k(x)\,e^{iky}}{A(k)}\right]\bigg|_{k=i\kappa} = -c_\kappa^2\varphi_{i\kappa}(x)\,e^{-\kappa y} \tag{10.66}$$

Substituting the definition in (10.43a) of $\varphi_{i\kappa}(x)$ into this equation yields the following expression for the integral on the left-hand side of (10.53):

$$\frac{1}{2\pi}\int_{-\infty}^{\infty} a(k)\psi_k(x)\,e^{iky}\,dk = -c_\kappa^2\,e^{-\kappa(x+y)} - \int_x^\infty K(x,s)c_\kappa^2\,e^{-\kappa(s+y)}\,ds \tag{10.67}$$

There is a corresponding contribution to this equation from every bound state of the potential. Thus, if there are N bound states, we need modify (10.67) only by including a summation over the individual states. Combining this with (10.53), and including the t-dependence of the various quantities, we obtain the integral equation satisfied by $\mathcal{K}$:

$$\mathcal{K}(x,y,t) + \mathcal{B}(x+y,t) + \int_x^\infty \mathcal{K}(x,s,t)\mathcal{B}(s+y,t)\,ds = 0 \tag{10.68}$$

where $\mathcal{B}(z,t)$ is now given by

$$\mathcal{B}(z,t) = \sum_{n=1}^{N} c_n^2\,e^{-\kappa_n z} + \frac{1}{2\pi}\int_{-\infty}^{\infty} b(k,t)\,e^{ikz}\,dk \tag{10.69}$$

The integral equation in (10.68) with the kernel given by (10.69) is known as the **Gel'fand–Levitan equation**. The solution $\mathcal{K}$ to this integral equation is related to the potential by (10.46):

$$u(x,t) = -2\frac{\partial}{\partial x}\mathcal{K}(x,x,t)$$

Notice that $\mathcal{B}$ is given entirely in terms of quantities that characterize the asymptotic behavior of both the bound and unbound states of the potential, namely, the eigenvalues, the normalization constants and the reflection coefficient.

The method of inverse scattering exhibits two important simplifications of the solution of the Korteweg–de Vries equation. First, both the Schrödinger equation (10.1) and the Gel'fand–Levitan equation (10.68) are *linear* equations for which methods of solution, both analytic and numerical, are abundant. Second, the variable t in the Korteweg–de Vries equation appears only as a parameter in Equations (10.68) and (10.69). Thus, once the scattering data have been determined for the initial potential, their evolution can be easily calculated using the relations derived in the preceding section, and the potential determined at any later time by solving (10.68) and (10.46). It is also noteworthy that the potential can be determined entirely from the eigenvalues and only the *asymptotic* behavior of the wavefunctions. All of these features conspire to make the inverse scattering method a flexible and intuitively appealing way to solve the initial-value problem for the Korteweg–de Vries equation. We now consider two examples.

Example 10.2. To illustrate the machinery of the inverse scattering procedure, we consider the particular case where the potential has a single bound state with the eigenvalue $\lambda = -\kappa^2$, normalization constant c, and with a vanishing reflection coefficient for all continuum states k. In this case $\mathcal{B}$ is given simply by

$$\mathcal{B}(z,t) = c^2(t)\,e^{-\kappa z} = c_0^2\,e^{8\kappa^3 t}\,e^{-\kappa z} \qquad (10.70)$$

where $c_0 = c(0)$. The Gel'fand–Levitan equation for $\mathcal{K}$ then reads

$$\mathcal{K}(x,y,t) + c_0^2\,e^{8\kappa^3 t}\,e^{-\kappa(x+y)}$$
$$+ c_0^2\,e^{8\kappa^3 t}\,e^{-\kappa y}\int_x^\infty \mathcal{K}(x,s,t)\,e^{-\kappa s}\,ds = 0 \qquad (10.71)$$

To solve this equation for $\mathcal{K}$, the x and y-dependence of $\mathcal{K}$ are separated as

$$\mathcal{K}(x,y;t) = f(x,t)\,e^{-\kappa y} \qquad (10.72)$$

The solution of an integral equation using the assumption (10.72) is called the method of separable kernels, in analogy with the method of separation of variables for partial differential equations (Tricomi, 1985). Substituting (10.72) into (10.71) and performing the integration over s produces an *algebraic* equation for f:

$$f(x,t)\,e^{-\kappa y} + c_0^2\,e^{8\kappa^3 t}\,e^{-\kappa(x+y)} + \frac{c_0^2}{2\kappa}\,e^{8\kappa^3 t}\,e^{-\kappa y}\,e^{-2\kappa x}f(x,t) = 0 \qquad (10.73)$$

This equation can be solved for $f(x,t)$ and, after a few simple manipulations, we obtain

$$f(x,t) = -\kappa \frac{e^{4\kappa^3 t} e^{\kappa x_0}}{\cosh[\kappa(x - x_0 - 4\kappa^2 t)]} \quad (10.74)$$

The quantity x_0 is defined by

$$x_0 = \frac{1}{2\kappa} \ln\left(\frac{c_0^2}{2\kappa}\right) \quad (10.75)$$

Substitution of (10.74) into (10.72) yields the solution $\mathcal{K}(x,y,t)$ of (10.71):

$$\mathcal{K}(x,y,t) = -\kappa \frac{e^{4\kappa^3 t} e^{-\kappa(y-x_0)}}{\cosh[\kappa(x - x_0 - 4\kappa^2 t)]} \quad (10.76)$$

The corresponding solution of the Korteweg–de Vries equation is obtained by applying (10.46) to (10.76):

$$u(x,t) = -2\kappa^2 \operatorname{sech}^2[\kappa(x - x_0 - 4\kappa^2 t)] \quad (10.77)$$

This solution is seen to correspond to a single soliton of amplitude $-2\kappa^2$, with speed $4\kappa^2$ (moving to the right), and centered initially at the point x_0.

The steps leading to Equation (10.77) can be generalized to include a reflectionless potential with N bound states (Problem 13). Thus, a reflectionless potential with N bound states corresponds, through the inverse scattering transform, to a pure N-soliton solution of the Korteweg–de Vries equation. The question now arises as to the contribution of the continuous part of the eigenvalue spectrum if the reflection coefficient does not vanish.

Example 10.3. The effect of the unbound states on solutions of the Korteweg–de Vries equation can be isolated by considering a potential with no bound states and with a nonvanishing reflection coefficient. An example is the initial-value problem with

$$u(x,0) = 2\operatorname{sech}^2 x \quad (10.78)$$

This potential represents a barrier localized around the origin (Figure 10.4), which produces a nonvanishing reflection coefficient and clearly cannot support any bound states. The time development of the solution from the initial condition in (10.78) can be determined by numerical integration by using a program given by Crandall (1991), and is shown in Figure 10.4.

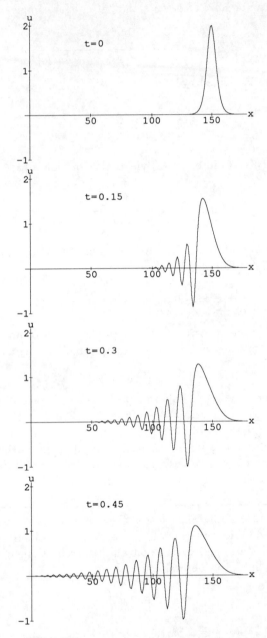

Figure 10.4 Solution of the initial-value problem (10.78) for the Korteweg–de Vries equation determined by numerical integration. The initially localized form decays into a decaying wave train moving to the left. The times are indicated in each plot and the x axis is labeled in units of the discretization interval of the numerical integration, $\Delta x = 0.2$. The time step in the integration is 0.002 units.

The initially localized wave form is seen to unfold into a wave train moving to the left with a decaying amplitude. This behavior is reminiscent of that found for solutions of the *linearized* Korteweg–de Vries equation (Problem 9.4), which also exhibits traveling wave solutions moving to the left, and will be seen to be a general feature of initial-value problems for this equation when the associated potentials have no bound states.

10.3 Solution of the Scattering Problem

Examples 10.2 and 10.3 showed how the discrete and continuous parts of the eigenvalue spectrum of Schrödinger's equation separately contribute to the solution of the initial-value problem of the Korteweg–de Vries equation. In this section, we will examine more closely the interplay between these two contributions by solving the scattering problem for the class of potentials given by

$$u(x) = U\,\text{sech}^2 x \tag{10.79}$$

where U is a constant. The behavior of the solutions will depend strongly on the value of U. In fact, we have already seen in Example 10.2 that if $U > 0$, then there are no bound states, and the solution evolves according to a decaying wave train. Alternatively, we can deduce from Example 10.1 that if $U = -2$, then there is a single bound state and the potential is reflectionless. We will see below that the solution of the scattering problem for this potential exhibits a number of additional intriguing features as U is varied, and we will examine how these features affect the solution of the Korteweg–de Vries equation.

We begin by transforming the Schrödinger equation

$$\frac{d^2\psi}{dx^2} + (\lambda - U\,\text{sech}^2 x)\psi = 0 \tag{10.80}$$

into a form that, for the potential in (10.79), is more amenable to solution. We change the independent variable from x to ξ by the transformation $\xi = \tanh x$ and introduce the function $\phi(\tanh x) = \psi(x)$. In making this variable change, we have changed the range of the independent variable from the entire real line, $-\infty < x < \infty$, to the bounded interval $-1 < \xi < 1$. In terms of the new independent variable ξ, the potential (10.79) becomes

$$U\,\text{sech}^2 x = U(1 - \tanh^2 x) = U(1 - \xi^2) \tag{10.81}$$

and the second derivative operator can be written in terms of ξ as

$$\frac{d^2}{dx^2} = \frac{d\xi}{dx}\frac{d}{d\xi}\left(\frac{d\xi}{dx}\frac{d}{d\xi}\right) = (1-\xi^2)\frac{d}{d\xi}\left[(1-\xi^2)\frac{d}{d\xi}\right] \tag{10.82}$$

Substitution of (10.81) and (10.82) into (10.80) yields the transformed Schrödinger equation:

$$(1 - \xi^2)\frac{d}{d\xi}\left[(1 - \xi^2)\frac{d\phi}{d\xi}\right] + [-U(1 - \xi^2) + \lambda]\phi = 0 \qquad (10.83)$$

Dividing this equation by $1 - \xi^2$, we obtain

$$\frac{d}{d\xi}\left[(1 - \xi^2)\frac{d\phi}{d\xi}\right] + \left[-U + \frac{\lambda}{1 - \xi^2}\right]\phi = 0 \qquad (10.84)$$

which is the generalized Legendre equation encountered in Section 6.2. This equation can be solved as in Section 6.3 by introducing the reduced function F through the substitution

$$\phi(\xi) = (1 - \xi^2)^{\kappa/2} F(\xi) \qquad (10.85)$$

In keeping with the notation used in the preceding section, we have used the notation $\lambda = -\kappa^2$. For bound states, $\lambda < 0$ and κ is a real number, while for continuum states $\lambda > 0$, so κ is purely imaginary. In the latter case, we replace κ by $-ik$, where k is real. By examining the asymptotic form of the solutions of (10.80), we see that κ is the quantity that describes the behavior of the solutions, namely, wave propagation for $\lambda > 0$ and exponential decay for $\lambda < 0$. Substitution of (10.85) into (10.84) yields the equation for F:

$$(1 - \xi^2)\frac{d^2 F}{d\xi^2} - 2\xi(\kappa + 1)\frac{dF}{d\xi} - [U + \kappa(\kappa + 1)]F = 0 \qquad (10.86)$$

Although we can now proceed as in Section 6.3 by solving (10.86) with the method of Frobenius, it is more convenient for treating both the bound and unbound states within the same framework to transform this equation with the change of variables $\zeta = \frac{1}{2}(1 - \xi)$ to obtain

$$\zeta(1 - \zeta)\frac{d^2 F}{d\zeta^2} + (\kappa + 1)(1 - 2\zeta)\frac{dF}{d\zeta} - [U + \kappa(\kappa + 1)]F = 0 \qquad (10.87)$$

This is the hypergeometric equation derived in Problem 5.14 and solved in Problem 5.15, using the method of Frobenius and solved in Section 8.3 using integral representations. The relationship between the hypergeometric function and the Legendre functions was established in Problem 6.15.

Comparing Equation (10.87) with the standard form of the hypergeometric equation,

$$x(1 - x)\frac{d^2 y}{dx^2} + [\gamma - (\alpha + \beta + 1)x]\frac{dy}{dx} - \alpha\beta y = 0 \qquad (10.88)$$

where α, β and γ are constants, allows these quantities to be identified, and the solution that is finite at the origin to be written in terms of the hypergeometric function. Particularly simple expressions for α, β and γ are obtained if we write $U = -\nu(\nu+1)$. Then,

$$\alpha = \kappa - \nu, \qquad \beta = \kappa + \nu + 1, \qquad \gamma = \kappa + 1 \qquad (10.89)$$

and the solution of Equation (10.84) can be written as

$$\phi(\xi) = (1-\xi^2)^{\kappa/2} F[\kappa - \nu, \kappa + \nu + 1; \kappa + 1; \tfrac{1}{2}(1-\xi)] \qquad (10.90)$$

This solution is valid both for $\lambda < 0$, where κ is real, and for $\lambda > 0$, where κ is imaginary.

To obtain the scattering data for the potential in (10.79), we must determine the asymptotic behavior of this solution as $x \to \pm\infty$. In particular, we must identify the values of κ that correspond to the bound states of the potential in (10.79) as a function of U. To proceed with this, we observe that bound states are expected only if $U < 0$. If $U = 0$, then the solution of the Korteweg–de Vries equation is the trivial solution $u(x,t) = 0$. If $U > 0$, then the only solutions that are obtained are those whose evolution is described according to a decaying wave train moving to the left, as shown in Example 10.3. Thus, in the following discussion, we will take $U < 0$, which means that $\nu > 0$.

Bound States. For bound states $\lambda < 0$, so κ is real. Since the standard normalization of the hypergeometric function is chosen by the requirement that $F(\alpha, \beta; \gamma; 0) = 1$, then as $x \to \infty$ ($\xi \to 1$) the asymptotic form of *any* function ψ given by (10.90) is

$$\lim_{x \to \infty} \psi(x) = [1 - \tanh^2 x]^{\kappa/2} = 2^\kappa\, e^{-\kappa x} \qquad (10.91)$$

Consider now the limit $x \to -\infty$, i.e. $\xi \to -1$. According to (10.90), this limit requires evaluating the hypergeometric function at the point $\zeta = 1$, where Equation (10.87) has a regular singular point. We showed in Section 1.4 that the series representation of the hypergeometric function converges at $\xi =$ if $\gamma - \alpha - \beta > 0$ and diverges otherwise. Applying this criterion to the quantities in (10.89), we find

$$\gamma - \alpha - \beta = -\kappa \qquad (10.92)$$

which is a negative number, since $\kappa > 0$. Thus, the hypergeometric function F diverges as $x \to -\infty$, which causes ψ to diverge as well. To remedy this situation, we use the result discussed in Section 8.3 that the hypergeometric function reduces to an nth-order polynomial if α is equal to a negative integer, i.e. $\alpha = -n$, where $n > 0$. Comparing this requirement with the value of α in Equation (10.89), we obtain $\kappa - \nu = -n$, or

$$\kappa = \nu - n \qquad (10.93)$$

Since ν is fixed, there are a finite number of values of n that can satisfy (10.93), while maintaining $\kappa > 0$. Thus, the larger the value of ν, that is, the deeper the potential, the greater the number of bound states in the potential.

Suppose first that ν is equal to an integer N. Then, according to (10.93), κ is also be an integer: $\kappa = N - n$. The allowed values of the eigenvalue $\lambda = -\kappa^2$ are thus given by

$$\lambda = -(N-n)^2, \qquad n = 0, 1, 2, \ldots, N-1 \qquad (10.94)$$

The case $n = N$ is excluded, since the zero eigenvalue does not correspond to a true bound state, as an examination of the associated eigenfunction shows (Problem 20).

Consider now the case where ν is not an integer. There is an integer N such that ν and N satisfy the relation $N < \nu < N+1$, i.e. $\nu = N + \delta$, where $0 < \delta < 1$. Thus, the condition (10.93) can be written as $\kappa = N + \delta - n$. In this case there are N bound states and the corresponding eigenvalues are

$$\lambda = -(\nu - n)^2, \qquad n = 0, 1, 2, \ldots, N-1 \qquad (10.95)$$

The case $n = N$ is again excluded for the reasons just mentioned. The results in Equations (10.94) and (10.95) together state that if $\nu = N + \delta$, with $0 \leq \delta < 1$, the potential $u(x) = -\nu(\nu+1)\operatorname{sech}^2 x$ has N bound states. In particular, if $0 < \nu < 1$, the potential does not support any bound states. The eigenvalues are given by Equation (10.95), which reduces to (10.94) if ν is equal to an integer.

Having found the eigenvalues of (10.80), we turn our attention to the eigenfunctions. Once we examine the scattering properties of these potentials, it will emerge that potentials for which $\nu = N$ will play a special role in solutions of the Korteweg–de Vries equation. Therefore, we will restrict our attention to this special case.

From the analysis of Section 8.3, the hypergeometric function with the condition (10.93) satisfied reduces to an nth-order polynomial. In fact, we can then make use of the connection between the hypergeometric function and the Legendre functions, $P_n^m(x)$, through the relation (Problem 6.15)

$$P_N^\kappa(\xi) = \frac{(N+\kappa)!}{2^\kappa (N-\kappa)!\kappa!}(1-\xi^2)^{\kappa/2} F[\kappa-N, \kappa+N+1; \kappa+1; \tfrac{1}{2}(1-\xi)] \qquad (10.96)$$

where κ and N are *positive* integers. The Legendre functions can also be obtained by solving (10.84) directly, as in Problem 21. The normalization condition for $\psi_N^\kappa(x) = P_N^\kappa(\tanh x)$ is determined by the integral (Problem 22)

$$\int_{-\infty}^{\infty} [\psi_n^\kappa(x)]^2 \, dx = \int_{-1}^{1} \frac{[P_N^\kappa(\xi)]^2}{1-\xi^2} \, d\xi = \frac{(N+\kappa)!}{(N-\kappa)!\kappa} \tag{10.97}$$

Combining (10.97) with (10.96), we obtain the asymptotic form for the discrete eigenfunctions as (Problem 23)

$$\lim_{x \to \infty} \left\{ \left[\frac{\kappa(N-\kappa)!}{(N+\kappa)!} \right]^{1/2} \psi_N^\kappa(x) \right\} = \frac{1}{\kappa!} \left[\frac{\kappa(N+\kappa)!}{(N-\kappa)!} \right]^{1/2} e^{-\kappa x} \tag{10.98}$$

From the definition in (10.23), we conclude that the constant, c_N^κ, for the state $\psi_N^\kappa(x)$ is given by

$$c_N^\kappa = \frac{1}{\kappa!} \left[\frac{\kappa(N+\kappa)!}{(N-\kappa)!} \right]^{1/2} \tag{10.99}$$

We consider a few examples. Suppose first that $\nu = 1$. Then, according to (10.94) there is a single bound state with $\kappa = 1$ and the normalized eigenfunction is

$$\psi_1^1(x) = \tfrac{1}{2}\sqrt{2}\,\text{sech}\,x \tag{10.100}$$

For the case $\nu = 2$, there are two bound states, one corresponding to $\kappa = 1$ and the other to $\kappa = 2$. The corresponding normalized eigenfunctions are

$$\psi_2^1(x) = \tfrac{1}{2}\sqrt{6}\,\tanh x \,\text{sech}\,x \qquad \psi_2^2(x) = \tfrac{1}{2}\sqrt{3}\,\text{sech}^2 x \tag{10.101}$$

Continuum States. To complete the characterization of the potential (10.79) for input into the Gel'fand–Levitan equation, we need to determine the reflection coefficient. For the unbound states, $\lambda > 0$, so κ is an imaginary number. The solution (10.90) is still valid, and we make the replacement $\kappa \to -ik$, to obtain

$$\phi(\xi) = (1-\xi^2)^{-ik/2} F[-ik-\nu, -ik+\nu+1; -ik+1; \tfrac{1}{2}(1-\xi)] \tag{10.102}$$

The asymptotic form of this solution as $x \to \infty$ ($\xi \to 1$) is given by an expression analogous to (10.91):

$$\lim_{x \to \infty} \psi(x) = 2^{-ik} e^{ikx} \tag{10.103}$$

where we again have used the fact that $F(\alpha, \beta; \gamma; 0) = 1$. To determine the transmission and reflection coefficients, we must express the wavefunction in (10.102) in the limit $x \to -\infty$ as the sum of an incident wave, e^{ikx}, and a reflected wave, $b(k)e^{-ikx}$. This can be accomplished with the formula (Problem 24)

$$F(\alpha, \beta; \gamma; x) = \frac{\Gamma(\gamma)\Gamma(\gamma - \alpha - \beta)}{\Gamma(\gamma - \alpha)\Gamma(\gamma - \beta)} F(\alpha, \beta; \alpha + \beta - \gamma + 1; 1 - x)$$

$$+ (1-x)^{\gamma-\alpha-\beta} \frac{\Gamma(\gamma)\Gamma(\alpha+\beta-\gamma)}{\Gamma(\alpha)\Gamma(\beta)} F(\gamma - \alpha, \gamma - \beta; 1 + \gamma - \alpha - \beta; 1 - x)$$

(10.104)

Substituting (10.104) into (10.102) and taking the limit $x \to -\infty$ ($\xi \to -1$), yields the asymptotic form

$$\lim_{x \to -\infty} \psi(x) = 2^{-ik} \frac{\Gamma(1-ik)\Gamma(ik)}{\Gamma(\nu+1)\Gamma(-\nu)} e^{-ikx}$$

$$+ 2^{-ik} \frac{\Gamma(1-ik)\Gamma(-ik)}{\Gamma(-ik-\nu)\Gamma(-ik+\nu+1)} e^{ikx} \qquad (10.105)$$

Thus, if we divide both sides of (10.102) by the coefficient of e^{ikx} in (10.105), then Equations (10.103) and (10.105) take the forms shown in (10.27), where the transmission and reflection coefficients are defined. This allows expressions for the transmission and reflection coefficients from the coefficients of the exponentials to be simply read off from the two equations, and we obtain

$$a^*(k) = \frac{\Gamma(-ik-\nu)\Gamma(-ik+\nu+1)}{\Gamma(1+ik)\Gamma(-ik)}$$

$$b^*(k) = \frac{\Gamma(-ik)\Gamma(-ik-\nu)\Gamma(-ik+\nu+1)}{\Gamma(\nu+1)\Gamma(-\nu)\Gamma(ik)}$$

(10.106)

The properties of these quantities are discussed below and in the problems at the end of the chapter. The corresponding continuum wavefunction (unnormalized, of course) is similarly obtained by dividing (10.102) by the coefficient of e^{ikx} in (10.105):

$$\psi_k(\xi) = 2^{ik} \frac{\Gamma(-ik-\nu)\Gamma(-ik+\nu+1)}{\Gamma(1-ik)\Gamma(-ik)} (1-\xi^2)^{-ik/2}$$

$$\times F[-ik - \nu, -ik + \nu + 1; -ik + 1; \tfrac{1}{2}(1-\xi)] \qquad (10.107)$$

There are several interesting features in the analytic structure of $a(k)$ and $b(k)$ that are revealed in (10.106). We recall in our definition of ν, immediately preceding Equation (10.89), that this quantity is required to be positive for a potential that supports bound states. Since for integer values of ν, the quantity $\Gamma(-\nu)$ is singular, *the reflection coefficient vanishes whenever ν is equal to an integer*. Physically, this means that the reflected waves from different parts of the potential interfere destructively. This allows all of the incident wave to be transmitted (Morse and Feshbach, 1953). The singularities of the Gamma function at negative integer values, combined with Equations (10.94) and (10.95), also show that both $a(k)$ and $b(k)$ have poles in the upper half of the complex k plane at points corresponding to the bound states of the potential.

Reflectionless potentials play a special role in the solution of the Korteweg–de Vries equation. A reflectionless potential with N bound states corresponds to a pure N-soliton solution; the initial function U evolves into a parade of N solitons moving to the right, with there being no contribution from the decaying wave train (Gardner et al., 1974). These potentials also provide one example for which a solution of the Gel'fand–Levitan equation can be obtained explicitly (Problem 13) and the corresponding solution of the Korteweg–de Vries equation expressed in closed form.

Example 10.4. The easiest way to see the effect of varying ν upon solutions of the Korteweg–de Vries equation, and in particular to verify the classification scheme we have just developed, is by direct numerical integration. We will again apply the method described by Crandall (1991) and display the solutions by using Plot. It is also instructive (and entertaining!) to animate these solutions. We consider initial values of the form in Equation (10.79) with (a) $U = -1$, (b) $U = -2$, (c) $U = -4$ and (d) $U = -6$. The time development of these solutions is shown in Figures 10.5, 10.6, 10.7 and 10.8, respectively.

(a) The case $U = -1$ corresponds to $\nu < 1$. According to the discussion following Equation (10.95) this potential does not support any bound states. The evolution of the solution is therefore seen to be qualitatively similar to that in Example 10.2. The initially localized profile decays into a decaying wave train moving to the left.

(b) The case $U = -2$ corresponds to $\nu = 1$. The potential has a single bound state and a vanishing reflection coefficient. The associated solution of the Korteweg–de Vries equation is simply a special case of the solution derived in Example 10.2 and is seen to correspond to a single soliton moving to the right with no decaying wave train. Thus, an initial condition of the Korteweg–de Vries equation corresponding to $\nu \leq 1$ does not lead to any qualitatively new behavior from that seen in Examples 10.2 and 10.3.

(c) The case $U = -4$ corresponds to $1 < \nu < 2$. The potential supports a single bound state, and the reflection coefficient is nonvanishing. Thus, there are contributions from *both* the discrete and continuous states to the function $\mathcal{B}$ in Equation (10.69). The effect of the different contributions is seen in the time development of the solution by the presence of both the decaying wave train, moving to the left, and the emergence of a single soliton moving to the right.

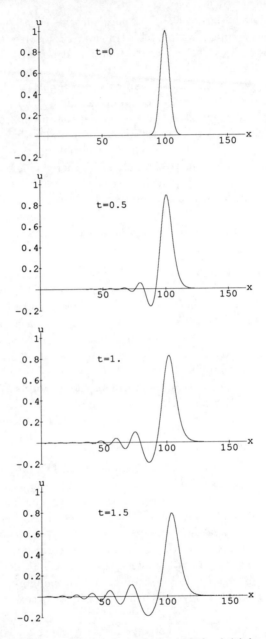

Figure 10.5 The *negative* of the solution of the initial-value problem for the Korteweg–de Vries equation (see the discussion accompanying Equations (9.15) and (9.16)) determined by numerical integration for case (a) in Example 10.4. The initial condition corresponds to a potential with no bound states and the solution evolves into a decaying wave train moving to the left. The times are indicated in each plot and the x axis is labeled in units of the discretization interval of the numerical integration, $\Delta x = 0.2$.

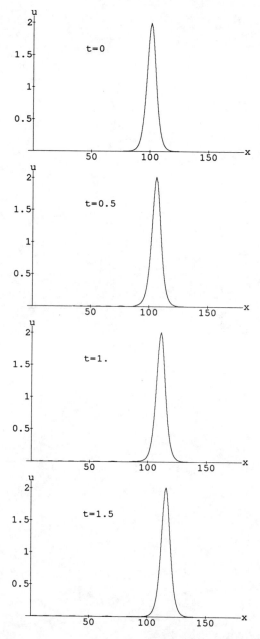

Figure 10.6 The *negative* of the solution of the initial-value problem for the Korteweg–de Vries equation determined by numerical integration for case (b) in Example 10.4. The initial condition corresponds to a potential with a single bound state and a vanishing reflection coefficient. The solution evolves with no change in form into a single soliton moving to the right with there being no decaying wave train. The labeling of the time sequence and the x axis are the same as in Figure 10.5.

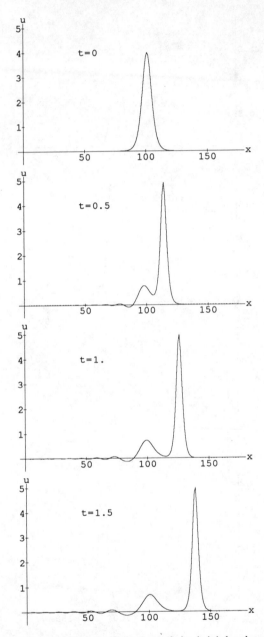

Figure 10.7 The *negative* of the solution of the initial-value problem for the Korteweg–de Vries equation determined by numerical integration for case (c) in Example 10.4. The initial condition corresponds to a potential with a single bound state and a nonvanishing reflection coefficient. The solution evolves into a decaying wave train moving to the left and a single soliton moving to the right. The labeling of the time sequence and the x axis are the same as in Figure 10.5.

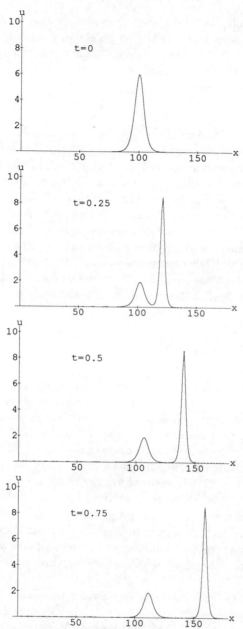

Figure 10.8 The *negative* of the solution of the initial-value problem for the Korteweg–de Vries equation determined by numerical integration for case (d) in Example 10.4. The initial condition corresponds to a potential with two bound states and a vanishing reflection coefficient. The solution evolves into two solitons moving to the right, with the larger soliton moving with the greater speed. The labeling of the time sequence and the x axis are the same as in Figure 10.5.

(d) Finally, the case $U = -6$ corresponds to $\nu = 2$. The potential has two bound states and a vanishing reflection coefficient. The associated solution of the Korteweg–de Vries equation is now seen to evolve into *two* solitons moving to the right.

The general features of the types of solutions obtained to the initial-value problem for the Korteweg–de Vries equation are now clear. The continuous part of the spectrum contributes to the solution whenever the reflection coefficient is nonvanishing, and leads to the appearance of a wave train moving to the left, by analogy with the solution of the linearized Korteweg–de Vries equation discussed in Problems 9.3 and 9.4. Any bound states of the potential lead asymptotically to the emergence of solitons moving to the right with a speed and amplitude determined by the eigenvalue. The method of inverse scattering has revealed many other remarkable properties of the Korteweg–de Vries equation, discussions of which may be found in the references provided at the end of the chapter.

10.4 Lax's Method

The success of the method of inverse scattering in solving the Korteweg–de Vries equation naturally raises the question of whether other nonlinear equations can be solved by analogous methods. The most surprising result of the inverse scattering procedure, and the central feature used for in generalizing the method, is the fact that the discrete eigenvalues of the Schrödinger equation do not change as the potential evolves according the Korteweg–de Vries equation. Lax (1968) approached the method of inverse scattering from a slightly different point of view based upon an abstract formulation of evolution equations and certain properties of operators, some of which are familiar in the context of quantum mechanics. Lax's argument has the virtue of associating certain nonlinear evolution equations with linear equations which are the analog of Schrödinger's equation for the Korteweg–de Vries equation.

To develop Lax's method, we consider two *linear* operators $\mathcal{L}$ and $\mathcal{B}$. The eigenvalue equation associated with the operator $\mathcal{L}$ corresponds to the Schrödinger equation for the Korteweg–de Vries equation. The general form of this eigenvalue equation is

$$\mathcal{L}\psi = \lambda\psi \tag{10.108}$$

where ψ is the eigenfunction and λ is the corresponding eigenvalue. The operator $\mathcal{B}$ characterizes the change of the eigenvalues with the parameter

t which in a nonlinear evolution equation usually corresponds to the time. The general form of this equation is

$$\psi_t = \mathcal{B}\psi \tag{10.109}$$

For the Korteweg–de Vries equation, this equation corresponds to (10.23). We now differentiate Equation (10.108) with respect to time,

$$\mathcal{L}_t\psi + \mathcal{L}\psi_t = \lambda_t\psi + \lambda\psi_t \tag{10.110}$$

and use Equation (10.109) to eliminate ψ_t:

$$\mathcal{L}_t\psi + \mathcal{L}\mathcal{B}\psi = \lambda_t\psi + \lambda\mathcal{B}\psi$$
$$= \lambda_t\psi + \mathcal{B}\lambda\psi$$
$$= \lambda_t\psi + \mathcal{B}\mathcal{L}\psi \tag{10.111}$$

This equation can be rearranged to produce

$$\lambda_t\psi = (\mathcal{L}_t + \mathcal{L}\mathcal{B} - \mathcal{B}\mathcal{L})\psi \tag{10.112}$$

Thus, for non-zero eigenfunctions the invariance of the eigenvalues with respect to changes in t, i.e. $\lambda_t = 0$, is obtained if and only if

$$\frac{\partial \mathcal{L}}{\partial t} = \mathcal{B}\mathcal{L} - \mathcal{L}\mathcal{B} \tag{10.113}$$

The derivative on the left-hand side of this equation is to be interpreted as the time derivative of the operator alone, *not* as the time derivative of the operator acting on a function. Equation (10.113) is called the **Lax equation** and the operators $\mathcal{B}$ and $\mathcal{L}$ are called a **Lax pair**. The problem is of course to find these operators for a given equation, and this is not always a straightforward task. In fact, no systematic procedure has been devised to determine whether a given nonlinear partial differential equation can be represented as in (10.113) and, if it can, to determine the Lax pair.

To see the connection between these ideas and inverse scattering, we can identify the operators $\mathcal{L}$ and $\mathcal{B}$ by comparing Equations (10.14) and (10.109) (Problem 26):

$$\mathcal{L} = -\frac{\partial^2}{\partial x^2} + u,$$
$$\mathcal{B} = -4\frac{\partial^3}{\partial x^3} + 6u\frac{\partial}{\partial x} + 3\frac{\partial u}{\partial x} \tag{10.114}$$

where $u = u(x,t)$. $\mathcal{L}$ is readily identified as the Schrödinger operator. Then, the required quantities are

$$\mathcal{BL}\psi = \left[-4\frac{\partial^3}{\partial x^3} + 6u\frac{\partial}{\partial x} + 3\frac{\partial u}{\partial x}\right]\left[-\frac{\partial^2}{\partial x^2} + u\right]\psi$$

$$= 4\frac{\partial^5 \psi}{\partial x^5} - 4\frac{\partial^3 u}{\partial x^3}\psi - 12\frac{\partial^2 u}{\partial x^2}\frac{\partial \psi}{\partial x} - 12\frac{\partial u}{\partial x}\frac{\partial^2 \psi}{\partial x^2} - 4u\frac{\partial^3 \psi}{\partial x^3}$$

$$- 6u\frac{\partial^3 \psi}{\partial x^3} + 6u\psi\frac{\partial u}{\partial x} + 6u^2\frac{\partial \psi}{\partial x} - 3\frac{\partial u}{\partial x}\frac{\partial^2 \psi}{\partial x^2} + 3u\psi\frac{\partial u}{\partial x} \quad (10.115a)$$

and

$$\mathcal{LB}\psi = \left[-\frac{\partial^2}{\partial x^2} + u\right]\left[-4\frac{\partial^3}{\partial x^3} + 6u\frac{\partial}{\partial x} + 3\frac{\partial u}{\partial x}\right]\psi$$

$$= 4\frac{\partial^5 \psi}{\partial x^5} - 6\frac{\partial^2 u}{\partial x^2}\frac{\partial \psi}{\partial x} - 12\frac{\partial u}{\partial x}\frac{\partial^2 \psi}{\partial x^2} - 6u\frac{\partial^3 \psi}{\partial x^3} - 3\frac{\partial^2 \psi}{\partial x^2}\frac{\partial u}{\partial x}$$

$$- 6\frac{\partial \psi}{\partial x}\frac{\partial^2 u}{\partial x^2} - 3\psi\frac{\partial^3 u}{\partial x^3} - 4u\frac{\partial^3 \psi}{\partial x^3} + 6u^2\frac{\partial \psi}{\partial x} + 3u\psi\frac{\partial u}{\partial x} \quad (10.115b)$$

By combining these expressions, we obtain

$$[\mathcal{BL} - \mathcal{LB}]\psi = -\frac{\partial^3 u}{\partial x^3}\psi + 6u\psi\frac{\partial u}{\partial x} = \left[-\frac{\partial^3 u}{\partial x^3} + 6u\frac{\partial u}{\partial x}\right]\psi \quad (10.116)$$

so that

$$\mathcal{BL} - \mathcal{LB} = -\frac{\partial^3 u}{\partial x^3} + 6u\frac{\partial u}{\partial x} \quad (10.117)$$

This determines the right-hand side of the equation of motion in (10.109). To determine the left-hand side, we must calculate the derivative $\mathcal{L}_t$. From Equation (10.114) this is seen to be

$$\mathcal{L}_t = u_t \quad (10.118)$$

Combining (10.117) with (10.118), we see that the relation $\mathcal{L}_t = \mathcal{BL} - \mathcal{LB}$ is satisfied provided that u satisfies the Korteweg–de Vries equation:

$$u_t = -u_{xxx} + 6uu_x \quad (10.119)$$

Lax's method clarifies the underlying structure that is responsible for the success of the inverse scattering method. The calculations that were carried out in a heuristic manner in Section 10.1 are seen to be the result of writing an evolution equation in the form (10.113). This method also helps to identify other pairs of operators that indicate a particular nonlinear equation is amenable to solution by inverse scattering. The problem is, of course, to determine the operators $\mathcal{L}$ and $\mathcal{B}$ which, for a given equation, involves calculating the quantities in (10.115). Although this procedure can be cumbersome if many trials are involved, *Mathematica* can be used to carry out these steps to identify quickly the evolution equation implied by a particular pair of operators, as the following example shows.

To carry out the algebraic steps in (10.115) with *Mathematica*, the operators $\mathcal{L}$ and $\mathcal{B}$ are first defined:

```
f:=Psi[x,y]

g:=u[x,y]

LaxL[f_]:=-D[f,{x,2}] + g f

LaxB[f_]:=-4 D[f,{x,3}] + 6 g D[f,x] + 3 f D[g,x]
```

The operations in equations (10.115) are now carried out with the command

```
Simplify[LaxB[LaxL[f]]-LaxL[LaxB[f]]-
    Dt[LaxL[f],y,Constants->{D[f,{x,2}],f,x}]]
```

```
                (0,1)                  (1,0)
-(Psi[x, y] (u      [x, y] - 6 u[x, y] u     [x, y] +

         (3,0)
        u     [x, y]))
```

which, when set equal to zero, yields (10.119). Notice that in calculating the left-hand side of this equation, the operation Dt was applied to $\mathcal{L}\psi$ and only u was considered as being t-dependent.

Further Reading

There are several books that provide good introductory discussions to the method of inverse scattering and its extensions, including those by Lamb (1980), Ablowitz and Segur (1981), Drazin (1983), Drazin and Johnson (1989) and Ablowitz and Clarkson (1991). An up-to-date compilation of equations that are solvable by the inverse scattering method is given by Ablowitz and Clarkson (1991). In addition the original papers by Gardner,

Greene, Kruskal and Miura are very readable, with Gardner *et al.* (1974) providing a useful summary and supporting discussions of many of the results presented in this chapter. Another useful summary of the inverse scattering method and of nonlinear wave equations has been given by Gibbon (1985). A comprehensive discussion of quantum mechanical inverse scattering methods in one dimension and in higher dimensions may be found in Chadan and Sabatier (1989). Crandall (1991) discusses the construction and graphical representation of pure soliton solutions of the Korteweg–de Vries equation, as well as the numerical integration of this equation, both within the framework of *Mathematica*.

References

Ablowitz M. J. and Clarkson P. A. (1991). *Solitons, Nonlinear Evolution Equations and Inverse Scattering*. Cambridge: Cambridge University Press.

Ablowitz M. J. and Segur H. (1981). *Solitons and the Inverse Scattering Transform*. Philadelphia: Society for Industrial and Applied Mathematics.

Abramowitz M. and Stegun I. A. (1965). *Handbook of Mathematical Functions*. New York: Dover.

Crandall R. E. (1991). *Mathematica for the Sciences*. Redwood City CA: Addison-Wesley.

Chadan K. and Sabatier P. C. (1989). *Inverse Problems in Quantum Scattering Theory*, 2nd edn. New York: Springer.

Drazin P. G. (1983). *Solitons*. Cambridge: Cambridge University Press.

Drazin P. G. and Johnson R. (1989). *Solitons: An Introduction*. Cambridge University Press.

Gardner C. S., Greene J. M., Kruskal M. D. and Miura R. M. (1974). Korteweg–de Vries equation and generalizations. VI. Methods for Exact Solution. *Commun. Pure Appl. Math.* **27**, 97–133.

Gibbon J. D. (1985). A survey of the origins and physical importance of soliton equations. *Phil. Trans. Roy. Soc. Lond. A* **315**, 335–65.

Lamb G. L. Jr. (1980). *Elements of Soliton Theory*. New York: Wiley.

Landau L. D. and Lifshitz E. M. (1965). *Quantum Mechanics*, 2nd edn. Oxford: Pergamon.

Lax P. D. (1968). Integrals of nonlinear equations of motion and solitary waves. *Commun. Pure Appl. Math.* **21**, 467–90.

Morse P. M. and Feshbach H. (1953). *Methods of Theoretical Physics* Vol. 2. New York: McGraw-Hill.

Tricomi F. G. (1985). *Integral Equations*. New York: Dover.

Problems

1. One of the equations comprising the Bäcklund transformation in Problem 9.16 is in terms of the function v that satisfies the *modified* Korteweg-de Vries equation:
$$u = v^2 + v_x$$
If we define the function ψ by $v = \psi_x/\psi$ then ψ is seen to satisfy the equation
$$\psi_{xx} - u\psi = 0$$
which has the appearance of a Schrödinger equation with eigenvalue zero. To proceed further, consider the quantity
$$u_\lambda = u - \lambda$$
where λ is a constant. Show that if $u(x,t)$ is a solution of the Korteweg–de Vries equation, then so is
$$u_\lambda(x,t) = u(x - 6\lambda t, t) - \lambda$$
Hence, by constructing
$$u - \lambda = v^2 + v_x$$
show that ψ satisfies the time-independent Schrödinger equation with the solution u of the Korteweg-de Vries equation as a potential and λ as the eigenvalue:
$$-\psi_{xx} + u\psi = \lambda\psi$$

2. Show that the Wronskian of two solutions $\psi(x)$ and $\phi(x)$ of Schrödinger's equation is independent of x:
$$\frac{d}{dx}\left[\psi\frac{d\phi}{dx} - \phi\frac{d\psi}{dx}\right] = 0$$
Determine the asymptotic form (where the effect of u can be neglected) of the two solutions of Schrödinger's equation corresponding to the eigenvalue $\lambda_n = -\kappa_n^2$ and show that these solutions are linearly independent.

3. Show that the Wronskian between the eigenfunction $\psi_k(x,t)$ of the Schrödinger equation in Equation (10.1) and the function $\Psi_k(x,t)$ in Equation (10.14) vanishes:
$$\psi_k\Psi_{k,x} - \Psi_k\psi_{k,x} = 0$$
In Problem 2 we showed that the Wronskian between two solutions of Schrödinger's equation with the same eigenvalue is independent of x. Thus, we can evaluate this quantity in the limit $x \to \pm\infty$ where $\psi_k(x,t)$ assumes the simple

forms in Equation (10.26). Show first that the requirement that the potential, $u(x,t)$, and its derivatives vanish in these limits implies that

$$\psi\Psi_x - \Psi\psi_x = \psi\psi_{xt} - 4k^2\psi\psi_{xx} - \psi_x\psi_t + 4k^2\psi_x^2$$

where the state label has been suppressed. Substitute the appropriate asymptotic forms for $\psi_k(x,t)$ by using (10.26) and obtain

$$\lim_{x\to\pm\infty}[\psi_k\Psi_{k,x} - \Psi_k\psi_{k,x}] = 0$$

By integrating this equation deduce that $\Psi_k(x,t)$ and $\psi_k(x,t)$ are proportional and that the proportionality constant is independent of x but may depend on t.

4. Another way (Gardner et al., 1974) to arrive at Equation (10.23) from (10.21) is to multiply the latter equation by ψ_n and integrate from $x=-\infty$ to $x=\infty$ to obtain

$$\int_{-\infty}^{\infty}\psi_{n,t}\psi_n\,dx + \int_{-\infty}^{\infty}[\psi_{n,xxx} - 3(u+\lambda_n)\psi_{n,x}]\psi_n\,dx = A_n\int_{-\infty}^{\infty}\psi_n^2\,dx$$

Then eliminate the potential from the second integral on the left-hand side and show that the resulting expression can be written as

$$\tfrac{1}{2}\int_{-\infty}^{\infty}(\psi_n^2)_t\,dx + \int_{-\infty}^{\infty}(\psi_n\psi_{n,xx} - 2\psi_{n,x}^2 - 3\lambda\psi_n^2)_x\,dx = A_n\int_{-\infty}^{\infty}\psi_n^2\,dx$$

Deduce from the fact that ψ_n is normalized and from the asymptotic form of a bound state that

$$A_n = 0$$

5. We can regard the Schrödinger equation,

$$\psi_{xx} + u\psi = \lambda\psi$$

where $u(x,t)$ is a solution of the Korteweg-de Vries equation and the equation satisfied by ψ_t in (10.21),

$$\psi_t = (\beta + u_x)\psi - (4\lambda + 2u)\psi_x$$

as a Bäcklund transformation. The quantity β in this equation is a Bäcklund parameter. Thus, differentiate the Schrödinger equation with respect to t, differentiate the evolution equation for ψ twice with respect to x and then use both equations to obtain expressions for ψ_{xxt} and ψ_{txx}. By requiring $\psi_{txx} = \psi_{xxt}$, obtain

$$u_t + 6uu_x + u_{xxx} = \lambda_t$$

Thus, u satisfies the Korteweg-de Vries equation, if and only if $\lambda_t = 0$.

6. Although the Jost solutions in (10.43) may at first glance appear unmotivated, their form is in fact a natural result of the method of variation of parameters for solving inhomogeneous ordinary differential equations.

Consider first the solution of the inhomogeneous equation

$$\frac{d^2 y}{dx^2} + k^2 y = V$$

where $V = V(x)$. The solution of the homogeneous equation can be written as

$$y(x) = A e^{ikx} + B e^{-ikx}$$

where A and B are constants. To obtain a particular solution of the inhomogeneous equation using the method of variation of parameters, assume a trial solution of the form,

$$y(x) = A(x) e^{ikx} + B(x) e^{-ikx}$$

where both A and B are now functions of x. Substitute this form into the inhomogeneous equation and show that the solutions satisfying the boundary condition (10.44) can be written as

$$y_{\pm k}(x) = e^{\pm ikx} + \frac{1}{2} \int_x^\infty F[\tfrac{1}{2}(x+s)] e^{\pm iks} \, ds$$

where

$$F(x) = \int_x^\infty V(s) e^{\mp iks} \, ds$$

In obtaining this result, you may assume that

$$\lim_{x \to \infty} V(x) = 0$$

If we now consider the inhomogeneous equation where $V(x)$ is replaced by $V(x) y(x)$,

$$\frac{d^2 y}{dx^2} + k^2 y = V y$$

show that a solution is obtained by making the corresponding replacement in the solution obtained for the first equation:

$$y_{\pm k}(x) = e^{\pm ikx} + \int_x^\infty e^{\pm iks} \left[\int_{(x+s)/2}^\infty V(t) y(t) e^{\mp ikt} \, dt \right] ds$$

which is an integral equation for y. This equation may solved at least formally by iterating, i.e. by writing the nth-order approximation to the solution as

$$y_{\pm k}^{(n)}(x) = e^{\pm ikx} + \int_x^\infty e^{\pm iks} \left[\int_{(x+s)/2}^\infty V(t) y_{\pm k}^{(n-1)}(t) e^{\mp ikt} \, dt \right] ds$$

together with the zeroth-order solution

$$y^{(0)}_{\pm k}(x) = e^{\pm ikx}$$

Show that this procedure leads to solutions that have the form of the Jost solutions in (10.43).

7. Consider solutions of the Schrödinger equation

$$\left[\frac{d^2}{dx^2} + (k^2 - u)\right]\varphi_k = 0$$

where $u = u(x,t)$, in the form

$$\varphi_k(x) = e^{ikx} + \int_x^\infty K(x,s)\, e^{iks}\, ds$$

$$\varphi_{-k}(x) = e^{-ikx} + \int_x^\infty K(x,s)\, e^{-iks}\, ds$$

By substituting $\varphi_{\pm k}(x)$ into the Schrödinger equation, show that K must satisfy an inhomogeneous wave equation

$$\frac{\partial^2 K}{\partial x^2} - \frac{\partial^2 K}{\partial s^2} = uK$$

and that K must be be related to the potential by

$$u(x,t) = -2\frac{dK(x,x)}{dx}$$

in addition to satisfying the following conditions:

$$\lim_{y\to\infty} K(x,y) = 0, \qquad \lim_{y\to\infty} K_y(x,y) = 0$$

8. Evaluate the quantities in Equation (10.58),

$$\left[\psi_k \frac{d\psi'_k}{dx} - \psi'_k \frac{d\psi_k}{dx}\right]\Bigg|_{-\infty}^{\infty} = -2k \int_{-\infty}^\infty \psi_k^2\, dx$$

using the representation of ψ_k in Equation (10.47)

$$\psi_k(x) = A(k)\varphi_{-k}(x) + B(k)\varphi_k(x)$$

where $\varphi_{\pm k}(x)$ are the Jost solutions in (10.43). By taking the required derivatives of $\psi_k(x)$, evaluate the resulting expression at the bound state $k = i\kappa$ and, observing that $A(i\kappa)$ must vanish (but $A'(i\kappa)$ need not), obtain the result in Equation (10.59)

$$iA'(i\kappa) = \frac{1}{B(i\kappa)} \int_{-\infty}^{\infty} \psi_{i\kappa}^2 \, dx$$

9. Using *Mathematica*, or otherwise, show that the kernel $\mathcal{K}(x,y,t)$ calculated in Example 10.2,

$$\mathcal{K}(x,y;t) = -\kappa \frac{\exp[4\kappa^3 t - \kappa(y - x_0)]}{\cosh[\kappa(x - x_0 - 4\kappa^2 t)]}$$

is a solution of

$$\mathcal{K}_{xx} - \mathcal{K}_{yy} - u\mathcal{K} = 0$$

where u is given by

$$u(x,t) = -2\frac{\partial}{\partial x}\mathcal{K}(x,x,t)$$
$$= -2\kappa^2 \operatorname{sech}^2[\kappa(x - x_0 - 4\kappa^2 t)]$$

10. Substitute the function $\mathcal{K}(x,y;t)$ calculated in Example 10.2 into equations (10.43) to obtain the following expressions for the Jost solutions for the Schrödinger equation with the potential given in (10.77):

$$\varphi_{\pm k}(x,t) = e^{\pm ikx} - \frac{\kappa}{\kappa \mp ik} \frac{\exp[4\kappa^3 t + \kappa x_0 - (\kappa \mp ik)x]}{\cosh[\kappa(x - x_0 - 4\kappa^2 t)]}$$

Using *Mathematica*, or otherwise, verify that these functions are solutions of Schrödinger's equation with the potential given by (10.77).

Show the Jost function corresponding to the bound state of this Schrödinger equation with eigenvalue $\lambda = -\kappa^2$ is given by

$$\varphi_{i\kappa}(x,t) = \frac{1}{2} \frac{\exp[-\kappa(x_0 + 4\kappa^2 t)]}{\cosh[\kappa(x - x_0 - 4\kappa^2 t)]}$$

Evaluate this expression in the limits $x \to \pm\infty$ to show that the asymptotic behavior is that expected for a bound state.

11. Show that the function $\varphi_{i\kappa}(x,0)$ calculated in Problem 10 is proportional to the wavefunction in (10.100). Obtained the normalized form of $\varphi_{i\kappa}(x,t)$ as

$$\varphi_{i\kappa}(x,t) = \tfrac{1}{2}\sqrt{2\kappa}\operatorname{sech}[\kappa(x - x_0 - 4\kappa^2 t)]$$

Evaluate this function in the limit $x \to \infty$ and show that

$$\lim_{x \to \infty} \varphi_{i\kappa}(x,t) = \sqrt{2\kappa}\exp[-\kappa(x - x_0 - 4\kappa^2 t)]$$

Hence, deduce that
$$c_\kappa(t) = c_\kappa(0) \exp(4\kappa^2 t)$$

12. Consider the eigenfunction of the potential
$$u(x,0) = -2\operatorname{sech}^2 x$$
with the eigenvalue $\kappa = 1$. Show from the results of Problem 11 that the normalized t-dependent eigenfunction is given by
$$\psi(x,t) = \tfrac{1}{2}\sqrt{2}\operatorname{sech}(x - 4t)$$
Using the Plot command of *Mathematica*, display both this normalized eigenfunction and the potential $u(x,t)$ in Equation (10.77) on the same graph and show how the two quantities change as a function of t.

Show that $\psi(x,t)$ satisfies Equation (10.23)
$$\psi_t + \psi_{xxx} - 3(u-1)\psi_x = 0$$
with the potential as given above.

13. The procedure described in Example 10.2 can be generalized to any potential with N bound states and with vanishing reflection coefficient (Gardner et al., 1974). Thus, suppose the bound states have parameters $\kappa_1, \ldots, \kappa_N$ and $c_1, \ldots, c_N$. Then the kernel of the Gel'fand–Levitan equation is given by
$$B(x+y;t) = \sum_{n=1}^{N} c_n^2(t) \exp[-\kappa_n(x+y)] \equiv \sum_{n=1}^{N} f_n(x,t) g_n(y)$$
where $f_n(x,t) = c_n^2(t)\exp(-\kappa_n x)$ and $g_n(y) = \exp(-\kappa_n y)$. Then, with $\mathcal{K}$ assumed to be of the form
$$\mathcal{K}(x,y,t) = \sum_{n=1}^{N} k_n(x,t) g_n(y)$$
substitute these expressions for B and $\mathcal{K}$ into the Gel'fand–Levitan equation to obtain
$$\sum_{n=1}^{N} k_n(x,t) g_n(y) + \sum_{n=1}^{N} f_n(x,t) g_n(y)$$
$$+ \int_x^\infty \left[\sum_{n=1}^{N} k_n(x,t) g_n(s) \sum_{n=1}^{N} f_n(s,t) g_n(y) \right] ds = 0$$

By equating the coefficients of $g_n(y)$ deduce

$$k_m + f_m + c_m^2 \sum_{n=1}^{N} k_n \frac{\exp[-(\kappa_m + \kappa_n)x]}{\kappa_m + \kappa_n} = 0$$

Write this equation in matrix form as

$$Mk + f = 0$$

where k and f are column vectors with entries k_n and f_n, respectively, and M is an $N \times N$ matrix with entries

$$M_{ij} = \delta_{ij} + c_i^2(t) \frac{\exp[-(\kappa_i + \kappa_j)x]}{\kappa_i + \kappa_j}$$

where δ_{ij} is the Kronecker delta:

$$\delta_{ij} = \begin{cases} 1 & \text{if } i = j \\ 0 & \text{if } i \neq j \end{cases}$$

Solve this matrix equation to obtain

$$K(x,x,t) = \sum_{n=1}^{N} k_n(x,t) g_n(x) = -g^T M^{-1} f$$

where g is the column vector with entries g_n and 'T' indicates transpose, so g^T is a row vector. Show that

$$\frac{\partial}{\partial x} M_{ij} = -f_i(x,t) g_j(x)$$

Then use the fact that $\ln \det M = \operatorname{Tr} \ln M$, where Tr means 'trace' and signifies the sum of the diagonal elements of a matrix, to obtain

$$K(x,x,t) = \operatorname{Tr}\left[M^{-1} \frac{\partial M}{\partial x}\right] = \frac{\partial}{\partial x} \ln \det M$$

and therefore

$$u(x,t) = -2 \frac{\partial^2}{\partial x^2} \ln \det M$$

14. Use the procedure outlined in Problem 13 to construct the two-soliton solution of the Korteweg–de Vries equation for the initial value given by

$$u(x,0) = -6 \operatorname{sech}^2 x$$

First use the eigenfunctions in Equation (10.101) to show that $\det M$ is given by the expression

$$\det M = 1 + 3 e^{8t - 2x} + 3 e^{64t - 4x} + e^{72t - 6x}$$

Then obtain $u(x,t)$ in the form

$$u(x,t) = -12\frac{3 + 4\cosh(2x - 8t) + \cosh(4x - 64t)}{[3\cosh(x - 28t) + \cosh(3x - 36t)]^2}$$

which is the solution obtained in Problem 9.24.

15. Use the procedure described in Problem 14 to construct the function $K(x,y,t)$ for the initial value of the Korteweg–de Vries equation given by

$$u(x,0) = -6\,\text{sech}^2 x$$

In particular, show that $K(x,y,t)$ can be written in the form

$$K(x,y,t) = f_1(x,t)\,e^{-y} + f_2(x,t)\,e^{-2y}$$

where $f_1(x,t)$ and $f_2(x,t)$ are given by

$$f_1(x,t) = -6\frac{e^{8t-x} - e^{72t-5x}}{1 + 3\,e^{8t-2x} + 3\,e^{64t-4x} + e^{72t-6x}}$$

$$f_2(x,t) = -12\frac{e^{64t-2x} + e^{72t-4x}}{1 + 3\,e^{8t-2x} + 3\,e^{64t-4x} + e^{72t-6x}}$$

Substitute this expression into the definition (10.43) for the Jost functions to obtain the solutions corresponding to the bound states at $\kappa=1$ and $\kappa=2$ as

$$\varphi_i(x,t) = \frac{e^{-x} - e^{64t-5x}}{1 + 3\,e^{8t-2x} + 3\,e^{64t-4x} + e^{72t-6x}}$$

$$\varphi_{2i}(x,t) = \frac{e^{-2x} + e^{8t-4x}}{1 + 3\,e^{8t-2x} + 3\,e^{64t-4x} + e^{72t-6x}}$$

Show that at $t=0$ these solutions are proportional to those in (10.101).

16. The Jost solutions obtained in Problem 15 can be used to obtain the wavefunctions corresponding to the potentials as $t \to \infty$. To carry out this analysis, first show that the two solutions $\varphi_i(x,t)$ and $\varphi_{2i}(x,t)$ can be written as

$$\varphi_i(x,t) = e^{-4t}\frac{\sinh(2x - 32t)}{\cosh(3x - 36t) + 3\cosh(x - 28t)}$$

$$\varphi_{2i}(x,t) = e^{-32t}\frac{\cosh(x - 4t)}{\cosh(3x - 36t) + 3\cosh(x - 28t)}$$

We now follow the steps taken in Problem 9.25. Introduce the variable $\xi_1 = x - 4t$ into $\varphi_i(x,t)$ and show that

$$\lim_{t \to \pm\infty} \varphi_i(\xi_1, t) = -\frac{e^{-4t}}{2\sqrt{3}}\text{sech}(\xi \pm \tfrac{1}{2}\ln 3)$$

What is the parity of this solution compared with the solution at $t=0$? Similarly, introduce the variable $\xi_2 = x - 16t$ into $\varphi_{2i}(x,t)$ and obtain

$$\lim_{t \to \pm\infty} \varphi_{2i}(\xi_2, t) = \frac{e^{-32t}}{2\sqrt{3}} \operatorname{sech}(2\xi_2 \mp \tfrac{1}{2} \ln 3)$$

17. Normalize the asymptotic forms of the solutions obtained in Problem 16 and use *Mathematica* to plot and animate the wavefunctions and the potentials for times $0 \leq t \leq 1$. In particular, observe how quickly both solutions evolve into their asymptotic forms. This is particularly striking in $\varphi_i(x,t)$, where the parity of the solution changes from being odd at $t=0$ to being even as $t \to \infty$.

18. The procedure described in Problem 14 can be used to construct the *three*-soliton solution of the Korteweg–de Vries equation for the initial value given by

$$u(x,0) = -12 \operatorname{sech}^2 x$$

First show that

$$\det M = 1 + e^{288t-12x} + 6 e^{280t-10x} + 15 e^{224t-8x} + 10 e^{72t-6x}$$
$$+ 10 e^{216t-6x} + 15 e^{64t-4x} + 6 e^{8t-2x}$$

Then, with the aid of *Mathematica* carry out the required derivatives and show that the resulting expression for the solution can be written in the form

$$u(x,t) = -24 \frac{N(x,t)}{[D(x,t)]^2}$$

where the numerator, $N(x,t)$, and the denominator, $D(x,t)$, are given by

$$N(x,t) = 126 + 25 \cosh(2x - 152t) + 135 \cosh(2x - 56t) + 50 \cosh(2x - 8t)$$
$$+ 40 \cosh(4x - 208t) + 80 \cosh(4x - 64t) + 30 \cosh(6x - 216t)$$
$$+ 15 \cosh(6x - 72t) + 10 \cosh(8x - 224t) + \cosh(10x - 280t)$$

$$D(x,t) = 10 \cosh(72t) + 15 \cosh(2x - 80t) + 6 \cosh(4x - 136t)$$
$$+ \cosh(6x - 144t)$$

Use *Mathematica* to verify that this expression and that in Problem 14 are solutions of the Korteweg–de Vries equation. Then use *Mathematica* to plot and animate these solutions.

19. The procedure derived in Problem 13 for obtaining the N-soliton solution of the Korteweg–de Vries equation with initial values of the form

$$u(x,0) = -N(N+1) \operatorname{sech}^2 x$$

can be carried out in *Mathematica* as a function of N with a few simple commands. Show from the discussion of Section 10.3 that the entries M_{ij} of the matrix M are given by

$$M_{ij} = \delta_{ij} + [c_N^i(t)]^2 \frac{e^{-(i+j)x}}{i+j}$$

where the $c_N^i(t)$ are given by combining Equations (10.25) and (10.99):

$$c_N^i(t) = \frac{1}{i!}\left[\frac{i(N+i)!}{(N-i)!}\right]^{1/2} \exp(4i^3 t)$$

The following sequence of commands in *Mathematica* constructs the matrix M from the M_{ij} and then differentiates the quantity $\ln \det M$ twice with respect to x to generate the N-soliton solution of the Korteweg–de Vries equation:

```
c[i_,j_]:=(1/i!)Sqrt[(i(j+i)!)/(j-i)!]Exp[4i^3t]

M[n_]:=IdentityMatrix[n]+
    Table[(c[i,n]^2/(i+j))Exp[-(i+j)x],{i,1,n},{j,1,n}]

u[n_]:=-2D[Log[Det[M[n]]],{x,2}]
```

Use these commands to construct expressions for N-soliton solutions for different values of N and to plot these expressions using the Plot and ContourPlot commands. A complementary discussion of this construction has been given by Crandall (1991).

20. If $\kappa = 0$ show that to within normalization the corresponding wavefunction is given by (*cf.* Problem 6.15)

$$\psi_0(x) = F[-N, N; 1; \tfrac{1}{2}(1-\tanh x)] = P_N(\tanh x)$$

Thus, deduce that

$$\lim_{x \to \infty} \psi_0(x) = 1$$

so that the zero eigenvalue does not correspond to a true bound state of the potential in Equation (10.79).

21. By substituting the quantities $U = -\nu(\nu+1)$ and $\lambda = -\kappa^2$ into Equation (10.84), this equation takes the form

$$\frac{d}{d\xi}\left[(1-\xi^2)\frac{d\phi}{d\xi}\right] + \left[\nu(\nu+1) - \frac{\kappa^2}{1-\xi^2}\right]\phi = 0$$

which is the same as the generalized Legendre equation in Equation (6.38):

$$\frac{d}{d\xi}\left[(1-\xi^2)\frac{d\Theta}{d\xi}\right] + \left[\ell(\ell+1) - \frac{m^2}{1-\xi^2}\right]\Theta = 0$$

We can use this analogy to obtain a solution of the first of these equations from the method of Frobenius by following the steps in Section 6.3. Thus, by introducing the reduced function $F(\xi)$ through the substitution in (6.55),

$$\phi(\xi) = (1-\xi^2)^{\kappa/2} F(\xi)$$

obtain the equation satisfied by $F(\xi)$ as

$$(1-\xi^2)\frac{d^2 F}{d\xi^2} - 2\xi(\kappa+1)\frac{dF}{d\xi} + \left[\nu(\nu+1) - \kappa(\kappa+1)\right] F = 0$$

which is equivalent to Equation (10.86). The recursion relations in Equation (6.60) can now be transcribed to obtain

$$a_n = \frac{(n+\nu+\kappa-1)(n-\nu+\kappa-2)}{n(n-1)} a_{n-2}$$

for $n \geq 2$. Then, following the reasoning in Section 6.3 show that solutions that are finite over the entire range of ξ can be obtained only by choosing ν and κ to satisfy the requirement $n-\nu+\kappa-2=0$ for some integer n, or

$$\nu - \kappa = n - 2$$

This shows that the condition for the existence of a bound state of the potential (10.79) is only that the *difference* of ν and κ be an integer. The quantities ν and κ are not required to be integers themselves. Complete the steps in Section 6.3 to obtain series representations of the solutions.

22. With a judicious use of recursion relations, or otherwise, establish the integral relation

$$\int_{-1}^{1} \frac{[P_n^m(x)]^2}{1-x^2} dx = \frac{(n+m)!}{(n-m)! m}$$

In fact, since equation (10.97) suggests that

$$\int_{-1}^{1} \frac{P_n^m(x) P_n^k(x)}{1-x^2} dx = \begin{cases} 0 & \text{if } m \neq k \\ \dfrac{(n+m)!}{(n-m)! m} & \text{if } m = k \end{cases}$$

so the orthogonality relation among the $\psi_m^n(x)$ leads naturally to an orthogonality relation among the Legendre functions that is different from the standard orthogonality relation obtained in Problem 6.45.

23. By writing the eigenfunctions for the bound states of the Schrödinger equation in (10.80) as $\psi_n^m(x) = P_n^m(\tanh x)$, show that Equations (10.96) and (10.97) imply that

$$\psi_n^m(x) = \frac{1}{2^m m!} \left[\frac{m(n+m)!}{(n-m)!}\right]^{1/2} (\cosh x)^{-m}$$

$$\times F[m-n, m+n+1; m+1; \tfrac{1}{2}(1-\tanh x)]$$

Hence, deduce that as $x \to \infty$, the form of $\psi_n^m(x)$ is

$$\lim_{x \to \infty} \psi_n^m(x) = \frac{1}{m!} \left[\frac{m(n+m)!}{(n-m)!} \right]^{1/2} e^{-mx}$$

This identifies the quantity c_n^m, as defined in Equation (10.24), as

$$c_n^m = \frac{1}{m!} \left[\frac{m(n+m)!}{(n-m)!} \right]^{1/2}$$

24. Equation (10.104) can be derived by considering the properties of solutions of the hypergeometric equation with respect to the point $x=1$, which is a regular singular point of the equation. Beginning with the standard form of the hypergeometric equation,

$$x(1-x)\frac{d^2y}{dx^2} + [\gamma - (\alpha+\beta+1)x]\frac{dy}{dx} - \alpha\beta y = 0$$

introduce the change of variables to $\xi = 1-x$ and deduce that the general solution can be written as a linear combination of the functions

$$F(\alpha,\beta;\alpha+\beta-\gamma+1;1-x)$$

and

$$(1-x)^{\gamma-\alpha-\beta} F(\gamma-\alpha,\gamma-\beta;1-\alpha-\beta+\gamma;1-x)$$

In particular, the solution of the hypergeometric equation expanded about the origin must be a linear combination of these two functions:

$$F(\alpha,\beta;\gamma;x) = AF(\alpha,\beta;\alpha+\beta-\gamma+1;1-x)$$
$$+ B(1-x)^{\gamma-\alpha-\beta} F(\gamma-\alpha,\gamma-\beta;1-\alpha-\beta+\gamma;1-x)$$

where A and B are constants. By evaluating this equation for $x=1$, and using the result of Problem 8.16,

$$F(\alpha,\beta;\gamma;1) = \frac{\Gamma(\gamma)\Gamma(\gamma-\alpha-\beta)}{\Gamma(\gamma-\alpha)\Gamma(\gamma-\beta)}$$

obtain

$$A = \frac{\Gamma(\gamma)\Gamma(\gamma-\alpha-\beta)}{\Gamma(\gamma-\alpha)\Gamma(\gamma-\beta)}$$

Similarly, by evaluating the equation at $x = 0$ and using the relation

$$\Gamma(x)\Gamma(1-x) = \frac{\pi}{\sin \pi x}$$

derived in Problem 6.23 to simplify the resulting expression, show that B is given by
$$B = \frac{\Gamma(\gamma)\Gamma(\alpha + \beta - \gamma)}{\Gamma(\alpha)\Gamma(\beta)}$$
Thus, the hypergeometric function can be represented as
$$F(\alpha, \beta; \gamma; x) = \frac{\Gamma(\gamma)\Gamma(\gamma - \alpha - \beta)}{\Gamma(\gamma - \alpha)\Gamma(\gamma - \beta)} F(\alpha, \beta; \alpha + \beta - \gamma + 1; 1 - x)$$
$$+ (1-x)^{\gamma-\alpha-\beta} \frac{\Gamma(\gamma)\Gamma(\alpha + \beta - \gamma)}{\Gamma(\alpha)\Gamma(\beta)} F(\gamma - \alpha, \gamma - \beta; 1 + \gamma - \alpha - \beta; 1 - x)$$

Other examples of transformations such as this are given by Abramowitz and Stegun (1965).

25. Using the relation
$$\Gamma(x)\Gamma(1-x) = \frac{\pi}{\sin \pi x}$$
derived in Problem 6.23, show that the quantities $T(k) = a(k)a^*(k)$ and $R(k) = b(k)b^*(k)$, where $a(k)$ and $b(k)$ are given in (10.106), can be written as
$$T(k) = \frac{\sinh^2(\pi k)}{\sin^2(\pi \nu) + \sinh^2(\pi k)}$$
$$R(k) = \frac{\sin^2(\pi \nu)}{\sin^2(\pi \nu) + \sinh^2(\pi k)}$$
which shows directly that $T(k) + R(k) = 1$. Suppose that ν is an integer. Obtain the values of $T(k)$ and $R(k)$ if k is considered to be real, i.e. for continuum states. Similarly, determine the poles of T and R in the *complex* k plane and verify that these correspond to the bound states of the potential.

26. By eliminating the eigenvalue in either of Equations (10.14) and comparing Equations (10.23) (with $\lambda_t = 0$) and (10.109), deduce that the operator B for the Korteweg-de Vries equation is given by
$$B = -4\frac{\partial^3}{\partial x^3} + 6u\frac{\partial}{\partial x} + 3\frac{\partial u}{\partial x}$$

27. Using *Mathematica*, or otherwise, verify that Equation (10.113) with the following Lax pair:
$$\mathcal{L} = \frac{\partial^3}{\partial x^3} + u\frac{\partial}{\partial x}$$
$$\mathcal{B} = 9\frac{\partial^5}{\partial x^5} + 15u\frac{\partial^3}{\partial x^3} + 15\frac{\partial u}{\partial x}\frac{\partial}{\partial x} + \left(5u^2 + 10\frac{\partial^2 u}{\partial x^2}\right)\frac{\partial}{\partial x}$$

implies the equation

$$\frac{\partial u}{\partial t} = \frac{\partial^5 u}{\partial x^5} + 5u\frac{\partial^3 u}{\partial x^3} + 5\frac{\partial u}{\partial x}\frac{\partial^2 u}{\partial x^2} + 5u^2\frac{\partial u}{\partial x}$$

28. Consider the Lax pairs

$$\mathcal{L} = i\begin{pmatrix} 1+s & 0 \\ 0 & 1-s \end{pmatrix}\frac{\partial}{\partial x} + \begin{pmatrix} 0 & u^* \\ u & 0 \end{pmatrix}$$

$$\mathcal{B} = is\begin{pmatrix} 1 & 0 \\ 0 & 1 \end{pmatrix}\frac{\partial^2}{\partial x^2} + \begin{pmatrix} -i|u|^2/(1+s) & u_x^* \\ -u_x & i|u|^2/(1-s) \end{pmatrix}$$

Show that the Lax equation

$$\mathcal{L}_t = \mathcal{BL} - \mathcal{LB}$$

implies that u satisfies the nonlinear Schrödinger equation:

$$iu_t + u_{xx} + \nu|u|^2 u = 0$$

where $\nu^{-1} = \frac{1}{2}(1-s^2)$ and $0 < s < 1$.

29. By suitably modifying the commands given in Section 10.4, derive the result in Problem 28 using matrix arithmetic in *Mathematica*.

Index

This index contains several types of entries. Most of these simply cite the occurrence of the word or phrase on the indicated page. An emboldened page number represents the definition of the entry. A pagenumber followed by a number enclosed in parentheses means that the entry occurs within the numbered problem on that page. Entries which refer either to a function or a command of *Mathematica* are typeset in a Courier font, e.g. Laplace.

A

Abs 341
absorbing boundary 136, 156(23), 161(32), 289–292, 295(5), 312(44, 46)
Airy equation 254(30)
Airy functions 390(4)
AiryAi 390(5)
analytic continuation 195, 251(24)
analytic function 5, **24**, 87, 102, 103, 144, 167, 168, 170–172, 176, 192, 218, 223, 261, 271, 318, 336, 415, 416, 428
animation 9, 10, 159(27), 161(31), 162(32), 382, 394(10), 429, 447(17, 18)
Apart 327, 334
ArcSin 331

ArcTan 394(10)
Arg 341
associated Laguerre equation 246(10), 318
   see also Laguerre equation
associated Laguerre polynomials
   expressions for 246(10)
   generating function for 345(5)
   integral representation for 345(5)
   relation to confluent hypergeometric function 246(11)
   Rodrigues' formula for 345(5)
   see also Laguerre polynomials
associated Legendre functions
   see Legendre functions
associated Legendre functions of second kind
   see Legendre functions of second kind, associated

453

454 *Index*

asymptotic formulae
  for Bessel functions 352(26, 27)
  for Hankel functions 352(26, 27)
  for eigenfunctions of Liouville
    equation 164(40), 165(41)
  for modified Bessel functions
    351(23), 352(26, 27), 354(32)
  for Weber–Hermite functions
    209, 211, 212, 243(3, 4)
auto-Bäcklund transformation **374**
  for Cauchy–Riemann equations
    395(14)
  for Korteweg–de Vries equation
    377, 378, 397(21), 398(23)
  and nonlinear superposition
    377–381
  for sine-Gordon equation 375, 376,
    395(17), 400(27)
  see also Bäcklund transformation

B

beta function 251(22, 23)
Bäcklund parameter 375, 381–385,
  449(5)
  see also Bäcklund transformation
Bäcklund transformation 356,
  371–377, 397(20), 439(1)
  for Burgers' equation 374, 395(15)
  integrability conditions for 374
  and inverse scattering 440(5)
  for Korteweg–de Vries equation
    377, 378, 385(16), 397(21)
  for Liouville equation 396(19)
  for modified Korteweg–de Vries
    equation 395(16)
  and solitons 378, 396(18),
    398(22, 23), 401(28)
  see also auto-Bäcklund
    transformation
base curve **84**, 85, 89, 92, 98, 99, 102,
  110(8)
  see also characteristic base curve
Bessel's equation 3, 162(34), 207, 228,
  343
  relation to other equations
    253(29, 30), 254(31), 255(35)
Bessel functions of first kind **230**,
  232–235, 253(27), 350(20),
  353(28)
  asymptotic form for 352(26, 27)
  generating function for 336
  integral representation for
    331–337, 349(18, 19), 350(21)

and Laplace transforms 298(12)
recursion relations for 253(28)
series for 686
series of 336, 337
Bessel functions of second kind **232**,
  233–235, 252(26), 253(29, 30),
  254(32), 256(35, 36), 340, 353(28)
  asymptotic form for 352(26, 27)
  integral representation for 353(28)
  recursion relations for 253(28)
  series for 232, 253(27)
Bessel functions of third kind **338**
  asymptotic form for 352(26)
  integral representation for
    338–342, 351(24), 353(29)
BesselJ 233, 234
BesselY 233
biharmonic equation 108(1)
boundary conditions
  Cauchy **91**, 94, 98, 99, 102, 103,
    104, 111(10, 11), 113(14, 15),
    115(22, 23), 116(23), 118(29),
    154(16)
  Dirichlet **91**, 99, 104, 105, 106
  for first-order nonlinear equations
    55, 62, 63, 75(29)
  for first-order quasi-linear equations
    6, 35, 38, 40, 45–51
  and Fourier series 126, 131–133,
    133–143, 312(46), 313(47)
  and Fourier transforms 295(5),
    312(44, 45)
  and Green's functions 261, 262,
    307(34), 312(44, 45), 312(46),
    313(47), 314(48, 49)
  and method of images 286–292,
    315(49)
  Neumann **91**, 99, 104
  for quantum mechanical scattering
    410
  and separation of variables 122–124
  see also absorbing boundary
  see also periodic boundary
    conditions
  see also reflecting boundary
  see also Sturm–Liouville problem
branch cut 271, 321, 342
Bromwich integral
  see Laplace transform, inverse of
Burgers' equation 356, 360, 372, 385,
  386
  Bäcklund transformation for 374,
    395(15)

Index   455

Hopf–Cole transformation 361–366
  initial-value problem 364–366
  shock-wave solutions 392(6–8)

C

Cauchy boundary conditions
  see boundary conditions, Cauchy
Cauchy–Kowalewski theorem 5, 85–89
  see also diffusion equation, series solution of
  see also telegraph equation, series solution of
  see also wave equation, series solution of
Cauchy–Riemann equations 395(14)
characteristic base curve **89**, 92, 93
  see also base curve
characteristics
  for diffusion equation 99
  for elliptic equations 101, 102
  for first-order equations 6, 35, 36–52, 56–60, 70(18–20), 71(22), 73(25), 110(8), 117(27, 28), 305(31), 376, 397(20)
  for hyperbolic equations 91–93
  for Laplace's equation 102, 103
  for parabolic equations 97–99
  second-order equations **89**, 108(4)
  for wave equation 94–97, 111(10), 113(14), 121, 135, 136, 137, 155(18)
  see also Charpit equations
Charpit equations **59**, 60, 62
  and strip condition **60**, 75
  and strip equations **60**
Chebyshev equation 162(34), 323
Coefficient 17, 83, 84, 87, 175, 179
Collect 175, 179
complementary error function 365, 366
complete integral **53**, 54, 55, 56, 60–63, 72(23), 73(26, 27), 74(28), 75(29)
complete solution
  see complete integral
completeness of **128**, 147, 240, 301(21)
confluence of singular points 201(17), 202(18, 19)
confluent hypergeometric equation 203(20), 244(7), 332
confluent hypergeometric function
  relation to associated Laguerre polynomials 246(11)
  relation to Hermite polynomials 245(7)
  relation to hypergeometric function 203(22)
  series for 203(21)
conservation equation 114(20), 276, 356, 388(1)
ContourPlot 97, 106, 113(14), 118(28), 155(18), 315(50), 341, 382, 392(8), 394(10), 400(26), 401(29), 448(19)
convolution theorem
  and diffusion equation 277, 278, 289, 304(28)
  for Fourier transforms **277**, 294(2), 299(16), 309(38)
  for Laplace transforms 286, 298(11), 309(38)
  and Poisson's equation 283
  and wave equation 286
  see also fundamental solutions
cosine
  Fourier series 126, 132
  Fourier transform 265, 266, 269, 296(5), 312(44)
curl 30(11), 32(13)
curvilinear coordinates
  cylindrical polar 29(8), 31, 226, 233
  elliptic cylinder 29(8)
  parabolic 29(8)
  parabolic cylinder 29(8), 31
  oblate spheroidal 29(8)
  orthogonal 17–25, 29(8), 113(16), 302(22, 23)
  polar 105, 108(3), 139, 259(21), 303(24)
  prolate spheroidal 29(8)
  spherical polar 18, 19, 20, 21, 23, 24, 25, 30(10), 160(29), 215, 218, 226, 236, 246(10), 255(35), 281, 308(27)
  see also curl
  see also divergence
  see also gradient
  see also Laplacian
Cos 12, 25, 117(25), 126, 129, 264, 267, 274, 333, 334

D

D 12, 24, 32(14), 33(17), 44, 45, 66(3), 67(7), 83, 87, 101, 109(6), 112(12), 175, 179, 185, 191, 238, 257(38), 394(10), 437, 448(19)

456 *Index*

d'Alembert's solution of wave equation
  95, 102, 111(10), 112(11), 116(23),
  135, 142, 143, 154(16–18), 155(19),
  286, 309(38), 311(42)
degree of a differential equation 3, 4
`Derivative` 66(3), 67(7), 84, 87
`Det` 12, 88, 448
delta function
  see Dirac delta function
diffusion equation **77**, 81, 98, 133,
  262, 356, 388(1)
  fundamental solution of 274–279,
    300(18), 303(25), 304(30)
  Green's functions for 289–292,
    312(44–46), 314(47, 48)
  inhomogeneous 303(26)
  relation to Burgers equation
    363–366, 372, 374, 392(6, 7),
    395(15)
  relation to Fokker–Planck equation
    78, 304(31), 306(32)
  series solution of 99–101, 113(17),
    114(17–19), 115(20, 21), 304(28),
    315(51)
  solution by separation of variables
    136–139, 156(21–23), 157(23–25),
    158(26, 27), 295(5)
  steady-state solutions of 155(20),
    159(28)
  three-dimensional 303(27)
  two-dimensional 161(32), 303(27)
diffusivity 98
digamma function 232, 251(25),
  252(26)
Dirac delta function 275, 278, 308(37),
  309(38), 365, 390(4), 415
  integral properties of 299(16),
    300(18, 19), 301(21)
  integral representation of 301(20)
  relation to Heaviside function
    300(17)
  and coordinate transformations
    302(22, 23)
directional derivative 90
distribution
  see generalized function
Dirichlet's kernel 152(10), 153(13)
Dirichlet boundary conditions
  see boundary conditions, Dirichlet
divergence
  theorem 104, 388(1)
  of a vector 23, 24, 25, 31(12), 32(13)
`Do` 17, 215, 234, 235, 282

E

eigenfunctions 5, 7, 8, 207, 208, 261,
  355
  and generating functions 235, 236
  and Green's functions 291, 292,
    314(48)
  and integral representations 318,
    322
  of Liouville equation 143–146, 147,
    162(33, 34), 163(35–38), 164(39),
    165(40, 41)
  see also individual special functions
    and orthogonal polynomials
eigenvalues of Liouville equation
  143–146
eikonal equation 74(29)
`Eliminate` 33(17), 65(3), 66(3), 109(6)
elliptic equations **82**, 83, 89, 101–104,
  108(4), 117(27)
  see also Laplace's equation
  see also Poisson's equation
elliptic function 369
  modulus of 369
elliptic integral 370
Euler constant 233, 252
Euler transform
  kernel for 323, 326
  for Bessel's equation 332
  for hypergeometric equation
    328–331
  for ordinary Legendre equation
    323–328
  see also integral representations
`EulerGamma` 234
even
  periodic extension 292
  function 132, 155(19), 265, 269,
    447(17)
`EvenQ` 213
`Exp` 12, 101, 190, 259(19), 257(38), 264,
  273, 279, 394(10), 448(19)
`Expand` 17, 25, 83, 331
exponents
  see Frobenius, method of

F

`Factor` 84, 179
Fejér's
  kernel 153(13, 14)
  theorem 153(14), 154(15)
Fermi–Pasta–Ulam problem 389(2)
`FindRoot` 159(27)

Fourier series 7, 119, 123, 126–133,
  147, 149(4), 150(5), 154(16), 207
  coefficients of 127, 134
  convergence of 128, 153(13, 14),
    154(15)
  complex 141–143, 151(9), 152(10),
    153(12)
  cosine 131, 132
  for diffusion equation 136–139,
    155(20, 21), 156(22, 23),
    157(24, 25), 158(26, 27), 159(28),
    161(32)
  and Gibbs' phenomenon 151(8)
  for Laplace's equation 139–143
  Parseval's theorem for 150(6, 7)
  and separation of variables
    133–143
  sine 131, 132, 138, 139, 149(3),
    151(8), 159(28)
  for wave equation 133–136,
    154(16–18), 155(19), 160(30),
    161(31)
Fourier transform 8, 123, 262–269,
  270, 271, 274, 275, 293, 294(1),
  295(4), 299(16), 317, 389(3),
  390(4), 404, 415
  convolution theorem for **277**,
    294(2), 299(16), 309(38)
  cosine 265–266, 296(5), 312(44)
  for diffusion equation 274–278,
    304(30)
  and Fourier integral formula 264
  Parseval's theorem for 295(3)
  for Poisson's equation 279–283,
    307(35)
  relation to Fourier series 263,
    264, 265
  sine 266, 296(5), 312(44)
  for wave equation 283–286,
    304(38)
Fourier's law 356, 388(1)
Frobenius, method of 7, 167, 171–180,
  196(1), 197(7), 198(7, 9, 10),
  201(15), 203(21), 204(24, 26),
  205(27)
  for double roots of indicial equation
    180–185
  indicial equation **173**, 174, 175,
    176, 177, 179, 180, 181, 182, 186,
    187, 188, 189, 191, 193, 198(9),
    203(23)
  irregular singular points **170**,
    199(10)

recursion relations **174**, 175, 176,
  177, 178, 179, 180, 181, 183, 184,
  186, 187, 188, 189, 191, 196(1, 3),
  199(10, 12), 203(23)
  for roots of indicial equation
    differing by an integer 186–191
  singular points 168–171, 191–194,
    196(2), 197(7), 198(10), 200(14),
    201(17), 202(17–19), 203(20, 23),
    205(27)
  see also series solutions of
    individual special functions
    and orthogonal polynomials
fundamental set of solutions
  for ordinary Legendre equation 326
  for Liouville equation 144, 163(36),
    164(39)
  for Schrödinger equation 414
fundamental solution 261, 262, 286, 287
  for diffusion equation 274–279,
    289, 290, 291, 292, 300(18),
    303(25–27),304(28–30), 314(47),
    362, 364, 365
  for Fokker–Planck equation 305(31),
    306(32)
  for Klein–Gordon equation 310(41)
  for Poisson's equation 279–283,
    288, 307(35, 36)
  for telegraph equation 311(42, 43)
  for wave equation 283–286,
    309(39, 40)
  see also Green's functions

G

Gamma 330, 333
gamma function 229, 234, 251(24),
  252(27), 428
  integral representation 249(18),
    259(19), 251(23)
  product representation 249(18),
    259(19)
  recursion relation for 249(18)
  relation to beta function 250(21),
    251(22)
  relation to digamma function 232,
    251(25)
Gauss equation
  see hypergeometric equation
Gauss' convergence test **15**, 16, 178,
  196(3), 220
Gel'fand–Levitan equation
  derivation of 412–419
  general form of 419

and soliton solutions 420, 421, 429, 444(13), 445(14, 15), 447(18, 19), 448(19)
generalized function 229(16)
  see also Dirac delta function
general solution
  of Bessel's equation 229, 232, 233, 253(29)
  of equations related to Bessel's equation 253(30), 254(31)
  of equations related to modified Bessel's equation 255(33.34)
  of first-order equations 38, **40**, 43, 44, 45, 47, 50, 52, 53, 65(1, 2), 66(4), 67(6), 68(10), 69(14), 71(21)
  of generalized Legendre equation 224
  of Helmholtz equation 256(35)
  of Hermite's equation 210, 211
  of Laplace's equation 102–104, 105, 106, 108(2)
  of modified Bessel's equation 254(32)
  of ordinary differential equations 5, 168, 171, 174, 177, 178, 179, 182, 184, 186, 191, 197(7), 198(9), 199(11), 202(18), 203(23)
  of ordinary Legendre equation 219, 220
  relation to complete integral 55, 56
  of Ricatti's equation 204(25, 26)
  of second-order equations 81, 84, 108(1, 4), 117(27), 119, 120, 270
  of wave equation 80, 94, 95, 111(10), 112(11), 134, 135, 309(38)
generating functions 7, 8, 208, 235, 236, 240, 318
  for associated Laguerre polynomials 345(5)
  for Bessel functions 335, 336, 349(18, 19), 350(20)
  for Hermite polynomials 257(39), 258(40), 322, 344(1)
  for Laguerre polynomials 344(2), 345(6)
  for Legendre polynomials 236–240, 258(42), 259(44), 328, 346(7)
  for modified Bessel functions of first kind 351(22)
geometrical optics 74(29)
Gibbs' phenomenon 130, 151(8), 261
Gram–Schmidt orthogonalization 125

Green's functions 5, 7, 8, 261, 262, 274, 278, 293, 355
  for boundary-value problems 286–292
  for diffusion equation 289–292, 312(44–46), 314(47, 48), 364
  for Poisson's equation 287–292, 307(34, 36), 308(37), 314(49)
  see also fundamental solutions

H

Hamilton–Jacobi equation 71(22)
Hankel functions
  see Bessel functions of third kind
harmonic oscillator
  classical 71(22), 245(8)
  quantum 208, 213, 245(8), 256(37)
heat equation
  see diffusion equation
Heaviside function **284**, 303(27), 309(39, 40), 310(41)
  integral representation for 298(15), 299(17)
  relation to Dirac delta function 300(17)
Helmholtz equation 226, 255(35)
Hermite's equation 162(34), 207, **210**, 211, 212, 219, 222, 244(6), 318, 323, 328, 242(1, 2)
Hermite functions 244(6)
Hermite polynomials
  generating function for 257(37, 38), 322, 344(1)
  integral representation of 321–323
  normalization 257(39)
  recursion relations for 257(39)
  relation to confluent hypergeometric function 245(7)
  relation to Weber–Hermite functions 213, 214, 245(8)
  series for 213, 243(5)
  series of 258(40, 41)
HermiteH 213
Hopf–Cole transformation 356, 361–366, 372, 374, 392(6, 7)
  see also Burgers' equation
Huygens' principle 285
hyperbolic equations **80**, 82, 83, 89, 91–97, 101, 108(4), 117(27), 118(28, 29), 358, 397(20)
  see also Klein–Gordon equation
  see also telegraph equation
  see also wave equation

hypergeometric equation 201(14), 323, 424
hypergeometric functions
  integral representation 328–331
  for quantum mechanics 424–428, 449(21, 23), 450(24)
  relation to confluent hypergeometric function 203(22)
  relation to elementary functions 201(16), 330, 331
  relation to Legendre functions of first kind 248(15)
  relation to Legendre functions of second kind 248(16)
  relation to Legendre polynomials 248(15)
  series for 15, 201(15)

I

IdentityMatrix 448(19)
If 51, 87, 112(12), 128, 129, 155(18), 180, 184, 213, 278
Im 34
indices
  see Frobenius, method of
indicial equation
  see Frobenius, method of
infinite series
  see power series
infinite products 149(4), 150(5)
Infinity 250(19, 282
initial strip
  see Charpit equations
initial-value problem
  see d'Alembert's solution of wave equation
  see diffusion equation
  see Laplace transform
integrability conditions
  see Bäcklund transformation
integral curves 38, 39, 40
  see also characteristics of first-order equations
integral representations 208, 236
  of associated Laguerre polynomials 345(5)
  of Bessel functions of first kind 333, 334–336, 340, 349(18, 19), 350(20, 21)
  of Bessel functions of second kind 341, 353(30)
  of Bessel functions of third kind 338–340, 351(24), 353(29)
  of Dirac delta function 301(20)
  of Gamma function 249(18), 250(19–21), 251(23, 24), 252(25)
  of Heaviside function 298(15), 299(17)
  of Hermite polynomials 322
  of hypergeometric functions 330, 331, 348(15, 16)
  of Laguerre polynomials 344(2)
  of Legendre functions 347(12)
  of Legendre functions of second kind 347(13)
  of Legendre polynomials 328
  of modified Bessel functions of first kind 351(22, 23)
  of modified Bessel functions of second kind 352(25), 354(31, 32)
  of spherical Bessel functions 333, 334
  see also Fourier transforms
  see also generating functions
  see also Green's functions
  see also Laplace transforms
Integrate 126, 129, 267, 282, 326, 330, 333
InverseSeries 29(7)
inverse scattering method
  see Jost solutions
  see Gel'fand–Levitan equation
  see Korteweg–de Vries equation
  see Lax pair
InverseLaplace 273, 274, 297(8)
InverseLaplace.m 273
irregular singular point 170, 199(10)
  see also Frobenius, method of

J

Jacobi elliptic function 369
Jacobian **18**, 21, 24, 32(15), 40, 42, 83, 117(28), 302(23)
JacobiSN 369
Jost solutions 413
  and inverse scattering 413–419, 442(8)
  derivation 441(6)
  examples of 443(10, 11), 446(15, 16)

K

kernel 317
  Dirichlet's 152(10, 11), 153(13)
  Euler's 323, 326
  Fejér's 153(13, 14)

Laplace's 323, 325
  of Gel'fand–Levitan equation
    413, 419, 420, 443(9), 444(13)
  see also Fourier transforms
  see also Gel'fand–Levitan equation
  see also integral representations
  see also Laplace transforms
Klein–Gordon equation
Korteweg–de Vries equation 4, 8, 360, 401(30),
  Bäcklund transformation for
    395(16), 397(21), 398(22, 23), 440(5)
  inverse scattering solution of 403, 404, 405–423, 425, 426, 428, 429–434
  linearized 371, 390(4, 5), 391
  N-soliton solution of 444(13), 447(19)
  and nonlinear superposition principle 377–381, 398(24)
  physical derivation of 360, 361
  relation to Fermi–Pasta–Ulam problem 389(2)
  and Schrödinger equation 439(1)
  single-soliton solution of 370–372, 378, 394(11), 420, 421, 443(9)
  three-soliton solution of 384, 385, 447(18)
  two-soliton solution of 381–383, 392(8), 399(24, 25), 400(26), 445(14)
  see also Lax pair
  see also modified Korteweg–de Vries equation

L

Laguerre equation 162(34), 218, 245(9), 318, 344(2)
  series solution of 246(9)
  see also associated Laguerre equation
Laguerre polynomials
  expressions for 345(4)
  generating function for 344(2)
  integral representation for 344(2), 345(3)
  normalization of 345(6)
  recursion relations for 346(6)
  Rodrigues' formula for 345(3)
  see also associated Laguerre polynomials
Laplace 273

Laplace's equation
  characteristics of 81, 82
  and generating function for Legendre polynomials 235–238
  Hadamard's example for 116(23)
  and Poisson's integral formula 141–143
  solution by separation of variables 139–141
  three-dimensional 208, 215, 216, 217, 222, 226, 233, 287
  two-dimensional 77, 78, 79, 95, 102–106, 107, 108(2, 3), 109(6), 115(22), 116(24), 117(25, 26), 133
  see also Poisson's equation
Laplace transform 8, 262, 269–274, 293, 296(6, 7), 297(8), 317, 354(32), 404
  convolution theorem for 298(11, 14)
  for ordinary differential equations 271–274, 297(9, 10)
  inverse of 270, 271, 298(12, 13, 15)
  see also fundamental solutions
  see also Green's functions
  see also kernel, Laplace's
Laplace.m 273
Laplacian 18, 215, 280
  in cylindrical coordinates 228
  in orthogonal curvilinear coordinates 22, 23, 32(13)
  in spherical polar coordinates 23–25, 215
Lax equation 435
Lax pair 435, 451(27)
  for Korteweg–de Vries equation 435–437
  for nonlinear Schrödinger equation 452(28, 29)
Legendre equation
  generalized 218, 219, 222, 223, 225, 424, 448(21)
  ordinary 162(34), 215, 219, 220, 221, 224, 225, 323
Legendre functions 207, 222, 247(13), 426
  normalization of 259(45), 449(22)
  relation to hypergeometric function 248(15), 424, 426, 449(23)
  relation to spherical harmonics 226, 227
  Rodrigues' formula for 259(44), 347(12)
  expressions for 225

Legendre functions of second kind
  integral representation for 326, 327, 328, 347(10)
  series for 248(14)
Legendre functions of second kind, associated 348(14)
  integral representation for 347(13)
  relation to hypergeometric function 249(16)
  Rodrigues' formula for 347(13)
Legendre polynomials 207, 215, 222, 315(49), 347(11)
  generating function for 236–239, 346(7)
  integral representation of 326, 328
  normalization of 259(42)
  recursion relations for 239, 240
  relation to hypergeometric function 248(15)
  Rodrigues' formula for 259(43), 328, 346(8)
  series for 221, 247(12)
LegendreP 239, 348(14)
Leibniz formula 259(44)
Liouville equation 120, **143**, 145, 146, 162(33, 34), 163(38), 164(39), 207, 221, 242(2), 259
Liouville equation, nonlinear 397(20)
  see also Sturm–Liouville problem
Liouville's normal form 165(40, 41)
ListPlot 159(7)
Log 185, 190, 234, 327, 331, 448(19)

M

Mathematica 2
  books about 26
  summary 8–10
  see also individual functions and commands
MatrixForm 12, 88
Miura transformation 395(16), 439(1)
Mod 155(18)
modified Bessel equation 254(32)
  relation to other equations 255(33, 34)
modified Bessel functions of first kind 255(32), 298(12, 13), 311(42, 43)
  asymptotic form of 351(23)
  generating function for 351(22)
  integral representation of 351(22, 23)
  series for 255(32)

modified Bessel functions of second kind 255(32)
  asymptotic form of 352(26)
  integral representation of 352(25), 354(21, 22)
modified Korteweg–de Vries equation
  Bäcklund transformation for 395(16)
  relation to Korteweg–de Vries equation 394(11), 395(16), 439(1)
  soliton solution of 394(11)
modulus, of elliptic function 369
Monge cone 57, 58, 72(24)

N

N 117(25), 234, 235
Nest 257(38)
Neumann boundary conditions **91**, 99, 104, 288, 289, 291
Neumann's formula 328, 347(10, 11), 348(11)
NIntegrate 117(25), 250(19), 264, 267, 278, 279, 282
nonlinear optics 355, 356
  see also nonlinear Schrödinger equation
nonlinear Schrödinger equation
  derivation of 356–358
  Lax pair for 452(28, 29)
  solitons of 392(9)
nonlinear superposition
  for Korteweg–de Vries equation 377–385, 398(24), 401(30)
  for sine-Gordon equation 400(27), 401(28, 29)
  see also auto-Bäcklund transformation
normalization
  of Hermite polynomials 257(39)
  of Laguerre polynomials 345(6)
  of Legendre functions 259(45)
  of Legendre polynomials 258(42)
  of spherical harmonics 260(45)
  see also standardization
numerical integration 9, 391, 421–423, 429–434, 438

O

odd
  periodic extension 292
  function 132, 134, 135, 136, 138, 155(19), 269, 292, 447(17)

order of a differential equation 2, 3, 4
orthogonal
    functions 124–126, 148(12)
    eigenfunctions of Liouville equation 145, 162(34), 163(35), 164(39)
    matrix 30(10)
    see also orthogonal polynomials
    see also Fourier series
    see also Sturm–Liouville problem
orthogonal curvilinear coordinates
    see curvilinear coordinates
orthogonal polynomials
    see associated Laguerre polynomials
    see Hermite polynomials
    see Laguerre polynomials
    see Legendre polynomials

P

parabolic equation **81**, 83, 89, 97–101, 108(4), 117(27)
    see also diffusion equation
    see also Fokker–Planck equation
ParametricPlot 33(18), 51, 52, 70(20), 108(3)
ParametricPlot3D 30(9), 34(20), 43, 44, 68(12), 69(13), 70(20), 74(28), 222, 227
Parseval's theorem
    for Fourier series 150(6, 7)
    for Fourier transforms 295(3, 4)
periodic
    boundary conditions 164(39)
    extension 134, 135, 136, 138, 154(16), 155(18), 262, 263, 292, 314(48)
    solutions of sine-Gordon equation 367, 369
Plot3D 34(20), 68(12), 74(28), 106, 161(31), 341, 382, 394(10)
Plus 327
plane curve 33(18)
    see also base curve
    see also characteristic base curve
Poisson integral formula 143
Poisson's equation 262, 262, 278
    fundamental solution of 279–283
    Green's functions for 293, 307(35), 308(37)
    reciprocity relation for 307(36)
    solution with Dirichlet boundary conditions 307(34)
    see also Laplace equation

PolyGamma 234
power series 7, 8, 9, 12, **13**
    difference of 16
    coefficients of 13
    Gauss test for convergence of **15**, 16, 178, 196(3), 220
    inverse of 29(7)
    partial sums **13**, 17, 29(5), 100, 129, 130, 131, 149(3), 150(5, 7), 151(8), 152(10), 153(14), 233, 234, 235, 258(41), 264
    product of two 17, 172
    radius of convergence **13**, 28(4), 29(5), 85, 87, 168, 173, 177, 183, 184, 198(7), 199(10), 205(27), 211, 220, 229, 247(14)
    ratio test for convergence of **14**, 178, 183, 211, 220, 224, 229
    sum of 16
    see also Cauchy-Kowalewski theorem
    see also diffusion equation, series solution of
    see also Fourier series
    see also Frobenius method of
    see also Laplace equation, series solution of
    see also telegraph equation, series solution of
    see also wave equation, series solution of
Print 17, 215, 234, 235, 282
Product 150(5), 250(19)
Protect 326, 327

Q

quasi-linear first-order equations 4, 6, 10, **35**, 52, 53, 56, 57, 58, 59, 63, 64, 65(2), 66(4, 5), 68(9, 10), 70(21), 71(22), 72(24), 73(25, 26)
    boundary conditions for 45–52
    general solution by method of characteristics 36–45
quasi-linear second-order equations **78**, 83
    boundary conditions for 89–91
    classification of 89
    solution by Cauchy-Kowalewski method 84–89
    see also elliptic equations
    see also hyperbolic equations
    see also parabolic equations

R

ratio test
   see power series, ratio test
     for convergence of
Real 341
reciprocity relation 307(36), 308(37)
recurrence formula
   see recursion relations
recursion relations
   for Bessel functions of first kind
     253(28), 256(36), 335
   for Bessel functions of second kind
     253(28)
   for digamma function 252(25)
   for gamma function 249(18),
     252(25)
   for Hermite polynomials 257(39)
   for Legendre polynomials 239, 240,
     346(6)
   for modified Bessel functions
     255(32)
   see also Frobenius, method of
reflecting boundary 156(23), 162(32),
   289, 290, 291, 292, 296(5), 312(45)
reflection coefficient **410**, 411, 412,
   414, 415, 420, 421, 427, 428, 429,
   434, 444(13)
reflectionless potential 421, 422, 429,
   431, 433, 444(13), 445(14),
   446(15, 16), 447(18, 19), 448(19)
regular point **168**, 169, 170, 171, 172,
   173, 176, 177, 180, 184, 192,
   196(2), 209, 210, 244(7)
   see also Frobenius, method of
regular singular point 167, **170**, 172,
   173, 176, 177, 179, 180, 182,
   188, 191, 192, 193, 194, 196(2),
   199(10), 200(13, 14), 201(14, 17),
   202(17), 203(20), 218, 219, 220,
   223, 228, 235, 323, 326, 329,
   425, 450(24)
   see also Frobenius, method of
renormalization group 71(21)
Ricatti equation 204(25, 26)
Rodrigues' formula
   for associated Laguerre
     polynomials 345(5)
   for Hermite polynomials 322
   for Legendre polynomials 259(43),
     328, 346(8), 347(10, 12)
   for Legendre functions 259(44),
     260(45)
   for Laguerre polynomials 345(3, 4)

S

scale factors **21**, 22, 23, 24, 29(8),
   302(22)
scattering data **405**, 406, 413, 414, 420,
   425
   see also Korteweg–de Vries
     equation
Schrödinger equation 22, 215, 226,
   255(35), 439(2)
   for harmonic oscillator 208
   for hydrogen atom 246(10)
   for hyperbolic secant potential
     422–428, 448(21), 449(23)
   and inverse scattering 403, 404,
     405–412, 428–434, 436, 439(3),
     440(5), 442(7)
   relation to Korteweg–de Vries
     equation 439(1)
   see also Jost solutions
separation constant **121**, 123, 124, 138,
   141, 143, 155(20), 217, 228, 242(1),
   256(35)
separation of variables 5, 6, 17, 113(13),
   119, 120–124, 133, 136, 139, 143,
   154(16), 160(30), 216, 242(1),
   256(35), 420
   see also diffusion equation, solution
     by separation of variables
   see also Laplace's equation,
     solution by separation of
     variables
   see also wave equation, solution
     by separation of variables
Series 16, 191
Simplify 12, 15, 32(14), 44, 45, 88,
   126, 175, 239, 257(38), 437
Sin 12, 24, 25, 117(25), 126, 264, 267,
   274, 282, 334
sine
   Fourier series 126, 131, 132, 134,
     138, 139, 149(3), 151(8), 153(12)
   Fourier integral 266, 269, 296(5),
     312(44)
sine-Gordon equation 358–360, 379
   anti-kink solution of 368, 369
   anti-soliton solution of 368, 369,
     394(12), 396(18)
   auto-Bäcklund transformation for
     375, 376, 396(18), 400(27)
   kink solution of 368, 369
   nonlinear superposition principle
     for 377, 400(27), 401(28, 29)
   periodic solutions of 367, 369

soliton solution of 368, 376, 377, 395(17), 396(18)
solutions with soliton pairs for 393(10), 401(28)
singular points
see irregular singular points
see regular singular points
solitons 8, 355, 356, **366**, 376
see also Korteweg–de Vries equation
see also nonlinear Schrödinger equation
see also sine-Gordon equation
solution surface
and Bäcklund transformations 372, 375, 397(20)
for first-order equations **36**, 37, 38, 56, 58, 40, 41, 43, 45, 47, 48, 49, 52, 54, 59, 68(12), 69(13, 14), 70(20), 72(24)
for Korteweg–de Vries equation 382
for second-order equations 84, 85, 90, 92
for sine-Gordon equation 394(10)
Solve 32(14), 67(6), 175, 273
space curve 19, 34(19), 56, 59, 84
SpaceCurve 30(9), 34(19), 72(22)
spherical Bessel functions 256(35, 36), 333, 334
spherical harmonics
expressions for 226
and Legendre functions 222–227
normalization of 226, 260(45)
and Unsöld's theorem 249(17)
spherical polar coordinates
see curvilinear coordinates, spherical polar
SphericalHarmonicY 249(17)
Sqrt 238, 279, 331, 333, 394(10), 448(19)
standardization
of associated Laguerre polynomials 246(10)
of Bessel functions of first kind 230, 333, 350(21)
of Bessel functions of second kind 232
of Bessel functions of third kind 338
of Hermite polynomials 212, 243(5), 322
of hypergeometric function 201(15), 330, 348(15)
of Laguerre polynomials 246(9)
of Legendre functions 225
of Legendre functions of second kind 248(14), 326
of Legendre polynomials 221, 328
of modified Bessel functions of first kind 255(32)
of modified Bessel function of second kind 353(25)
of spherical Bessel functions 256(35)
see also normalization
Sturm–Liouville problem 7, 120, 143–146, 207
eigenfunctions of 145, 162(33), 163(36–38), 164(39, 40), 165(41)
eigenvalues of 144, 145, 146, 162(33), 165(41)
periodic boundary conditions for 164(39)
see also Liouville equation
Subtract 327
Sum 17, 24, 101, 112(12), 129, 179, 184, 190, 213, 233, 234, 264

T

Table 12, 24, 87, 448(19)
telegraph equation 35
fundamental solution of 311(42)
and separation of variables 112(13)
series solution of 112(12, 13)
solution of initial-value problem 311(43)
thermal diffusivity 98, 388(1)
Times 327
Together 238, 240, 394(10)
transmission coefficient 404, **410**, 411, 412, 414, 416, 417, 418, 427, 428
True 44, 240

U

uniqueness of solutions 2, 5, 6, 13
of first-order equations 46, 47
of second-order equations 85, 87, 93, 94, 104, 110(8), 111(10)
Unprotect 326, 327
Unsöld's theorem 249(17)

V

van der Pol equation 3
variation of parameters 441(6)

vector field **37**, 38, 39, 40, 41, 58
  see also integral curves
vector operations in curvilinear
    coordinates
  see curl
  see divergence
  see gradient
  see Laplacian
VectorAnalysis.m 24
VectorField 42, 341

W

wave equation 22, 32(14), 77, 78, 79, 80, 126, 142, 143, 255(35), 261, 262, 287, 311(42, 43), 357, 389
  characteristics of 94
  d'Alembert's solution of 95, 96, 97, 111(10), 113(14, 15), 135, 136, 137, 154(18, 19), 309(28)
  fundamental solution of 283–286, 310(40)
  general solution of 94
  and Huygens' principle 285
  series solution of 111(11)
  solution by separation of variables 120–123, 133–135, 136, 137, 154(16, 17)
  two-dimensional 160(30), 161(31), 309(39)
  three-dimensional 160(29), 309(39)
Weber–Hermite functions 213, 214, 245(8)
  see also Hermite functions
  see also Hermite polynomials
weight function **145**, 162(34), 163(35), 212, 221, 258(39)
  see also Sturm–Liouville problem
Wronskian **11**, 12, 28(1–3), 123, 124, 164(38), 197(4), 198(8), 410, 439(2, 3)